Studies in Probability and Ergodic Theory

ADVANCES IN MATHEMATICS
SUPPLEMENTARY STUDIES, VOLUME 2

ADVANCES IN
Mathematics
SUPPLEMENTARY STUDIES

EDITED BY Gian-Carlo Rota

EDITORIAL BOARD:

Michael F. Atiyah
Lipman Bers
Raoul Bott
Felix Browder
A. P. Calderón
S. S. Chern
J. Dieudonné
J. L. Doob
Samuel Eilenberg
Paul Erdös
Adriano Garsia
Marshall Hall, Jr.
Lars Hörmander
Konrad Jacobs
Nathan Jacobson
Mark Kac
Richard V. Kadison
Shizuo Kakutani
Samuel Karlin
Donald Knuth
K. Kodaira
J. J. Kohn
Bertram Kostant
Peter D. Lax
C. C. Lin
John Milnor
Calvin C. Moore
D. S. Ornstein
Claudio Procesi
Gerald E. Sacks
M. Schutzenberger
J. T. Schwartz
I. M. Singer
D. C. Spencer
Guido Stampacchia
Oscar Zariski

Studies in Probability and Ergodic Theory

ADVANCES IN MATHEMATICS
SUPPLEMENTARY STUDIES, VOLUME 2

EDITED BY

Gian-Carlo Rota

Department of Mathematics
Massachusetts Institute of Technology
Cambridge, Massachusetts

With the Editorial Board
of *Advances in Mathematics*

ACADEMIC PRESS New York San Francisco London 1978
A Subsidiary of Harcourt Brace Jovanovich, Publishers

ACADEMIC PRESS, INC.
111 Fifth Avenue, New York, New York 10003

United Kingdom Edition published by
ACADEMIC PRESS, INC. (LONDON) LTD.
24/28 Oval Road, London NW1 7DX

Library of Congress Cataloging in Publication Data
Main entry under title:

Studies in probability and ergodic theory.

(Advances in mathematics : Supplementary studies; v.2)
Includes bibliographies.
1. Probabilities--Addresses, essays, lectures.
2. Ergodic theory--Addresses, essays, lectures.
I. Rota, Gian Carlo, Date II. Series.
QA273.18.S75 519.2 78-13074
ISBN 0-12-599102-9

PRINTED IN THE UNITED STATES OF AMERICA

Contents

Coupling Methods for Markov Processes

David Griffeath

On Fluctuations of Sums of Random Variables

Lajos Takács

Almost-Sure Invariance Principle for Branching Brownian Motion

L. G. Gorostiza and A. R. Moncayo

On Operator Inequalities and Projections

M. Haseeb Rizvi and R. W. Shorrock

Boundary Behavior of Laplace–Stieltjes Transforms with Applications to Uniformly Distributed Sequences

Jeffrey D. Vaaler

Regularities of Distribution

Leonard Shapiro

Strong Liftings on Topological Measured Spaces

Richard J. Maher

Mixing Transformations in an Infinite Measure Space

Nathaniel A. Friedman

On Eventually Weakly Wandering Sequences

Martin H. Ellis and Nathaniel A. Friedman

Gap Sequences and Eventually Weakly Wandering Sequences

Martin H. Ellis and Nathaniel A. Friedman

Measurable Transformations on Homogeneous Spaces

J. R. Choksi and R. R. Simha

Ergodic Transformations of Lebesgue Spaces

Anthony Lo Bello

List of Contributors

Numbers in parentheses indicate the pages on which the authors' contributions begin.

J. R. CHOKSI (269), Department of Mathematics, McGill University, Montreal, Quebec, Canada

MARTIN H. ELLIS (185, 195), Department of Mathematics, State University of New York, Albany, New York 12222

NATHANIEL A. FRIEDMAN (167, 185, 195), Department of Mathematics, State University of New York, Albany, New York 12222

L. G. GOROSTIZA (95), Departamento de Matemáticas, Centro de Investigación del IPN, Mexico City, Mexico

DAVID GRIFFEATH† (1), Department of Mathematics, Cornell University, Ithaca, New York 14853

J. C. KIEFFER (251), Department of Mathematics, University of Missouri—Rolla, Rolla, Missouri 65401

ANTHONY LO BELLO (287), Department of Mathematics, Allegheny College, Meadville, Pennsylvania 16335

RICHARD J. MAHER (155), Department of Mathematics, Loyola University of Chicago, Chicago, Illinois 60626

G. MILES (207, 219, 231), Department of Mathematics, Birkbeck College, London, England

A. R. MONCAYO (95), Universidad Autonoma Metropolitana (Iztapalapa), Mexico City, Mexico

M. HASEEB RIZVI‡ (113), Department of Statistics, Stanford University, Stanford, California 94305, and Department of Mathematics, Sir George Williams University, Montreal, Canada

LEONARD SHAPIRO (135), Mathematical Sciences Department, North Dakota State University, Fargo, North Dakota

R. W. SHORROCK§ (113), Centre de Recherches Mathématiques, Université de Montréal, Montréal, Canada

R. R. SIMHA (269), School of Mathematics, Tata Institute of Fundamental Research, Bombay, India

† Present address: Department of Mathematics, University of Wisconsin at Madison, Madison, Wisconsin 53706.

‡ Present address: Committee on National Statistics, National Academy of Sciences, Washington, D.C. 20418.

§ Present address: Management Sciences Division, Bell Canada, Montreal, Canada.

LAJOS TAKÁCS (45), Department of Mathematics and Statistics, Case Western Reserve University, Cleveland, Ohio 44106

R. K. THOMAS (207, 219, 231), Department of Mathematics, Birkbeck College, London, England

JEFFREY D. VAALER (119), Department of Mathematics, University of Texas at Austin, Austin, Texas 78712

Preface

The supplementary volumes of the journal *Advances in Mathematics* are issued from time to time to facilitate publication of papers already accepted for publication in the journal. The volumes will deal in general—but not always—with papers on related subjects, such as algebra, topology, foundations, etc., and are available individually and independently of the journal.

Coupling Methods for Markov Processes[†]

DAVID GRIFFEATH[‡]

Department of Mathematics, Cornell University, Ithaca, New York

DEDICATED TO KIYOSI ITÔ, A GOOD FRIEND AND A GREAT TEACHER

Contents

INTRODUCTION

Probably the most familiar limit law for Markov chains is the ergodic theorem for regular finite chains. Recall that a homogeneous chain ξ_n, $n = 0, 1, \ldots,$ taking on values in $S = \{0, 1, \ldots, N\}$ is said to be *regular* if

$$\min_{i,\, j \in S} p_{m_0}(i, j) = \epsilon_0 > 0 \qquad \text{for some} \quad m_0 > 0, \tag{$*$}$$

where $p_m(i, j)$ denotes the probability that the process is in state j after m steps starting from i. The result we refer to is

[†] This paper is based on the author's Ph.D. thesis, Cornell University, 1976.

[‡] Present address: Department of Mathematics, University of Wisconsin, Madison, Wisconsin.

ISBN 0-12-599102-9

THEOREM 0 (Markov–Frechet). *If* (ξ_n) *is a regular Markov chain on* $S = \{0, 1, \ldots, N\}$, *then*

$$\lim_{n\to\infty} p_n(i, j) = \mu_j, \qquad i, j \in S,$$

where (μ_j) *is a strictly positive invariant probability distribution.*

In a 1938 paper [13], Doeblin gave a proof of Theorem 0 which may be viewed as the starting point for the present study. His argument went like this: Consider two *independent* copies of the given chain, (ξ_n^1) and (ξ_n^2), and observe them simultaneously as a bivariate process $(\tilde{\xi}_n) = (\xi_n^1, \xi_n^2)$ with state space $S \times S$. We claim that under the regularity hypothesis, if the copies start in any states i and j, then they will eventually reach the same state k at the same time m with probability one. To see this, choose m_0 as in $(*)$. Then the probability, starting from (i, j) that $\tilde{\xi}_{m_0} = (k, k)$ for some k is

$$\sum_{k=0}^{N} p_{m_0}(i, k) p_{m_0}(j, k) \geqslant N\epsilon_0^2 = \epsilon > 0.$$

Similarly, if $\xi_{m_0}^1 \neq \xi_{m_0}^2$ then the probability that $\tilde{\xi}_{2m_0} = (k, k)$ for some k exceeds ϵ, and so forth. Consequently the probability that (ξ_n^1) and (ξ_n^2) meet by time mm_0 is at least $1 - (1-\epsilon)^m$, which tends to 1 as $m \to \infty$. Doeblin now proposes that we modify $(\tilde{\xi}_n)$ by giving the two copies the same random trajectory once they meet, so that $\xi_n^1 = \xi_n^2$ from that time on. If we do this, and let $\tilde{P}_{(i,j)}$ denote the resultant joint probability law,

$$\begin{aligned} |p_m(i, k) - p_m(j, k)| &= |\tilde{P}_{(i,j)}(\xi_m^1 = k, \xi_m^2 \neq k) - \tilde{P}_{(i,j)}(\xi_m^1 \neq k, \xi_m^2 = k)| \\ &\leqslant \tilde{P}_{(i,j)}((\xi_n^1) \text{ and } (\xi_n^2) \text{ first meet after time } m) \\ &\to 0 \quad \text{as} \quad m \to \infty \qquad \text{for all} \quad i, j, k \in S. \end{aligned}$$

The rest of the proof is easy, since $\max_i p_m(i, k)$ decreases and $\min_i p_m(i, k)$ increases for each k as m increases.

Our goal in these notes is to isolate the key ingredients in Doeblin's proof and use them to derive ergodic theorems for much more general Markov processes on a typically uncountable state space. The present work extends the author's papers [15–17] which treated denumerable chains, at the same time providing a unified exposition of numerous recent results. For our purposes, the essentials are as follows: (ξ_n) is a discrete time Markov process on a quite general measurable space S. P_ι is the probability law when (ξ_n) starts with initial distribution ι. P_ι is a measure on the path space $\Omega = S^{\mathbb{N}}$ with the appropriate σ-algebra $\mathscr{B}$. In $S \times S$, the *diagonal* is $\Delta = \{(s, s) : s \in S\}$. We consider a collection $(\tilde{P}_{(\mu,\nu)})$ of probability laws for a bivariate stochastic process $(\tilde{\xi}_n) = (\xi_n{}^1, \xi_n{}^2)$ such that the marginal processes are copies of the given (ξ_n) starting with initial distributions μ and ν, respectively. In addition we require that $\tilde{P}_{(\mu,\nu)}(\xi_n{}^1 = \xi_n{}^2 \text{ for all } n \geqslant \tau_\Delta) = 1$, where $\tau_\Delta = \min\{n : \tilde{\xi}_n \in \Delta\}$ ($= \infty$ if no such n exists) is the *hitting time to the diagonal*. Such a collection of measures is called a *coupling* for (ξ_n). The object of the coupling method is to construct $\tilde{P}_{(\mu,\nu)}$ with the additional property that $\tau_\Delta < \infty$ almost surely. (Such a coupling is called "successful.") Just as in Doeblin's proof, this guarantees ergodic theorems. But our generalized formulation is much more powerful since we do not insist that the marginals be independent until τ_Δ. In fact, some of the most useful applications of the coupling method employ processes $(\tilde{\xi}_n)$, which are not even Markovian. Such couplings are more powerful than Markovian ones, but there is a price to pay: the joint sample path behavior may well become unmanageable.

Section 1 develops the coupling method in detail. After the necessary notation has been introduced in Section 1.1, Sections 1.2 and 1.3 present various formulations of "loss of memory" for Markov processes, the kinds of properties guaranteed by successful coupling. Since our arguments even apply in some cases when $\tau_\Delta < \infty$ with a positive probability less than one, interest centers around the tail σ-algebra $\mathscr{T}$ for (ξ_n). Roughly, the atomic structure of $\mathscr{T}$ measures the degree of ergodicity: the simpler the structure, the more forgetful the process. Section 1.4 defines the notion of coupling precisely, and details the kind of information which successful or partially successful coupling provides. Section 1.5 presents examples where $(\tilde{\xi}_n)$ is Markovian, while Section 1.6 is devoted to the construction of a non-Markovian "maximal" coupling for which $(\tilde{\xi}_n)$ hits the diagonal Δ as efficiently as possible.

In Section 2 we present some applications of coupling to nonhomogeneous Markov processes. The first three sections use maximal coupling to obtain a general structure theorem for the tail σ-algebra $\mathscr{T}$, an equivalence theorem for various formulations of loss of memory, and related results. Section 2.4 deals with two types of uniform ergodicity, and ergodic coefficients.

Fundamental ergodic laws in the homogeneous setting, namely Orey's theorem for φ-recurrent Markov processes and the corresponding result for aperiodic random walk, are proved by means of coupling in the first two sections of Section 3. The final section contains a brief discussion of rates of convergence for certain φ-recurrent processes.

In an Appendix we mention how most of the theory developed in Sections 1–3 may be applied to continuous time Markov processes.

For the sake of exposition, citations in the text have been kept to a minimum. A separate section at the end contains a list of bibliographical notes which I hope is reasonably accurate and complete, together with various remarks. This is followed by the references.

1. The Coupling Method

1.1. *Preliminaries*

Let S be a polish space (i.e., a topological space homeomorphic to a complete separable metric space), and let $\mathscr{S}$ be the topological σ-algebra on S generated by open sets. S will be called the *state space*. The *time parameter set* is $\mathbb{N} = \{0, 1, \ldots\}$ (discrete time). Put $\Omega = S^{\mathbb{N}}$ = the *path space* of sequences $\omega : \mathbb{N} \to S$, $\omega = (\omega_n)_{n \in \mathbb{N}}$. The *nth coordinate map* $\xi_n : \Omega \to S$ is given by $\xi_n(\omega) = \omega_n$. Denote by $\mathscr{B} = \sigma[(\xi_n)_{n \in \mathbb{N}}]$ the σ-algebra generated by the sequence (ξ_n), $\mathscr{B}_m{}^n = \sigma[(\xi_k)_{m \leqslant k \leqslant n}]$, $\mathscr{B}_m{}^\infty = \sigma[(\xi_k)_{m \leqslant k \leqslant \infty}]$, and abbreviate $\mathscr{B}_m = \mathscr{B}_m{}^m$; here $m, n \in \mathbb{N}$. The *tail σ-algebra* is $\mathscr{T} = \bigcap_{m \in N} \mathscr{B}_m{}^\infty$.

Let $\mathscr{P}$ be the collection of probability measures on $(S, \mathscr{S})$. A map $\pi : S \times \mathscr{S} \to [0, 1]$ is called a *transition function* (t.f.) iff (i) $\pi(x, \cdot) \in \mathscr{P}$ for each $x \in S$, and (ii) $\pi(\cdot, E)$ is $\mathscr{S}$-measurable for each $E \in \mathscr{S}$. A collection $(\pi^m)_{m \in \mathbb{N}}$ of transition functions will be termed a *transition probability system* (t.p.s.). Let (π^m) be a given t.p.s. If $\iota \in \mathscr{P}$, the *Markov measure P_ι with initial distribution* ι (and t.p.s. (π^m)) is the unique measure on $(\Omega, \mathscr{B})$ satisfying $P_\iota(\xi_0 \in E) = \iota(E)$, and

$$P_\iota(\xi_{m+1} \in E \,|\, \mathscr{B}_0{}^m)(\omega) = P_\iota(\xi_{m+1} \in E \,|\, \mathscr{B}_m)(\omega) = \pi^m(x, E)$$

when $\omega \in \{\xi_m = x\} = \{\omega : \xi_m(\omega) = x\}$, $E \in \mathscr{S}$, $m \in \mathbb{N}$. For our purposes, a (discrete time) *Markov process on* $(\Omega, \mathscr{B}, P_\iota)$ is the coordinate process $(\xi_n)_{n \in \mathbb{N}}$ for some P_ι manufactured from a t.p.s. (π^m) according to the above recipe. There is no loss of generality here when S is polish, by the existence theorem for regular conditional probabilities and the Kolmogorov extension theorem. It will be convenient to speak of a Markov process on $(\Omega, \mathscr{B})$ as the coordinate process (ξ_n) governed by any member of the family $(P_\iota)_{\iota \in \mathscr{P}}$ induced by (π^m). It should be stressed that the underlying t.p.s. (π^m) will always be regarded as fixed.

The Markov process (ξ_n) is *homogeneous* iff $\pi^m(x, E) = \pi(x, E)$, independently of m. We call (ξ_n) a *Markov chain* iff S is countable; in this case we will always take $\mathscr{S} = 2^S$ = all subsets of S, and generally assume for notational convenience that $S = \mathbb{N}$. When $S = \mathbb{R} = (-\infty, \infty)$, $\mathscr{S}$ = the Borel σ-algebra, and $\pi^m(x, E) \equiv \pi(E - x)$ ($E - x$ means translation of E by $-x$) for some Borel measure π on $\mathbb{R}$, then (ξ_n) will be referred to as a *random walk*.

For $m \in \mathbb{N}$, the *m-shift* $\theta^m : \Omega \to \Omega$ is given by $\xi_n \circ \theta^m = \xi_{m+n}$, $n \in \mathbb{N}$. Viewing θ^m as a set map,

$$\theta^m B = \{\theta^m \omega : \omega \in B\}, \qquad \theta^{-m} B = \{\omega : \theta^m \omega \in B\}, \qquad B \in \mathscr{B}.$$

(These sets are in $\mathscr{B}$, though in general $\theta^{-m}\theta^m B \neq B$ unless $B \in \mathscr{B}_m{}^\infty$.) If P_ι is a Markov measure with t.p.s. (π^m), define the modified measure $P_\mu{}^m$, $\mu \in \mathscr{P}$,

$m \in \mathbb{N}$, on $(\Omega, \mathscr{B})$ as the Markov measure with initial distribution μ and $t.p.s.$ $({}^m\pi^n)_{n\in\mathbb{N}}$, where ${}^m\pi^n = \pi^{m+n}$. Let E_ι be the expectation operator for P_ι, $E_\mu{}^m$ the expectation operator for $P_\mu{}^m$. Also, write $p_{mn}(\mu, E) = P_\mu{}^m(\xi_{n-m} \in E)$, $p_n(\iota, E) = p_{0n}(\iota, E), m \leqslant n, E \in \mathscr{S}$. Let $\iota^n \in \mathscr{P}$ be given by $(\iota^n)(\cdot) = p_n(\iota, \cdot)$. When ι or $\mu = \epsilon_x =$ the initial distribution concentrated at $x \in S$, we will always replace ϵ_x by x in subscripts and arguments. Note in particular that we put $x^n = \epsilon_x{}^n$.

The conditional probability measures $P^m_{\xi_m}:\Omega \to \mathscr{P}$ are defined as

$$P^m_{\xi_m(\omega)} = P_x{}^m \qquad \text{on} \quad \{\xi_m = x\}.$$

Throughout the discussion, analogous notation will be used to describe various quantities depending on $\xi_m(\omega)$. Moreover, it will often be convenient to consider expressions such as $P^m_{\xi_m}(B_{\xi_m})$, where B_{ξ_m} is some (random) event determined by ξ_m. Though somewhat ambiguous, this will always mean

$$P^m_{\xi_m(\omega)}(B_{\xi_m(\omega)}) = P_x{}^m(B_x) \qquad \text{on} \quad \{\xi_m = x\}$$

(where, of course, $B_x \in \mathscr{B}$ for each x). With the above notation, the Markov property for (ξ_n) becomes

$$P_\iota(B\,|\,\mathscr{B}_0{}^m) = P^m_{\xi_m}(\theta^m B), \qquad \iota \in \mathscr{P}, \quad m \in N, \quad B \in \mathscr{B}_m{}^\infty.$$

If $f:S \times \mathbb{N} \to \mathbb{R}$ is bounded and $\mathscr{S} \otimes 2^{\mathbb{N}}$-measurable, we write

$$(M_n f)(x, m) = E_x{}^m[f(\xi_{n-m}, n-m)] = \int_S p_{mn}(x, dy) f(y, n)$$

$m; n \in \mathbb{N}, m \leqslant n$. f is called *space–time harmonic* iff $M_n f = f$ for every n.

For $\mu, \nu \in \mathscr{P}$, $\mu - \nu$ may be viewed as a signed measure on $(S, \mathscr{S})$ with total measure 0. Letting $\mu - \nu = (\mu - \nu)^+ - (\mu - \nu)^-$ be the Jordan decomposition, we can find Hahn sets H^+, H^- such that $(\mu - \nu)^+(E) = \mu(H^+ \cap E) - \nu(H^+ \cap E) = \sup_{F\in\mathscr{S}}\{\mu(F \cap E) - \nu(F \cap E)\} = (\mu - \nu)^-(E) = \mu(H^- \cap E) - \nu(H^- \cap E) = -\inf_{F\in\mathscr{S}}\{\mu(F \cap E) - \nu(F \cap E)\}; E \in \mathscr{S}$. Thus the total variation measure satisfies $|\mu - \nu| = 2(\mu - \nu)^+$, and it is convenient to introduce the norm

$$\|\mu - \nu\| = \tfrac{1}{2}|\mu - \nu|(S) = \sup_{E\in\mathscr{S}}\{\mu(E) - \nu(E)\}$$

$(=(\mu - \nu)^+(S) = (\mu - \nu)^-(S) = \tfrac{1}{2}$ the total variation norm of $\mu - \nu)$. Also, we set $\mu \wedge \nu = \mu - (\mu - \nu)^+$, remarking that

$$(\mu \wedge \nu)(S) = 1 - \|\mu - \nu\|,$$

and also that for any μ_1, μ_2, and ν,

$$(\mu_1 \wedge \mu_2) + \nu = (\mu_1 + \nu) \wedge (\mu_2 + \nu).$$

When $a, b \in \mathbb{R}$, set $a \wedge b = \min\{a, b\}$, $a \vee b = \max\{a, b\}$, $a^+ = a \vee 0$, $[a] =$ the greatest integer in a. Then $0 \leqslant (\mu \wedge \nu)(E) \leqslant \mu(E) \wedge \nu(E)$, $E \in \mathscr{S}$, and when S is countable we have

$$\|\mu - \nu\| = \tfrac{1}{2} \sum_{x \in S} |\mu(x) - \nu(x)| = 1 - \sum_{x \in S} (\mu(x) \wedge \nu(x))$$
$$= 1 - (\mu \wedge \nu)(S).$$

(*Note*: Here and below we write x for $\{x\}$ whenever convenient.)

Finally, a central object of study for us will be the signed measure

$$p_{mn}(\mu, \cdot) - p_{mn}(\nu, \cdot) \qquad \text{for given} \quad \mu, \nu \in \mathscr{P}, \quad m \leqslant n.$$

It will be useful to denote by $H_{mn}^+(\mu, \nu)$ a positive Hahn set for the Jordan decomposition of the above measure, and to set $H_{mn}^-(\mu, \nu)$ to be the complementary set in S. When S is countable one can take

$$H_{mn}^+(\mu, \nu) = \{x \in S : p_{mn}(\mu, x) \geqslant p_{mn}(\nu, x)\}.$$

1.2. P_ι-*Atomic Decomposition of* $\mathscr{T}$

Let P_ι be a Markov measure on $(\Omega, \mathscr{B})$. The set $A \in \mathscr{T}$ is a P_ι-*atom* of $\mathscr{T}$ iff $P_\iota(A) > 0$ *and* whenever $B \subset A$ and $B \in \mathscr{T}$ then $P_\iota(B) = 0$ or $P_\iota(B) = P_\iota(A)$. For each P_ι, Ω admits the decomposition

$$\Omega = F + \sum_{r \in I} A_r, \qquad F, A_r \in \mathscr{T}, \tag{1}$$

into a disjoint sum of P_ι-atoms A_r, and a fully nonatomic set F. The index set I may be empty, finite, or countable, and the partition is unique modulo P_ι-null sets. $\mathscr{T}$ is P_ι-*atomic* iff F may be taken as $\varnothing$, in which case it is P_ι-*finite* when $|I| < \infty$ and P_ι-*trivial* when $|I| = 1$. Clearly all these notions depend fundamentally on ι, and in general $|I|$ will vary as P_ι runs over the family of measures corresponding to a given t.p.s. (π^m). When P_ι is understood, write $B_1 \cong B_2$ iff $P_\iota(B_1 \triangle B_2) = 0$ $(B_1, B_2 \in \mathscr{B})$, and $\mathscr{B}_1 \cong \mathscr{B}_2$ iff the P_ι-completions of the σ-algebras $\mathscr{B}_1$ and $\mathscr{B}_2$ are equivalent.

We mention here a construction of the atomic sets, due in essence to Blackwell, and an important corollary for finite Markov chains.

PROPOSITION 1. *Let* (1) *be the* P_ι-*atomic decomposition of* $\mathscr{T}$. *There exist sets* $E^n, E_r^n \in \mathscr{S}$; $n \in \mathbb{N}$, $r \in I$, *such that*

(i) *for each fixed* n, $\{E^n, E_r^n : r \in I\}$ *is a partition of* S, *and*

(ii) $F \cong \{\xi_n \in E_r^n \text{ f.o. for every } r\}$

$A_r \cong \{\xi_n \in E_r^n \text{ ev.}\}, \qquad r \in I$

(*f.o.* and *ev.* abbreviate "*for only finitely many n*" and "*for all sufficiently large n*," *respectively*.)

Proof. Set $E_r{}^n = \{x: P_x{}^n(A_r) > \frac{1}{2}\}$, $r \in I$, $E^n = S - \sum_{r \in I} E_r{}^n$. For each n, the $(E_r{}^n)_{r \in I}$ are clearly disjoint, so (i) holds. By martingale convergence, $\lim_{n\to\infty} P^n_{\xi_n}(A_r) = \lim_{n\to\infty} P_\iota(A_r | \mathscr{B}_0{}^n) = 1_{A_r}$ a.s. Hence $A_R \cong \{\omega: \lim_{n\to\infty} P^n_{\xi_n(\omega)}(A_r) > \frac{1}{2}\} = \lim_{n\to\infty}\{\xi_n \in E_r{}^n\}$, and (ii) follows. ∎

COROLLARY 1 (Cohn–Senchenko). *If* $|S| = N < \infty$, *then* $\mathscr{T}$ *is finite and has at most* N *atoms w.r.t. every* P_ι.

Proof. Let $(E_r{}^n)$ be the sets constructed above. If $|I| > N$, then we can find $r_1, \ldots, r_{N+1} \in I$ such that for some large m one has $P_\iota(\xi_m \in E^m_{r_k}) > 0$, $1 \leqslant k \leqslant N+1$, contradicting the fact that $|S| = N$. Hence $|I| \leqslant N$, and it now follows that

$$F \cong \left\{\xi_n \in \bigcup_{r\in I} E_r{}^n \text{ f.o.}\right\} = \{\xi_n \in S \text{ f.o.}\} = \varnothing. \quad ∎$$

The proof of the last result is due to Cohn [10]; earlier arguments were much more complicated.

Our principal objective in these notes is to study the partition (1) of $\mathscr{T}$ for Markov processes, and to derive results guaranteeing a relatively simple structure. Roughly speaking, the simpler the ingredients of $\mathscr{T}$, the more (ξ_n) tends to lose track of its past history. Several precise formulations of loss of memory will be given in the next section, but first we present some examples illustrating various possibilities for the atomic decomposition.

EXAMPLE 1. Let (ξ_n) be a random walk with t.p.s. (π^m) such that $\pi^m(x, E) \equiv \pi(E - x)$. If π has an absolutely continuous part, then $\mathscr{T}$ is P_ι-trivial for every $\iota \in \mathscr{P}$. If $\pi(1) = \pi(-1) = \frac{1}{2}$, then $F \cong \varnothing$, $I = \{0, 1\}$, and $A_0 \cong \{\xi_0$ is even$\}$, $A_1 \cong \{\xi_0$ is odd$\}$ whenever ι gives positive measure to both of these sets. These facts will be proved in Section 3, where other possibilities for random walk are also discussed.

EXAMPLE 2. $S = \{e^{i\theta}: \theta \in [0, 2\pi)\}$, $\iota = \epsilon_1$. (ξ_n) is homogeneous with $\pi(1, 1) = \pi(1, e^{i\theta_0}) = \frac{1}{2}$, and $\pi(x, e^{i\theta_0}x) = 1$ when $x \neq 1$, for some fixed $\theta_0 \in [0, 2\pi)$. If θ_0 is irrational, then $F \cong \varnothing$, $I = \mathbb{N}$, and $A_r \cong \{\omega^r\}$, where ω^r is the path such that $\omega_n{}^r = e^{i(n-r)+\theta_0}$. When θ_0 is rational (ξ_n) reduces to a finite Markov chain, and $\mathscr{T}$ is trivial.

EXAMPLE 3. $S = \mathbb{N}$, $\iota = \epsilon_0$, $\pi^n(x, x) = \pi^n(x, x + 2^n) = \frac{1}{2}$, $x \in S$, $n \in \mathbb{N}$. It is not hard to see that $\xi_n(\omega)$ determines $\xi_0(\omega), \xi_1(\omega), \ldots, \xi_{n-1}(\omega)$ uniquely, so that $\mathscr{B}_0{}^n \cong \mathscr{B}_n$. It follows that $\mathscr{T} \cong \mathscr{B}$ and $F \cong \Omega$, i.e., the tail σ-algebra for this Markov *chain* is full and fully nonatomic.

1.3. λ-*Loss of Memory*

Throughout this section, let (ξ_n) be a Markov process on $(\Omega, \mathscr{B}, P_\iota)$ with given t.p.s. (π^m). We are going to introduce various expressions of asymptotic partial loss of memory for (ξ_n). These conditions are indexed by a parameter $\lambda \in (0, 1]$; intuitively: the larger λ the more forgetful (ξ_n). The most important case $\lambda = 1$, will give formulations for total loss of memory of the chain's past history.

First, say that a 0–λ *law holds for* (ξ_n) (w.r.t. P_ι) iff

$$P_\iota(B) = 0 \qquad \text{or} \qquad P_\iota(B) \geqslant \lambda \qquad \text{whenever} \quad B \in \mathscr{T}. \tag{2}$$

This clearly implies that $\mathscr{T}$ is finite with at most $[1/\lambda]$ atoms, while any Markov process with finite tail σ-algebra satisfies (2) with $\lambda = \min_{r \in I} P_\iota(A_r)$. The following simple fact will have curious consequences later on (cf. Corollary 2.3):

Proposition 2. *If a* 0–λ *law holds for* (ξ_n) *with* $\lambda > \frac{1}{2}$, *then a* 0–1 *law holds (i.e.,* $\mathscr{T}$ *is trivial).*

Proof. Immediate since B, $B^c \in \mathscr{T}$ cannot satisfy $P_\iota(B) > \frac{1}{2}$ *and* $P_\iota(B^c) > \frac{1}{2}$. ∎

Next, for $\mu, \nu \in \mathscr{P}$, $m, n \in \mathbb{N}$ with $m \leqslant n$, write

$$\gamma_m{}^n(\mu, \nu) = \sup_{B \in \mathscr{B}_{n-m}^{\infty}} \{P_\mu{}^m(B) - P_\nu{}^m(B)\},$$

$$\delta_m{}^n(\mu, \nu) = \|p_{mn}(\mu, \cdot) - p_{mn}(\nu, \cdot)\|,$$

$$\gamma_m(\mu, \nu) = \lim_{n \to \infty} \gamma_m{}^n(\mu, \nu), \qquad \delta_m(\mu, \nu) = \lim_{n \to \infty} \delta_m{}^n(\mu, \nu)$$

$(\gamma_m{}^n \downarrow \gamma_m,\ \delta_m{}^n \downarrow \delta_m)$. We call (ξ_n) λ-*mixing* (w.r.t. P_ι) iff

$$\gamma_m(\xi_m, \iota^m) \leqslant 1 - \lambda \quad \text{a.s.} \qquad \text{for each} \quad m \in \mathbb{N} \tag{3}$$

and λ-*ergodic* (w.r.t. P_ι) iff

$$\delta_m(\xi_m, \iota^m) \leqslant 1 - \lambda \quad \text{a.s.} \quad \text{for each} \quad m \in \mathbb{N}. \tag{4}$$

Here $\gamma_m(\xi_m, \iota^m) = \gamma_m(\epsilon_x, \iota^m)$ when $\omega \in \{\xi_m = x\}$, and similarly for δ_m. Note that by the Markov property

$$\gamma_m(\xi_m, \iota^m) = \lim_{n \to \infty} \sup_{B \in \mathscr{B}_n{}^\infty} \{P_\iota(B \mid \mathscr{B}_0{}^m) - P_\iota(B)\}. \tag{5}$$

Evidently, λ-mixing expresses near independence of events in $\mathscr{B}_0{}^m$ and $\mathscr{B}_n{}^\infty$ for each fixed m when n is sufficiently large. Similarly λ-ergodicity asserts that (ξ_n) tends to lose track of the value of ξ_m as $n \to \infty$, to a degree indexed by λ.

Finally, we say that (ξ_n) has *λ-constant space–time harmonics* (w.r.t. P_ι) iff the following property holds: For any $\mathscr{S} \otimes 2$ -measurable space–time harmonic $f: S \times T \to \mathbb{R}$ such that $0 \leqslant f \leqslant 1$,

$$|f(\xi_m, m) - E_\iota[f(\xi_0, 0)]| \leqslant 1 - \lambda \qquad \text{a.s.} \tag{6}$$

for every $m \in \mathbb{N}$.

Eventually we will prove that (2), (3), (4), and (6) are all *equivalent*, but for the moment we establish the implications in one direction only.

LEMMA 1. *Let P_ι be a Markov measure.*

(a)
$$1_A - P_\iota(A) \leqslant \liminf_{m \to \infty} \sup_{B \in \mathscr{T}} \{P_\iota(B \mid \mathscr{B}_0{}^m) - P_\iota(B)\}$$
$$\leqslant \liminf_{m \to \infty} \gamma_m(\xi_m, \iota^m) \qquad \text{a.s.}$$

for any $A \in \mathscr{T}$.

(b) *If $0 < \lambda \leqslant 1$, $B \in \mathscr{T}$ and $1_B - P_\iota(B) \leqslant 1 - \lambda$ a.s. on B, then $P_\iota(B) = 0$ or $P_\iota(B) \geqslant \lambda$.*

(Here $1_B(\omega)$ denotes the indicator of B.)

Proof. (a) The first inequality follows from the martingale convergence theorem; the second is trivial by (5) since $\mathscr{T} \subset \mathscr{B}_n{}^\infty$ for every n.

(b) Integrate over B to get $P_\iota(B) - [P_\iota(B)]^2 \leqslant (1 - \lambda)P_\iota(B)$. ∎

PROPOSITION 3. *Let (ξ_n) be a Markov process on $(\Omega, \mathscr{B}, P_\iota)$, and suppose $0 < \lambda \leqslant 1$. Then λ-mixing $\Rightarrow$ λ-ergodicity $\Rightarrow$ λ-constant space–time harmonics $\Rightarrow$ a 0–λ law for (ξ_n).*

Proof. $\delta_m{}^n(\mu, \nu)$ may be written in the form of $\gamma_m{}^n$ with $\mathscr{B}_{n-m}^\infty$ replaced by $\mathscr{B}_{n-m} = \mathscr{S}$; this shows that

$$\delta_m{}^n(\mu, \nu) \leqslant \gamma_m{}^n(\mu, \nu), \tag{7}$$

and hence (3) $\Rightarrow$ (4). If f satisfies the hypotheses of (6), then

$$\begin{aligned} |f(\xi_m, m) - E_\iota[f(\xi_0, 0)]| &= |(M_n f)(\xi_m, m) - E_\iota[f(\xi_n, n)]| \\ &= \left| \int_S [p_{mn}(\xi_m, dy) - \iota^n(dy)] f(y, n) \right| \\ &\leqslant \int_S |p_{mn}(\xi_m, \cdot) - \iota^n(\cdot)| = \delta_m{}^n(\xi_m, \iota^m) \end{aligned}$$

for every $n \geqslant m$. Letting $n \to \infty$ shows that (4) $\Rightarrow$ (6). Now put $f(x, m) = P_x^{\ m}(B)$ for fixed $B \in \mathscr{T}$. This f satisfies the hypotheses of (6), and the conclusion becomes $P(B \mid \mathscr{B}_0^{\ m}) - P(B) \leqslant 1 - \lambda$ a.s. Applying Lemma 1 yields property (2), as desired. ∎

1.4. *Coupling: The General Formulation*

In order to present the general definition of a coupling for a Markov process, some additional notation is required. Put $\tilde{S} = S \times S$, and give $\tilde{S}$ the product topology. $\tilde{S}$ is polish since S is, and the topological σ-algebra on $\tilde{S}$ coincides with the product σ-algebra $\tilde{\mathscr{S}} = \mathscr{S} \otimes \mathscr{S}$ generated by the semialgebra of rectangles $\mathscr{S} \times \mathscr{S} = \{(B^1, B^2): B^1, B^2 \in \mathscr{S}\}$. Set $\tilde{\Omega} = \tilde{S}^{\mathbb{N}}$, and let $\tilde{\xi}_n = (\xi_n^{\ 1}, \xi_n^{\ 2})$ be the bivariate coordinate map sending $\tilde{\omega} = ((\omega_0^{\ 1}, \omega_0^{\ 2}), (\omega_1^{\ 1}, \omega_1^{\ 2}), \ldots) \in \tilde{\Omega}$ to $\tilde{\omega}_n = (\omega_n^{\ 1}, \omega_n^{\ 2})$. Write $\tilde{\mathscr{B}} = \sigma[(\tilde{\xi}_n)_{n \in \mathbb{N}}]$. It will sometimes be convenient to think of $\tilde{\Omega}$ as $S^{\mathbb{N}} \times S^{\mathbb{N}}$, with $\tilde{\omega} = (\omega^1, \omega^2) \in \tilde{\Omega}$. The *diagonal* of $\tilde{S}$ is $\Delta = \{(x, x): x \in S\}$. Because $\tilde{S}$ is polish, $\Delta \in \tilde{\mathscr{S}}$ and $\tilde{S} - \Delta$ is a countable union of rectangles. According to the usual extension theorems, a unique probability measure $\tilde{\mu}$ on $(\tilde{S}, \tilde{\mathscr{S}})$ is determined by any consistent prescription on $\mathscr{S} \times \mathscr{S}$, and any consistent prescription on cylinders of $\tilde{\Omega}$ extends to a unique probability measure $\tilde{P}$ on $(\tilde{\Omega}, \tilde{\mathscr{B}})$. Also $\tilde{\mu}(\Delta) = 1$ if and only if $\tilde{\mu}(E^1, E^2) = 0$ whenever $E^1, E^2 \in \mathscr{S}$ and $E^1 \cap E^2 = \varnothing$. The *hitting time to the diagonal* is $\tau_\Delta(\tilde{\omega}) = \min\{n \in \mathbb{N}: \tilde{\omega}_n \in \Delta\}$ $(= \infty$ if $\tilde{\omega}_n \notin \Delta$ for all $n)$. We will make repeated use of the partition:

$$\begin{aligned} \tilde{\Omega} = \tilde{\Omega}^* \qquad & (= \{\tilde{\omega}: \tau_\Delta(\tilde{\omega}) < \infty \text{ and } \tilde{\omega}_n \in \Delta \text{ for all } n > \tau_\Delta(\tilde{\omega})\}) \\ + \tilde{\Omega}^0 \qquad & (= \{\tilde{\omega}: \tau_\Delta(\tilde{\omega}) < \infty \text{ and } \tilde{\omega}_n \notin \Delta \text{ for some } n > \tau_\Delta(\tilde{\omega})\}) \\ + \tilde{\Omega}^\infty \qquad & (= \{\tilde{\omega}: \tau_\Delta(\tilde{\omega}) = \infty\}). \end{aligned}$$

Now let $(P_\iota)_{\iota \in \mathscr{P}}$ be the collection of Markov measures on $(\Omega, \mathscr{B})$ with a given t.p.s. (π^m), and let (ξ_n) be the coordinate process. A *coupling* for (ξ_n) is a collection of measures $\tilde{P}^m_{(\mu,\nu)}$, $\mu, \nu \in \mathscr{P}$, $m \in \mathbb{N}$, on $(\tilde{\Omega}, \tilde{\mathscr{B}})$ which satisfies

$$\tilde{P}^m_{(\mu,\nu)}(\cdot, \Omega) = P_\mu^{\ m}(\cdot), \qquad \tilde{P}^m_{(\mu,\nu)}(\Omega, \cdot) = P_\nu^{\ m}(\cdot), \tag{8}$$

and

$$\tilde{P}^m_{(\mu,\nu)}(\tilde{\Omega}^0) = 0. \tag{9}$$

In this context, $(\tilde{\xi}_n)$ is called the *coupled process* for (ξ_n) (corresponding to the coupling $(\tilde{P}^m_{(\mu,\nu)})$). Condition (8) states that when $(\tilde{\xi}_n)$ is governed by $\tilde{P}^m_{(\mu,\nu)}$, the marginal processes $(\xi_n^{\ 1})$ and $(\xi_n^{\ 2})$ are copies of the given Markov process (ξ_n) starting from initial distributions μ and ν, respectively; thus we may think of $(\tilde{\xi}_n)$ as two simultaneously evolving copies of (ξ_n). The property (9) requires that $(\tilde{\xi}_n)$ remain on the diagonal Δ at all times after τ_Δ; we do not, however, rule out the possibility that τ_Δ might be infinite.

The coupling method for Markov processes is based on the next key inequality.

LEMMA 2. *If* $(\tilde{P}^m_{(\mu,\nu)})$ *is a coupling for* (ξ_n), *then*

$$\tilde{P}^m_{(\mu,\nu)}(\tilde{\Omega}^*) \leqslant 1 - \gamma_m(\mu, \nu).$$

Proof. If $B \in \mathscr{B}_n{}^\infty$, then (9) implies that $\{\tau_\Delta > n\} \supset \{(B, B^c)\}$ modulo a $\tilde{P}^m_{(\mu,\nu)}$-null set for each fixed $\mu, \nu \in \mathscr{P}$; $m, n \in \mathbb{N}$. Hence (8) yields

$$\tilde{P}^m_{(\mu,\nu)}(\tau_\Delta > n) \geqslant \sup_{B \in \mathscr{B}_n{}^\infty} \{P_\mu{}^m(B) - P_\nu{}^m(B)\} = \gamma_m^{m+n}(\mu, \nu). \tag{10}$$

Let $n \to \infty$ to complete the proof. ∎

Our next two results show how couplings provide information about the atomic structure of $\mathscr{T}$. The first gives a sufficient condition for λ-loss of memory as formulated in the previous section, while the second identifies the decomposition (1) explicitly in an important special case. Examples of couplings will constitute the remainder of this first section, and subsequent sections will contain numerous applications of the coupling method.

THEOREM 1. *Let* $(\tilde{P}^m_{(\mu,\nu)})$ *be a coupling for* (ξ_n). *If for some* $\iota \in \mathscr{P}$ *and* $\lambda \in (0, 1]$, *one has*

$$\tilde{P}^m_{(\xi_m{}^1, \iota^m)}(\tilde{\Omega}^*) \geqslant \lambda \quad \text{a.s.} \qquad \text{for every} \quad m \in \mathbb{N}, \tag{11}$$

then λ*-mixing,* λ*-ergodicity,* λ*-constant space–time harmonics, and a* 0–λ *law all hold for* (ξ_n) *w.r.t.* ι.

Proof. Assuming (11), Lemma 2 shows that $\lambda \leqslant \tilde{P}^m_{(\xi_m{}^1, \iota^m)}(\tilde{\Omega}) \leqslant 1 - \gamma_m(\xi_m, \iota^m)$ a.s. for each m, so (3) holds. The remainder follows from Proposition 3. ∎

THEOREM 2. *Let* I *be a finite or countable index set. Suppose there are sets* $E_r{}^n$; $r \in I$, $n \in \mathbb{N}$, *such that*:

(i) *for each fixed* n, $\{E_r{}^n; r \in I\}$ *is a partition of* S,
(ii) $\pi^n(x, E_r^{n+1}) = 1$ *for all* $x \in E_r{}^n$, *and*
(iii) $\tilde{P}^m_{(x,y)}(\tilde{\Omega}^*) = 1$ *whenever* $x, y \in E_r{}^m$ *for some* r.

Then for each $\iota \in \mathscr{P}$, *the* P_ι*-atomic decomposition of* $\mathscr{T}$ *is equivalent to*

$$\Omega = \sum_{r:\, \iota(E_r{}^0) > 0} \{\xi_0 \in E_r{}^0\} \qquad (\text{mod } P_\iota\text{-null sets}). \tag{12}$$

Thus, if $\iota(E_r{}^0) = 1$ *for some* r, *then* $\mathscr{T}$ *is* P_ι*-trivial. In particular, if* $\tilde{P}^m_{(x,y)}(\Omega^*) = 1$ *for every* $x, y \in S$, *then* $\mathscr{T}$ *is* P_ι*-trivial w.r.t. every* $\iota \in \mathscr{P}$.

Proof. Suppose first that $\iota(E_r^0) = 1$ for some r. To show $\mathscr{T}$ P_ι-trivial, it suffices by Proposition 3 to check that $\gamma_m(\xi_m, \iota^m) = 0$ a.s. for each m. But by Fatou, and since (ii) implies that $\iota^m(E_r^m) = 1$,

$$\gamma_m(\xi_m, \iota^m) \leqslant \int_S \gamma_m(\xi_m, y)\iota^m(dy) = \int_{E_r^m} \gamma_m(\xi_m, y)\iota^m(dy).$$

To complete the check we note that $\xi_m \in E_r^m$ a.s., again by (ii), while (iii) and Lemma 2 imply that $\gamma_m(x, y) = 0$ for all $x, y \in E_r^m$. When $x, y \in E_r^0$, taking ι to be the mixture $\frac{1}{2}x + \frac{1}{2}y$ shows that for any $B \in \mathscr{T}$, $P_x(B) = P_y(B) = 0$ *or* 1. It follows that for arbitrary $\iota \in \mathscr{P}$, $B \in \mathscr{T}$,

$$P_\iota(B) = \int_{x:\, P_x(B)=1} P_x(B)\iota(dx) = \sum_{\substack{r:\, P_x(B)=1 \\ \forall x \in E_r^0, \\ \iota(E_r^0) > 0}} P_\iota(\xi_0 \in E_r^0).$$

Thus B is equivalent to a sum of events from the partition (12), and if $B \subset \{\xi_0 \in E_r^0\}$ then $P_\iota(B) = 0$ or $P_\iota(\xi_0 \in E_r^0)$. Moreover, (i) and (ii) imply that $\{\xi_0 \in E_r^0\} \cong \{\xi_n \in E_r^n \text{ ev.}\} \in \mathscr{T}$, so that (12) is indeed an atomic decomposition. This completes the proof except for the last statement, which is immediate if we take $I = \{0\}$ and $E_0^m = S$ for all m. ▮

We remark that under the hypotheses of Theorem 2, the E_r^n such that $P_\iota(E_r^0) > 0$ satisfy the requirements of the E_r^n guaranteed by Proposition 1.

1.5. *Markovian Couplings*

Let (ξ_m) be a Markov process on $(\Omega, \mathscr{B})$ with t.p.s. (π^m). In this section we introduce three examples of coupled processes for (ξ_n) which are themselves Markov processes on the state space $\tilde{S}$. Such couplings are completely determined by a t.p.s. $(\tilde{\pi}^m)$, $\tilde{\pi}^m : \tilde{S} \times \tilde{\mathscr{S}} \to [0, 1]$. Moreover, $\tilde{\pi}^m$ is uniquely determined by consistent prescriptions for

$$\{\tilde{\pi}^m((x, y), (E^1, E^2)) : x, y \in S, E^1, E^2 \in \mathscr{S}\}.$$

According to (8) and (9) these must satisfy

$$\tilde{\pi}^m((x, y), (E, S)) = \pi^m(x, E), \qquad \tilde{\pi}^m((x, y), (S, E)) = \pi^m(y, E) \tag{13}$$

and

$$\tilde{\pi}^m((x, x), (E^1, E^2)) = \pi^m(x, E^1 \cap E^2). \tag{14}$$

Given such a $(\tilde{\pi}^m)$, if we choose $\tilde{P}^m_{(\mu,\nu)}$ to be the Markov measure on $(\tilde{\Omega}, \tilde{\mathscr{B}})$ with t.p.s. $(\tilde{\pi}^{m+n})_{n \in \mathbb{N}}$ and initial distribution $\mu \times \nu$, then $(\tilde{P}^m_{(\mu,\nu)})$ is a coupling for (ξ_n).

(i) *The classical coupling.* Let $(\tilde{\xi}_n)$ be a Markov process on $(\tilde{\Omega}, \tilde{\mathscr{B}})$ with t.p.s. $(\tilde{\pi}^m)$ satisfying (14) and

$$\tilde{\pi}^m((x, y), (E^1, E^2)) = \pi^m(x, E^1)\pi^m(y, E^2), \qquad x \neq y.$$

It is a simple matter to check that cylinder prescriptions are consistent, satisfy (13), and hence induce a coupled process $(\tilde{\xi}_n)$. The coupled process evolves as two independent copies of (ξ_n) until these copies reach a common state; thereafter the marginal processes use the *same* transition mechanism. This was surely the first known coupling, dating back as far as the 1938 paper by Doeblin discussed in the introduction; we call it the *classical coupling*.

(ii) *Ornstein's coupling.* Let (ξ_n) be a random walk on $\mathbb{R}$ with $\pi^m(x, E) \equiv \pi(E - x)$, and let K be a positive real number. Here is an intuitive description of a Markovian coupling for (ξ_n): To determine the progress of $(\tilde{\xi}_n)$ from time m to time $m + 1$ we observe the difference in the displacement of two independent copies of our original process (ξ_n) at this step. If the absolute value of this difference does not exceed K we let the copies move to their new states, while if the difference exceeds K we instead translate both processes by the *same* value taken on by one of the two independent copies. These heuristics lead to the *Ornstein coupling*, with t.p.s. $(\tilde{\pi}^m)$ determined by (14) and

$$\begin{aligned}\tilde{\pi}^m((x, y), (x + E^1, y + E^2)) &= \iint\limits_{\substack{w \in E^1,\, z \in E^2: \\ |w - z| \leqslant K}} \pi(dw)\pi(dz) \\ &\quad + \int_{w \in E^1 \cap E^2} \left\{\pi(dw)\left[1 - \int_{z \in \mathbb{R}: |w - z| \leqslant K} \pi(dz)\right]\right\}, \qquad x \neq y. \qquad (15)\end{aligned}$$

Once again, a straightforward check shows that these specifications give rise to a t.p.s. $(\tilde{\pi}^m)$ satisfying (13), hence to a well-defined coupling. The key feature of the construction is that until τ_Δ, the difference $\xi_n{}^2 - \xi_n{}^1$ in the marginals is a random walk with bounded symmetric increments. This fact will be exploited in Section 3.

(iii) *Vasershtein's coupling.* In this example $(\tilde{\xi}_n)$ is a Markov process on $(\tilde{\Omega}, \tilde{\mathscr{B}})$ with $(\tilde{\pi}^m)$ determined by

$$\begin{aligned}\tilde{\pi}^m((x, y), (E^1, E^2)) &= ((\pi^m(x, \cdot) \wedge \pi^m(y, \cdot))(E^1 \cap E^2) \\ &\quad + \frac{(\pi^m(x, \cdot) - \pi^m(y, \cdot))^+(E^1)(\pi^m(y, \cdot) - \pi^m(x, \cdot))^+(E^2)}{\delta_m^{m+1}(x, y)} \qquad (16)\end{aligned}$$

(omit the last term if $\delta_m^{m+1}(x, y) = 0$). Once again, straightforward computations show that (16) leads to a coupling, the version for Markov processes of a more general construction due to Vasershtein; call it the *Vasershtein coupling*. Note that the classical coupled process is "weaker" than the Vasershtein coupled process since the former moves from $(x, y) \notin \Delta$ to Δ with probability $\int \pi(x, \{z\})\pi(y, dz)$, while under the latter the probability of this transition is $\int (\pi(x, \cdot) \wedge \pi(y, \cdot))(dz)$, and the second integral dominates the first. In fact, the Vasershtein coupling is characterized by the property that at each step it sends us to the diagonal as efficiently as any *Markovian* coupling can.

Numerous applications of the three Markovian couplings just described will be presented in Sections 2 and 3.

1.6. *The Maximal Coupling*

Equations (7) and (10) combined show that

$$\tilde{P}^m_{(\mu,\nu)}(\tau_\Delta > n) \geqslant \gamma_m^{m+n}(\mu, \nu) \geqslant \delta_m^{m+n}(\mu, \nu)$$

for any coupling. Our objective in this section is to construct, for an arbitrary Markov process (ξ_n), a coupling which satisfies these relations with equality. This will be called the *maximal coupling*, since it sends us to the diagonal Δ as rapidly as *any* coupling. The construction is tedious because $(\tilde{\xi}_n)$ is necessarily non-Markovian. Once we have shown existence, though, the maximality will be used to advantage in applications (cf. Section 2).

For notational convenience, set

$$\alpha_m{}^n(\mu, \nu) = (p_{mn}(\mu, \cdot) \wedge p_{mn}(\nu, \cdot))(S) = 1 - \delta_m{}^n(\mu, \nu),$$
$$\alpha_m(\mu, \nu) = \lim_{n \to \infty} \alpha_m{}^n(\mu, \nu) = 1 - \delta_m(\mu, \nu).$$

THEOREM 3. *Let (ξ_n) be any Markov process on $(\Omega, \mathscr{B})$. There is a maximal coupling $(\tilde{P}^m_{(\mu,\nu)})$ for (ξ_n) such that*

$$\tilde{P}^m_{(\mu,\nu)}(\tilde{\xi}_n \in (E^1, E^2), \tau_\Delta \leqslant n) = (p_{m\,m+n}(\mu, \cdot) \wedge p_{m\,m+n}(\nu, \cdot))(E^1 \cap E^2), \quad (17)$$
$$m, n \in \mathbb{N}, \quad \mu, \nu \in \mathscr{P}, \quad E^1, E^2 \in \mathscr{S}.$$

In particular,

$$\tilde{P}^m_{(\mu,\nu)}(\tau_\Delta \leqslant n) = \alpha_m^{m+n}(\mu, \nu). \quad (18)$$

Proof. We will construct $\tilde{P}^0_{(\mu,\nu)}$; shifting by m and applying the same procedure yields $\tilde{P}^m_{(\mu,\nu)}$. Thus, put $m = 0$, fix $\mu, \nu \in \mathscr{P}$, and to simplify notation write the desired $\tilde{P}^0_{(\mu,\nu)}$ simply as $\tilde{P}$. By the Kolmogorov extension theorem, it suffices to prescribe consistent values of $\tilde{P}$ on cylinders of the form:

$$\{\xi_0 \in (E_0{}^1, E_0{}^2), \xi_1 \in (E_1{}^1, E_1{}^2), \dots, \xi_n \in (E_n{}^1, E_n{}^2)\}, \qquad n \in \mathbb{N}, \quad (19)$$

and check that they satisfy (8), (9), and (17). For brevity's sake we adopt the following notational conventions throughout the proof. The symbol $E_n{}^1$ will be reserved for a Borel set in the first coordinate of $\tilde{\mathscr{S}}$ at time n, $E_n{}^2$ for the second coordinate at time n. Abbreviate:

$$[(E_n{}^1, E_n{}^2), (E^1_{n'}, E^2_{n'})] \quad \text{for} \quad (E_n{}^1, E_n{}^2), (E^1_{n+1}, E^2_{n+1}), \dots, (E^1_{n'}, E^2_{n'}),$$
$$[E_n{}^1, E^1_{n'}] \quad \text{for} \quad E_n{}^1, E^1_{n+1}, \dots, E^1_{n'},$$

$E_k{}^1, E_k{}^2 \in \mathscr{S}$, and so forth. Any event of the form $\{\tilde{\xi}_{n_k} \in (E^1_{n_k}, E^2_{n_k}); 1 \leqslant k \leqslant l\}$ will be written simply as $\{(E^1_{n_k}, E^2_{n_k}); 1 \leqslant k \leqslant l\}$ and similarly for marginal events. Thus, for example, the $\tilde{P}$ measure of the event (19) will be denoted as $\tilde{P}([(E_0{}^1, E_0{}^2), (E_n{}^1, E_n{}^2)])$.

The required $\tilde{P}$ will now be built quite formally (some motivation, in the Markov chain case, may be found in [15]). First, for each $n \in \mathbb{N}$ define a measure $\psi_n{}^1$ on $(\tilde{S}, \tilde{\mathscr{S}})$ by means of its specification on the semi-algebra $\mathscr{S} \times \mathscr{S}$ as

$$\psi_n{}^1(dx, dy) = (p_n(\mu, \cdot) - p_n(\nu, \cdot))^+(dx)\pi^n(x, dy). \tag{20}$$

Next, let $y \mapsto \theta_n{}^1(E, y)$ be a regular conditional probability distribution for $\psi_n{}^1$ w.r.t. the map $(x, y) \mapsto x$, i.e., versions

$$\theta_n{}^1(E, y) = \frac{d\psi_n{}^1(E, \cdot)}{d\psi_n{}^1(S, \cdot)}(y) \tag{21}$$

of the Radon–Nikodym derivatives such that $\theta_n{}^1(\cdot, y) \in \mathscr{P}$ for all $y \in S$ and $y \mapsto \theta_n{}^1(E, y)$ is $\mathscr{S}$-measurable for each $E \in \mathscr{S}$. Now for $n \in \mathbb{N}$, define $\rho_n{}^1 : \mathscr{S}^{n+1} \times S \to [0, 1]$ inductively by putting $\rho_0{}^1(dx, y) = \theta_1{}^1(dx, y)$, and

$$\rho^1_{n+1}([E_0, E_n], dx, y) = \rho_n{}^1([E_0, E_n], x)\theta^1_{n+1}(dx, y). \tag{22}$$

Define $\psi_n{}^2$ by switching μ and ν in (20), and obtain $\theta_n{}^2$ and $\rho_n{}^2$ by replacing 1 with 2 in the superscripts of (21) and (22). It follows that for every n and fixed y, $\rho_n{}^j([E_0, E_n], y)$ $(j = 1, 2)$ are countably additive in each coordinate k $(0 \leqslant k \leqslant n)$, and equal 1 if $E_k = S$ for all k.

Specify $\tilde{P}$ on cylinders restricted to $\{\tau_\Delta = n\}$, $n \in \mathbb{N}$, as

$$\begin{aligned}\tilde{P}([(E_0{}^1, E_0{}^2), (E^1_{n'}, E^2_{n'})], \tau_\Delta = 0)\\ = \int_{E_0{}^1 \cap E_0{}^2} (\mu \wedge \nu)(dx) P_x([E_1{}^1 \cap E_1{}^2, E^1_{n'} \cap E^2_{n'}]), \qquad n' \geqslant 1, \quad (23)\end{aligned}$$

and for $n' > n+1$,

$$\begin{aligned}
&\tilde{P}([(E_0{}^1, E_0{}^2), (E^1_{n'}, E^2_{n'})], \tau_\Delta = n+1) \\
&\quad = \int_{E^1_{n+1} \cap E^2_{n+1}} (\psi_n{}^1(S, \cdot) \wedge (\psi_n{}^2(S, \cdot))(dx)\rho_n{}^1([E_0{}^1, E_n{}^1], x) \\
&\qquad \times \rho_n{}^2([E_0{}^2, E_n{}^2], x) P_x^{n+1}([E^1_{n+2} \cap E^2_{n+2}, E^1_{n'} \cap E^2_{n'}]).
\end{aligned} \tag{24}$$

We claim that (17) and the following two properties are satisfied:

$$\tilde{P}([E_0{}^1, E_n{}^1], \tau_\Delta < \infty) \leqslant P_\mu([E_0{}^1, E_n{}^1])$$

and (25)

$$\tilde{P}([E_0{}^2, E_n{}^2], \tau_\Delta < \infty) \leqslant P_\nu([E_0{}^2, E_n{}^2]);$$

If $n' \geqslant n$, then

$$\tilde{P}((E^1_{n'}, E^2_{n'}), \tau_\Delta = n) = 0 \qquad \text{whenever} \quad E^1_{n'} \cap E^2_{n'} = \varnothing. \tag{26}$$

The proof of (17) is by induction. For $n = 0$ the claim follows from (23). Assuming (17) for n we have

$$\begin{aligned}
&\tilde{P}((E^1_{n+1}, E^2_{n+1}), \tau_\Delta \leqslant n+1) \\
&\quad = \tilde{P}((E^1_{n+1}, E^2_{n+1}), \tau_\Delta \leqslant n) + \tilde{P}((E^1_{n+1}, E^2_{n+1}), \tau_\Delta = n+1) \\
&\quad = \qquad \tilde{P}_1 \qquad + \qquad \tilde{P}_2.
\end{aligned}$$

For $k \leqslant n$ we see from (24) that

$$\tilde{P}((E^1_{n+1}, E^2_{n+1}), \tau_\Delta = k) = \int_{x \in S} \tilde{P}(\xi_n{}^1 \in dx, \tau_\Delta = k)\pi^n(x, E^1_{n+1} \cap E^2_{n+1}),$$

and hence, using the induction hypothesis,

$$\tilde{P}_1 = \int_{x \in S} (p_n(\mu, \cdot) \wedge p_n(\nu, \cdot))(dx)\pi^n(x, E^1_{n+1} \cap E^2_{n+1}).$$

By (24),

$$\tilde{P}_2 = (\psi_n{}^1(S, \cdot) \wedge \psi_n{}^2(S, \cdot))(E^1_{n+1} \cap E^2_{n+1}).$$

The observation that

$$\begin{aligned}
&(p_{n+1}(\mu, \cdot) \wedge p_{n+1}(\nu, \cdot)(dy) \\
&\quad = (\psi_n{}^1(S, \cdot\,) \wedge \psi_n{}^2(S, \cdot\,))(dy) + \int_{x \in S} (p_n(\mu, \cdot) \wedge p_n(\nu, \cdot)(dx)\pi^n(x, dy),
\end{aligned}$$

and integration over $y \in E^1_{n+1} \cap E^2_{n+1}$, establishes the claim for $n+1$. The verification of (25) is more involved. First, we show inductively that for all $n \in \mathbb{N}$,

$$\begin{aligned}
&\tilde{P}([E_0{}^1, E^1_{n-1}], \xi_n{}^1 \in dx, \tau_\Delta \leqslant n) \\
&\quad = P_\mu([E_0{}^1, E^1_{n-1}], \xi_n \in dx) - (p_n(\mu, \cdot) - p_n(\nu, \cdot))^+(dx)\rho^1_{n-1}([E_0, E_{n-1}], x)
\end{aligned}$$

(by convention, omit $[E_0{}^1, E_{n-1}^1]$ if $n = 0$, and set $\rho_{-1}^1(x) = 1$). For $n = 0$ this is clear from (23). Assume (27) for n, and write

$$\begin{aligned}
&\tilde{P}([E_0{}^1, E_{n+1}^1], \tau_\Delta \leqslant n+1) \\
&\quad = \tilde{P}([E_0{}^1, E_{n+1}^1], \tau_\Delta \leqslant n) + \tilde{P}([E_0{}^1, E_{n+1}^1], \tau_\Delta = n+1) \\
&\quad = \qquad \tilde{P}_3 \qquad + \qquad \tilde{P}_4 .
\end{aligned}$$

We compute

$$\begin{aligned}
\tilde{P}_3 &= \int_{x \in E_n{}^1} \tilde{P}([E_0{}^1, E_{n-1}^1], \xi_n{}^1 \in dx, \tau_\Delta \leqslant n)\pi^n(x, E_{n+1}^1) && ((23) \text{ and } (24)) \\
&= \int_{x \in E_n{}^1} [P_\mu([E_0{}^1, E_{n-1}^1], \xi_n \in dx) - (p_n(\mu, \cdot) \\
&\quad - p_n(\nu, \cdot))^+(dx)\rho_{n-1}^1([E_0{}^1, E_{n-1}^1], x)] \cdot \pi^n(x, E_{n+1}^1) \\
& && (\text{induction hypothesis}) \\
&= P_\mu([E_0{}^1, E_{n+1}^1]) - \int_{x \in E_n{}^1} \psi_n{}^1(dx, E_{n+1}^1)\rho_{n-1}^1([E_0{}^1, E_{n-1}^1], x) && ((20)).
\end{aligned}$$

The last integral may be rewritten as

$$\begin{aligned}
&\int_{x \in E_n{}^1} \int_{y \in E_{n+1}^1} \psi_n{}^1(S, dy)\theta_n{}^1(dx, y)\rho_{n-1}^1([E_0{}^1, E_{n-1}^1], x) && ((21)) \\
&\quad = \int_{y \in E_{n+1}^1} \psi_n{}^1(S, dy)\rho_n{}^1([E_0{}^1, E_n{}^1], y) && (\text{Fubini and } (22)).
\end{aligned}$$

Also, by (24),

$$\tilde{P}_4 = \int_{y \in E_{n+1}^1} (\psi_n{}^1(S, \cdot) \wedge \psi_n{}^2(S, \cdot))(dy)\rho_n{}^1([E_0{}^1, E_n{}^1], y).$$

Combining calculations,

$$\begin{aligned}
\tilde{P}_3 + \tilde{P}_4 &= P_\mu([E_0{}^1, E_{n+1}^1]) - \int_{y \in E_{n+1}^n} (\psi_n{}^1(S, \cdot) \\
&\quad - \psi_n{}^2(S, \cdot))^+(dy)\rho_n{}^1([E_0{}^1, E_n{}^1], y) \\
&= P_\mu([E_0{}^1, E_{n+1}^1]) - \int_{y \in E_{n+1}^n} (p_{n+1}(\mu, \cdot) \\
&\quad - p_{n+1}(\nu, \cdot))^+(dy)\rho_n{}^1([E_0{}^1, E_n{}^1], y),
\end{aligned}$$

this last since

$$\begin{aligned}
&\psi_n{}^1(S, dy) - \psi_n{}^2(S, dy) \\
&\quad = (p_{n+1}(\mu, dy) - \int_{x \in S} (p_n(\mu, \cdot) \wedge p_n(\nu, \cdot))(dx)\pi^n(x, dy)) \\
&\qquad - (p_{n+1}(\nu, dy) - \int_{x \in S} (p_n(\mu, \cdot) \wedge p_n(\nu, \cdot))(dx)\pi^n(x, dy)).
\end{aligned}$$

We have proved (27), which implies that

$$\tilde{P}([E_0{}^1, E_n{}^1], \tau_\Delta \leqslant n) \leqslant P_\mu([E_0{}^1, E_n{}^1]), \qquad n \in \mathbb{N}.$$

But for $n' > n$,

$$\tilde{P}([E_0{}^1, E_n{}^1], \tau_\Delta \leqslant n') = \tilde{P}([E_0{}^1, E_n{}^1][S_{n+1}, S_{n'}], \tau_\Delta \leqslant n') \leqslant P_\mu([E_0{}^1, E_n{}^1]).$$

Letting $n' \to \infty$ yields the first inequality of (25). Since an analogous argument clearly applies to the second marginal, the verification of (25) is complete. Property (26) is immediate from (23) and (24).

To finish the construction of $\tilde{P}$, note that by (17), $\tilde{P}(\tau_\Delta < \infty) = \alpha_0(\mu, \nu) \leqslant 1$. If $\tilde{P}(\tau_\Delta < \infty) = 1$ we must have equality in (25) in all cases. Then since (23) and (24) are consistent on $(\mathscr{S} \times \mathscr{S})^{\mathbb{N}}$, the unique extension to $\tilde{P}$ on $(\tilde{\Omega}, \tilde{\mathscr{B}})$ satisfies (8) and (9) as a consequence of (25) and (26), respectively. If $\tilde{P}(\tau_\Delta < \infty) < 1$, we put

$$\tilde{P}([(E_0{}^1, E_0{}^2), (E_n{}^1, E_n{}^2)], \tau_\Delta = \infty)$$
$$= \frac{[P_\mu([E_0{}^1, E_n{}^1]) - \tilde{P}([E_0{}^1, E_n{}^1], \tau_\Delta < \infty)] \times [P_\nu([E_0{}^2, E_n{}^2]) - \tilde{P}([E_0{}^2, E_n{}^2], \tau_\Delta < \infty)]}{1 - \tilde{P}(\tau_\Delta < \infty)}.$$

Then $\tilde{P}(\Omega^\infty) = 1 - \tilde{P}(\tau_\Delta < \infty)$, $\tilde{P}([E_0{}^1, E_n{}^1], \tau_\Delta = \infty) = P_\mu([E_0{}^1, E_n{}^1]) - \tilde{P}([E_0{}^1, E_n{}^1], \tau_\Delta < \infty)$, and similarly for the second marginal. Again, the cylinder specifications determined by (23), (24) and (28) extend to a coupling for $\tilde{P}$ which satisfies (17). This completes the proof of the theorem. ∎

COROLLARY 2. *Let (ξ_n) be a Markov process on $(\Omega, \mathscr{B})$. Then*

$$\gamma_m{}^n(\mu, \nu) = \delta_m{}^n(\mu, \nu) \qquad \text{for all} \quad \mu, \nu \in \mathscr{P}, \quad m \leqslant n.$$

Proof. As already mentioned, if $(\tilde{P}^m_{(\mu,\nu)})$ is the maximal coupling for (ξ_n), we find that

$$\delta_m^{m+n}(\mu, \nu) = 1 - \alpha_m^{m+n}(\mu, \nu) = \tilde{P}^m_{(\mu,\nu)}(\tau_\Delta > n)$$
$$\geqslant \gamma_m^{m+n}(\mu, \nu) \geqslant \delta_m^{m+n}(\mu, \nu)$$

by combining (7), (10), and (18). ∎

2. APPLICATIONS TO NONHOMOGENEOUS PROCESSES

2.1. *A Structure Theorem for $\mathscr{T}$*

In the first three subsections of this section we use maximal coupling to study asymptotic properties of general nonhomogeneous Markov processes. Our immediate objective is a means of determining the atomic structure of $\mathscr{T}$.

Recall the definition of $H^{\pm}_{mn}(\mu, \nu)$ from the end of Section 1.1, and introduce

$$\begin{aligned} C_m{}^{+}(\mu, \nu) &= \{\omega : \omega_n \in H^{+}_{m\,m+n}(\mu, \nu) \text{ i.o.}\} \in \mathscr{T}, \\ C_m{}^{-}(\mu, \nu) &= \{\omega : \omega_n \in H^{-}_{m\,m+n}(\mu, \nu) \text{ i.o.}\} \in \mathscr{T}, \\ \tilde{C}_m{}^{1}(\mu, \nu) &= \{\tilde{\omega} : \omega_n{}^{1} \in H^{-}_{m\,m+n}(\mu, \nu) \text{ i.o.}\} \in \tilde{\mathscr{B}}, \\ \tilde{C}_m{}^{2}(\mu, \nu) &= \{\tilde{\omega} : \omega_n{}^{2} \in H^{+}_{m\,m+n}(\mu, \nu) \text{ i.o.}\} \in \tilde{\mathscr{B}}; \end{aligned}$$

$m \in \mathbb{N}$, $\mu, \nu \in \mathscr{P}$ ("i.o." abbreviates "for infinitely many n"). These events will play a central role in the derivation. Until further notice (ξ_n) is a Markov process on $(\Omega, \mathscr{B})$ and $(\tilde{P}^m_{(\mu,\nu)})$ is the *maximal coupling* for (ξ_n).

PROPOSITION 1. *For any $\mu, \upsilon \in \mathscr{P}$ and $m \in \mathbb{N}$,*

$$\tilde{P}^m_{(\mu,\nu)}(\tau_\Delta < \infty) \geqslant P^m(C_m{}^{-}(\mu, \nu)) \vee P^m(C_m{}^{+}(\mu, \nu)).$$

Proof. By properties (1.9) and (1.17) of the maximal coupling,

$$\begin{aligned} \tilde{P}^m_{(\mu,\nu)}(\tilde{C}_m{}^{1}(\mu, \nu), \tau_\Delta = \infty) &\leqslant \sum_{n>m} \tilde{P}^m_{(\mu,\nu)}(\xi^1_{n-m} \in H^{-}_{mn}(\mu, \nu), \tau_\Delta = \infty) \\ &\leqslant \sum_{n>m} \tilde{P}^m_{(\mu,\nu)}(\xi^1_{n-m} \in H^{-}_{mn}(\mu, \nu), \tau_\Delta > n-m) \\ &= \sum_{n>m} (p_{mn}(\mu, \cdot) - p_{mn}(\nu, \cdot))^{+}(H^{-}_{mn}(\mu, \nu)) = 0. \end{aligned}$$

Thus $\tilde{C}_m{}^{1}(\mu, \nu) \subset \{\tau_\Delta < \infty\}$ modulo a $\tilde{P}^m_{(\mu,\nu)}$-null set. Similarly, $\tilde{C}_m{}^{2}(\mu, \nu) \subset \{\tau_\Delta < \infty\}$ $\tilde{P}^m_{(\mu,\nu)}$-almost surely. But by (1.8), $\tilde{P}^m_{(\mu,\nu)}(\tilde{C}_m{}^{1}(\mu, \nu)) = P_\mu{}^m(C_m{}^{-}(\mu, \nu))$ and $\tilde{P}^m_{(\mu,\nu)}(\tilde{C}_m{}^{2}(\mu, \nu)) = P_\nu{}^m(C_m{}^{+}(\mu, \nu))$. The claim follows. ∎

PROPOSITION 2. *If A_r is a P_ι-atom of $\mathscr{T}$, then*

$$\liminf_{m \to \infty} \tilde{P}^m_{(\xi_m, \iota^m)}(\tau_\Delta < \infty) \geqslant P_\iota(A_r) \quad P_\iota\text{-a.s.} \qquad \textit{on} \quad A_r.$$

Proof. In our customary notation, put $C_m{}^{+}(\xi_m(\omega), \iota^m) = C_m{}^{+}(x, \iota^m)$ when $\omega \in \{\xi_m = x\}$, similarly for $C_m{}^{-}$, and to simplify matters, temporarily abbreviate $C_m{}^{\pm}(\omega) = C_m{}^{\pm}(\xi_m(\omega), \iota^m)$. Fixing A_r, set

$$\begin{aligned} \mathbb{N}_0(\omega) &= \{m \in \mathbb{N} : P_\iota(A_r \cap \theta^{-m} C_m{}^{-}(\omega)) = 0\}, \\ \mathbb{N}_1(\omega) &= \{m \in \mathbb{N} : P_\iota(A_r \cap \theta^{-m} C_m{}^{-}(\omega)) = P_\iota(A_r)\}. \end{aligned}$$

$\mathbb{N}_0(\omega) \cup \mathbb{N}_1(\omega) = \mathbb{N}$ for each ω since $C_m{}^{-}(\omega) \in \mathscr{T}$ and A_r is an atom of $\mathscr{T}$. If $m \in \mathbb{N}_0(\omega)$, then $P_\iota(A_r \cap \theta^{-m} C_m{}^{+}(\omega)) = P_\iota(A_r)$ because $\theta^{-m} C_m{}^{+}(\omega) \cup \theta^{-m} C_m{}^{-}(\omega) = \Omega$ for all ω; hence

$$P^m_{\iota^m}(C_m{}^{+}(\omega)) \geqslant P_\iota(A_r), \qquad m \in \mathbb{N}_0(\omega). \tag{1}$$

If $m \in \mathbb{N}_1(\omega)$, then $P^m_{\xi_m(\omega)}(\theta^m A_r \cap C_m^-(\omega)) = P_\iota(A_r \cap \theta^{-m} C_m^-(\omega) | \mathscr{B}_0{}^m)(\omega)$. By martingale convergence this last term tends to 1 as $m \to \infty$, $m \in \mathbb{N}_1(\omega)$, for P_ι-almost all $\omega \in A_r$. Thus

$$\lim_{\substack{m\to\infty \\ m \in \mathbb{N}_1(\omega)}} P^m_{\xi_m(\omega)}(C_m^-(\omega)) = 1 \tag{2}$$

for P_ι-almost all $\omega \in A_r$ such that $\mathbb{N}_1(\omega)$ is infinite. Together, (1) and (2) show that

$$\liminf_{m\to\infty} P^m_{\xi_m}(C_m^-) \vee P^m_{\iota^m}(C_m^+) \geqslant P_\iota(A_r) \quad P_\iota\text{-a.s.} \quad \text{on} \quad A_r.$$

An application of Proposition 1 with $\mu = \iota^m$ and $v = x$ on $\{\xi_m = x\}$ completes the proof. ∎

We now state and prove the structure theorem for $\mathscr{T}$.

THEOREM 1. *$\alpha(\omega) = \lim_{m\to\infty} \alpha_m(\xi_m(\omega), \iota^m)$ exists P_ι-almost surely, and if (1.1) is the P_ι-atomic decomposition of $\mathscr{T}$, then*

$$\alpha(\omega) = \begin{cases} 0 & P_\iota\text{-a.s.} \quad \text{on} \quad F \\ P_\iota(A_r) & P_\iota\text{-a.s.} \quad \text{on} \quad A_r \end{cases}.$$

Proof. Write $\bar{\alpha}(\omega) = \limsup_{m\to\infty} \alpha_m(\xi_m(\omega), \iota^m)$, $D_k = \{\omega : \bar{\alpha}(\omega) \geqslant 1/k\}$. Clearly $D_k \in \mathscr{T}$. Suppose $B \in \mathscr{T}$, $B \subset D_k$, and $P_\iota(B) > 0$. Corollary 1.2 shows that $\bar{\alpha}(\omega) = 1 - \liminf_{m\to\infty} \gamma_m(\xi_m(\omega), \iota^m)$, so according to Lemma 1.1(a) we have $1_B - P_\iota(B) \leqslant 1 - (1/k)P_\iota$-a.s. on B. Now part (b) of the same lemma shows that $P_\iota(B) \geqslant 1/k$. Thus D_k contains only atoms, at most k in number. Hence $D = \{\omega : \tilde{\alpha}(\omega) > 0\}$ contains only atoms, which shows that $\lim_{m\to\infty} \alpha_m(\xi_m(\omega), i^m) = \bar{\alpha}(\omega) = 0$ P_i-a.s. on F. Next, fix A_r, a P_i-atom of $\mathscr{T}$. Using the above corollary and lemma once more,

$$\bar{\alpha}(\omega) \leqslant 1 - \{1_{A_r} - P_\iota(A_r)\} = P_\iota(A_r) \qquad P_\iota\text{-a.s.} \quad \text{on} \quad A_r. \tag{3}$$

Letting $\underline{\alpha}(\omega) = \liminf_{m\to\infty} \alpha_m(\xi_m(\omega), \iota^m)$, Theorem 1.3 and Proposition 2 yield

$$\underline{\alpha}(\omega) = \liminf_{m\to\infty} \tilde{P}^m_{(\xi_m(\omega), \iota^m)}(\tau_\Delta < \infty) \geqslant P_\iota(A_r) \qquad P_\iota\text{-a.s.} \quad \text{on} \quad A_r. \tag{4}$$

In combination, (3) and (4) complete the proof. ∎

As an immediate consequence we obtain

COROLLARY 1. $\mathscr{T}$ is

(a) *P_ι-trivial* *iff* $\alpha(\omega) = 1$ *P_ι-a.s.;*
(b) *P_ι-finite* *iff* $\alpha(\omega) \geqslant \lambda$ *P_ι-a.s. for some* $\lambda > 0$;
(c) *P_ι-atomic* *iff* $\alpha(\omega) > 0$ *P_ι-a.s.;*
(d) *P_ι-fully nonatomic* *iff* $\alpha(\omega) = 0$ *P_ι-a.s.*

Another byproduct of Theorem 1 is the fact that the P_ι-atomic structure of $\mathcal{T}$ is completely determined by any subsequential Markov process $(\zeta_k) = (\xi_{n_k})$.

Corollary 2. *The P_ι-atomic decomposition* (1.1) *of $\mathcal{T}$ is equivalent to the P_ι-atomic decomposition of*

$$\bigcap_{m=1}^{\infty} \sigma\langle \xi_{n_k}; k \geqslant m \rangle, \tag{5}$$

where (n_k) is any strictly increasing sequence in $\mathbb{N}$.

Proof. Fix (n_k), and let $\mathcal{T}'$ be the σ-algebra (5). If A_r is a P_ι-atom of $\mathcal{T}$, then since $\mathcal{T} \subset \mathcal{T}'$, $A_r = F_0' + \sum_{q \in I_0} A_q'$, where F_0' is fully nonatomic and A_q' are P_ι-atoms *in* $\mathcal{T}'$. Theorem 1 says that $\lim_{n\to\infty} \alpha_n(\omega) = P_\iota(A_r) > 0$ on A_r, and for the subsequential process that $\lim_{k\to\infty} \alpha_{n_k}(\omega) = 0$ on F_0', $P_\iota(A_q')$ on A_q'. Evidently the only possibility is $|I_0| = 1$ and $F_0' \cong \varnothing$. Hence A_r is equivalent to some atom of $\mathcal{T}'$. A similar argument shows that the nonatomic set F of $\mathcal{T}$ is equivalent to a nonatomic set in $\mathcal{T}'$. ∎

2.2. *λ-loss of Memory (Continued)*

As an application of the methods of the last subsection, we now complete the proof of an equivalence theorem for the various expressions of asymptotic partial loss of memory introduced in Section 1.3.

Theorem 2. *Let (ξ_n) be a Markov process on $(\Omega, \mathcal{B}, P_\iota)$. Consider the following four conditions for fixed $\lambda \in (0, 1]$:*

(i) $P_{\iota m}^{m}(C^{m-}(\xi_m, \iota^m)) \vee P_{\iota m}^{m}(C_m{}^{+}(\xi_m, \iota^m)) \geqslant \lambda$ *i.o. a.s.,*
(ii) $\gamma_m(\xi_m, \iota^m) \leqslant 1 - \lambda$ *i.o. a.s.,*
(iii) $\delta_m(\xi_m, \iota^m) \leqslant 1 - \lambda$ *i.o. a.s.,*
(iv) *For any $\mathcal{S} \otimes 2^{\mathbb{N}}$-measurable space–time harmonic $f: S \times \mathbb{N} \to \mathbb{R}$ such that $0 \leqslant f \leqslant 1$,*

$$|f(\xi_m, m) - E_\iota[f(\xi_0, 0)]| \leqslant 1 - \lambda \qquad \text{i.o.} \quad \text{a.s.}$$

(In each case "i.o." abbreviates "for infinitely many m.") If any of (i)–(iv) *holds, then (ξ_n) satisfies a 0–λ law w.r.t. P_ι. More generally, if $(\sigma_m)_{m \in \mathbb{N}}$ is a strictly increasing sequence of P_ι-a.s. finite stopping times, and if any of* (i)–(iv) *holds with m replaced by σ_m, then* (1.2) *follows. Conversely, if a 0–λ law holds for (ξ_n) w.r.t. P_ι, and if* (i′)–(iv′) *denote the above conditions with "i.o." replaced*

by "for every m" then (i′)–(iv′) *all hold. In particular,* (1.2), (1.3), (1.4), *and* (1.6) *are all equivalent.*

Proof. (ii′)–(iv′) are equivalent to (1.3), (1.4), and (1.6), respectively, since $\mathbb{N}$ is countable. (i) ⇒ (ii) ⇔ (iii), and (i′) ⇒ (ii′) ⇔ (iii′) by Proposition 1, (1.18), and Corollary 1.2. It was proved in Proposition 1.3 that (iii′) ⇒ (iv′) ⇒ (1.2); the proof that (iii) ⇒ (iv) ⇒ (1.2) is strictly analogous. The stopping time versions of all of these results are routine extensions. Hence it remains only to show (1.2) ⇒ (i′). To this end, fix $m \in \mathbb{N}$ and temporarily write

$$\Omega_1 = \{\omega: P^m_{\xi_m(\omega)}(C_m^-(\xi_m(\omega), \iota^m)) = 1\},$$
$$\Omega_2 = \{\omega: P^m_{\xi_m(\omega)}(C_m^+(\xi_m(\omega), \iota^m)) > 0\}.$$

The inequality in (i′) clearly holds if $\omega \in \Omega_1$, and otherwise $\omega \in \Omega_2$ because $C_m^+ \cup C_m^- = \Omega$. Since $P^m_{\xi_m}(C_m^+(\xi_m, \iota^m)) = P_\iota(\theta^{-m} C_m^+(\xi_m, \iota^m) \mid \mathscr{B}_0{}^m)$, it follows that $P_\iota(\theta^{-m} C_m^+(\xi_m, \iota^m)) > 0$ P_ι-a.s. on Ω_2. But by (1.2) this implies $P^m_{\iota m}(C_m^+(\xi_m, \iota^m)) \geqslant \lambda$ P_ι-a.s. on Ω_2, so the desired inequality again holds. (i′) is proved, since m is arbitrary. ▌

By combining the last theorem with Proposition 1.2 we obtain the following result, which seems impossible to verify directly:

COROLLARY 3. *If any of* (i)–(iv) *holds with* $\lambda > \frac{1}{2}$, *then all of* (i)–(iv) *hold with* $\lambda = 1$.

When combined with Theorem 1.3, Theorem 2 also shows that (1.11) is *necessary* for λ-loss of memory w.r.t. P_ι if $\tilde{P}$ is the maximal coupling. In Theorem 1.2, however, it is generally not quite necessary that

$$\tilde{P}^m_{(x,y)}(\Omega^*) = 1 \qquad \text{for all} \quad x, y \in S, \quad m \in \mathbb{N}, \tag{6}$$

to ensure total loss of memory w.r.t. every P_ι, even when $\tilde{P}$ is maximal. The next proposition identifies a necessary and sufficient condition.

PROPOSITION 3. *$\mathscr{T}$ is P_ι-trivial w.r.t. every $\iota \in \mathscr{P}$ if and only if*

$$\int_{x \in S} \int_{y \in S} \delta_m(x, y) p_m(y_0, dy) p_m(x_0, dx) = 0 \qquad \text{for all} \quad m \in \mathbb{N}, \quad x_0, y_0 \in S. \tag{7}$$

Proof. By Theorem 2, $\mathscr{T}$ is P_ι-trivial w.r.t. every $\iota \in \mathscr{P}$ if and only if $\delta_m(\xi_m, \iota^m) = 0$ P_ι-as for every $m \in \mathbb{N}$, $\iota \in \mathscr{P}$, or equivalently if and only if

$$\int_{x \in S} \delta_m(x, \iota^m) p_m(\iota, dx) = 0 \qquad \text{for all} \quad m \in \mathbb{N}, \quad \iota \in \mathscr{P}. \tag{8}$$

Assuming (7), and using Fatou's lemma (as in Theorem 1.2), we see that (8) is majorized by

$$\int_{x \in S} \int_{y \in S} \delta_m(x, y) p_m(\iota, dy) p_m(\iota, dx)$$
$$= \int_{x_0 \in S} \int_{y_0 \in S} \left[\int_{x \in S} \int_{y \in S} \delta_m(x, y) p_m(y_0, dy) p_m(x_0, dx) \right] \iota(dy_0) \iota(dx_0) = 0.$$

Conversely, suppose that (8) holds. Taking $m = 0$ and ι to be the convex combination $\frac{1}{2}x_0 + \frac{1}{2}y_0$ yields $\delta_0(x_0, y_0) = 0$ for each pair $x_0, y_0 \in S$. Since $\delta_m(x, y) \leqslant \delta_m(x, x_0{}^m) + \delta_m(x_0{}^m, y_0{}^m) + \delta_m(y, y_0{}^m)$ and $\delta_m(x_0{}^m, y_0{}^m) = \delta_0(x_0, y_0) = 0$, to establish (7) it suffices to show

$$\int_{x \in S} \delta_m(x, x_0{}^m) p_m(x_0, dx) = 0 \qquad \text{for all} \quad m \in \mathbb{N}, \quad x_0 \in S.$$

But this is simply (8) with $\iota = x_0$. ∎

COROLLARY 4. *If* (ξ_n) *is homogeneous and* $(\tilde{P}_{(\mu,\nu)})$ *is the maximal coupling, then* (6) *is necessary and sufficient for total loss of memory w.r.t. every* P_ι.

Proof. In the homogeneous case $\delta_m(x, y) = \delta_0(x, y)$ for every m, while (7) implies that $\delta_0(x, y) = 0$. The claim follows since $\tilde{P}_{(x,y)}(\Omega^*) = 1 - \delta_0(x, y)$. ∎

2.3. *Additional Applications of the Maximal Coupling*

LEMMA. $\qquad \alpha_m(\xi_m, \iota^m) \geqslant p_m(\iota, \xi_m), \qquad m \in \mathbb{N}, \quad \iota \in \mathscr{P}.$

(*Here* $p_m(\iota, \xi_m) = p_m(\iota, \{\xi_m\})$.)

Proof. If $(\tilde{P}^m_{(\mu,\nu)})$ is the maximal coupling, then

$$\alpha_m(\xi_m, \iota^m) = \tilde{P}^m_{(\xi_m, \iota^m)}(\tau_\Delta < \infty) \geqslant \tilde{P}_{(\xi_m, \iota^m)}(\tau_\Delta = 0) = p_m(\iota, \xi_m). \quad ∎$$

The above estimate yields simple sufficient conditions for trivial, finite, or atomic $\mathscr{T}$ w.r.t. P_ι. It should be noted, though, that the results which follow are of interest primarily in the Markov chain case, since otherwise $p_m(\iota, \xi_m)$ is typically identically 0.

PROPOSITION 4. *If there is a sequence* $(\sigma_m)_{m \in \mathbb{N}}$ *of strictly increasing* P_ι*-a.s. finite stopping times such that*

$$p_{\sigma_m}(\iota, \xi_{\sigma_m}) \geqslant \lambda \qquad \text{i.o.} \quad P_\iota\text{-a.s.}$$

for a fixed $\lambda \in (0, 1]$, *then a* 0–λ *law holds for* (ξ_n) *w.r.t.* P_ι. *If*

$$P_\iota(p_{\sigma_m}(\iota, \xi_{\sigma_m}) \geqslant \lambda \text{ i.o. for some } \lambda > 0) = 1,$$

then $\mathscr{T}$ *is* P_ι*-atomic*. (*Again*, "*i.o.*" = "*for infinitely many m*.")

Proof. The first assertion is a consequence of the Lemma, the stopping time version of (iii) in Theorem 2, and the fact that $\alpha = 1 - \delta$. Also, by Theorem 1 and the Lemma,

$$P_\iota(\alpha(\omega) > 0) = P_\iota\left(\limsup_{m\to\infty} \alpha_{\sigma_m}(\omega) \geqslant \lambda \text{ for some } \lambda > 0\right)$$

$$\geqslant P_\iota(p_{\sigma_m}(\iota, \xi_{\sigma_m}) \geqslant \lambda \text{ i.o. for some } \lambda > 0).$$

The hypothesis and Corollary 1 imply the second assertion. ∎

As special cases of this last proposition we have:

COROLLARY 5. *If* $\limsup_{m\to\infty} P_\iota(p_m(\iota, \xi_m) \geqslant \lambda) = 1$ *for fixed* $\lambda \in (0, 1]$, *then a* 0–λ *law holds for* (ξ_n) *w.r.t.* P_ι. $\mathscr{T}$ *is* P_ι*-atomic if*

$$\lim_{\lambda\to 0} \limsup_{m\to\infty} P_\iota(p_m(\iota, \xi_m) \geqslant \lambda) = 1.$$

Proof. The expressions in the hypotheses are majorized by $P_\iota(p_m(\iota, \xi_m) \geqslant \lambda \text{ i.o.})$ and $P_\iota(p_m(\iota, \xi_m) \geqslant \lambda \text{ i.o. for some } \lambda > 0)$, respectively. Apply Proposition 4 with $\sigma_m(\omega) \equiv m$. ∎

COROLLARY 6. *For each n, let* E_n *be a finite subset of S, and suppose that* $P_\iota(\xi_n \in E_n$ *for infinitely many* $n) = 1$. *If* $\liminf_{n\to\infty} \inf_{x\in E_n} p_n(\iota, x) \geqslant \lambda > 0$, *then a* 0–λ *law holds for* (ξ_n) *w.r.t.* P_ι. (*Note*: $E_n = \varnothing$ *for some n is allowed*; *the empty* inf *is* $+\infty$).

Proof. Apply Proposition 4 with σ_m = the mth time that $\xi_n \in E_n$. ∎

2.4. *Uniform Coupling*

In this section we discuss two types of *uniform* ergodicity, stronger forms of total loss of memory than the $\lambda = 1$ case of (1.4). To introduce these notions, define

$$\underline{\alpha}_m{}^n = \inf_{x,\, y\in S} \alpha_m{}^n(x, y); \qquad \overline{\delta}_m{}^n = \sup_{x,\, y\in S} \delta_m{}^n(x, y);$$

$$\underline{\underline{\alpha}}_l = \inf_{m\in\mathbb{N}} \underline{\alpha}_m^{m+l}; \qquad \overline{\overline{\delta}}_l = \sup_{m\in\mathbb{N}} \overline{\delta}_m^{m+l}.$$

Note that $\underline{\alpha}_m{}^n = 1 - \overline{\delta}_m{}^n$, $\underline{\underline{\alpha}}_l = 1 - \overline{\overline{\delta}}_l$, that $\overline{\delta}_m{}^n$ decreases as $n \to \infty$ for each fixed m, and that $\overline{\overline{\delta}}_l$ decreases as $l \to \infty$. A Markov process (ξ_n) on $(\Omega, \mathscr{B})$ is *S-uniformly ergodic* iff

$$\lim_{n\to\infty} \overline{\delta}_m{}^n = 0 \qquad \text{for all} \quad m \in \mathbb{N}, \tag{9}$$

and *ST-uniformly ergodic* iff

$$\lim_{l\to\infty} \overline{\overline{\delta}}_l = 0. \tag{10}$$

Condition (9) asserts that (ξ_n) tends to lose track of its past uniformly over the state space S, while (10) demands that this forgetfulness occur uniformly with respect to *both* S and the time parameter set $\mathbb{N}$. Clearly (10) $\Rightarrow$ (9), and (9) $\Rightarrow$ (1.4) with $\lambda = 1$ for every $\iota \in \mathscr{P}$ by Theorem 1.2. Note that (9) and (10) are equivalent when (ξ_n) is homogeneous.

A method of proving conditions for uniform ergodicity will now be presented. This is accomplished by introducing an auxiliary $\{0, 1\}$-valued process (η_n). Thus, let $\hat{S} = \tilde{S} \times \{0, 1\} = S \times S \times \{0, 1\}$, $\hat{\mathscr{S}} = \tilde{\mathscr{S}} \otimes 2^{\{0,1\}}$, $\hat{\Omega} = \hat{S}^{\mathbb{N}} = \tilde{S}^{\mathbb{N}} \times \{0, 1\}^{\mathbb{N}}$, write $\kappa_n = (\tilde{\xi}_n, \eta_n) = (\xi_n{}^1, \xi_n{}^2, \eta_n)$ for the coordinate process on $\hat{\Omega}$, and $\hat{\mathscr{B}}$ for the σ-algebra generated by (κ_n). The uniform coupling technique is formalized as follows.

Proposition 5. *Let (ξ_n) be a Markov process on $(\Omega, \mathscr{B})$. Suppose there is a collection of measures $(\hat{P}^m_{(\mu,\nu)}; m \in \mathbb{N}, \mu, \nu \in \mathscr{P})$ on $(\hat{\Omega}, \hat{\mathscr{B}})$ such that*

(i) *$(\hat{P}^m_{(\mu,\nu)}(\cdot, \cdot, \{0, 1\}^{\mathbb{N}}); m \in \mathbb{N}, \mu, \nu \in \mathscr{P})$ is a coupling for (ξ_n), and*
(ii) *For every $m \in \mathbb{N}$, $\mu, \nu \in \mathscr{P}$,*

$$\hat{P}^m_{(\mu,\nu)}(\tilde{\xi}_n \in \Delta \text{ whenever } \eta_n = 0) = 1,$$

and

$$\hat{P}^m_{(\mu,\nu)}(\eta_n = 1) = \epsilon(m, n) \to 0 \qquad \text{as} \quad n \to \infty.$$

Then (ξ_n) is S-uniformly ergodic. If, in addition,
(iii) *$\epsilon(m, n) \equiv \epsilon(n)$ is independent of m, then (ξ_n) is ST-uniformly ergodic.*

Proof. Properties (i) and (ii) yield

$$\hat{P}^m_{(x,y)}(\tilde{\xi}_n \in \Delta) \leqslant \alpha_m^{m+n}(x, y) \qquad (\text{cf. } (1.10)) \tag{11}$$

and

$$\hat{P}^m_{(x,y)}(\tilde{\xi}_n \in \Delta) \geqslant P^m_{(x,y)}(\eta_n = 0) \to 1 \tag{12}$$

uniformly in x and y as $n \to \infty$, respectively. Hence $\underline{\alpha}_m^{m+n} \to 1$ for each m as $n \to \infty$. If (iii) holds, then the convergence in (12) is also uniform in m, so $\underline{\underline{\alpha}}_n \to 1$ as $n \to \infty$. This completes the proof since $\bar{\delta} = 1 - \underline{\alpha}$, $\bar{\bar{\delta}} = 1 - \underline{\underline{\alpha}}$. ∎

With the aid of the Vasershtein coupling and the last proposition, we now derive sufficient criteria for S- and ST-uniform ergodicity.

Theorem 3 (Markov–Dobrushin–Iosifescu). *Let (ξ_n) be a Markov process on $(\Omega, \mathscr{B})$.*

(a) *If $\sum_{m=0}^{\infty} \underline{\alpha}_m^{m+1} = \infty$, then (ξ_n) is S-uniformly ergodic;*
(b) *If $\underline{\underline{\alpha}}_1 > 0$, then (ξ_n) is ST-uniformly ergodic.*

Proof. Measures $\hat{P}$ satisfying the hypotheses of Proposition 5 will be constructed. To prove (a), let $\hat{P}^m_{(\mu,\nu)}$ be a Markov measure on $(\hat{\Omega}, \hat{\mathscr{B}})$ with initial distribution $\mu \times \nu \times \epsilon_1$ and t.p.s. $(\hat{\pi}^m)$, with $\hat{\pi}^m : \hat{S} \times \hat{\mathscr{S}} \to [0, 1]$ prescribed in terms of the $\alpha_m{}^n(x, y)$, $\underline{\alpha}_m{}^n$, and the t.p.s. $(\tilde{\pi}^m)$ of the *Vasershtein coupling*, as follows:

$$\hat{\pi}^m((x, y, 0), (E^1, E^2, 0)) = \tilde{\pi}^m((x, y), (E^1, E^2));$$

$$\hat{\pi}^m((x, y, 1), (E^1, E^2, 0)) = \chi_m(x, y)\tilde{\pi}^m((x, y), (E^1 \cap E^2, E^1 \cap E^2));$$

$$\hat{\pi}^m((x, y, 1), (E^1, E^2, 1)) = (1 - \chi_m(x, y))\tilde{\pi}((x, y), (E^1 \cap E^2, E^1 \cap E^2)) + \tilde{\pi}^m((x, y), (E^1 \backslash E^2, E^2 \backslash E^1));$$

$m \in \mathbb{N}$, $x, y \in S$, E^1, $E^2 \in \mathscr{S}$, where $\chi_m(x, y) = \underline{\alpha}_m^{m+1}/\alpha_m^{m+1}(x, y)$ $(= 0$ if $\alpha_m^{m+1}(x, y) = 0)$, and $E \backslash F = E \cap F^c$. Then $\hat{P}^m_{(\mu,\nu)}$ is well defined, and $(\tilde{\xi}_n)$ is the Vasershtein coupled process. Hence (i) of Proposition 5 holds. Now (η_n) is a $\{0, 1\}$-valued Markov chain, with one-step transition matrix

$$\begin{pmatrix} 1 & 0 \\ \underline{\alpha}_{m+n}^{m+n+1} & 1 - \underline{\alpha}_{m+n}^{m+n+1} \end{pmatrix}$$

at time n when governed by $\hat{P}^m_{(\mu,\nu)}$. Thus

$$\hat{P}^m_{(\mu,\nu)}(\eta_n = 1) = \prod_{k=0}^{n-1} (1 - \alpha_{m+k}^{m+k+1}) = \epsilon(m, n), \qquad 0 \leqslant m \leqslant n < \infty. \tag{13}$$

If the condition in (a) holds, then the above product tends to 0 as $n \to \infty$. Also, $\eta_0 = 1$ and inspection of $(\hat{\pi}^m)$ shows that with probability one transition to $\eta_n = 0$ occurs only when $\tilde{\xi}_n \in \Delta$. These remarks establish condition (ii) of Proposition 5. Thus (ξ_n) is S-uniformly ergodic. To show (b), redefine $\chi_m(x, y) = \underline{\underline{\alpha}}_1/\alpha_m^{m+1}(x, y)$. This has no effect on the law governing $(\tilde{\xi}_n)$, while (η_n) now has transition matrix

$$\begin{pmatrix} 1 & 0 \\ \underline{\underline{\alpha}}_1 & 1 - \underline{\underline{\alpha}}_1 \end{pmatrix}$$

at every time, independently of m. Hence

$$\hat{P}^m_{(\mu,\nu)}(\eta_n = 1) = (1 - \underline{\underline{\alpha}}_1)^n = \epsilon(n). \tag{14}$$

Evidently (iii) of Proposition 5 holds, and if $\underline{\underline{\alpha}}_1 > 0$, then so does (ii). It follows that (ξ_n) is ST-uniformly ergodic. This completes the proof; we remark in passing that (11)–(14) give rates of ergodicity in terms of $\underline{\alpha}$ and $\underline{\underline{\alpha}}$. ∎

If π is an arbitrary transition function, the *ergodic coefficient* of π is given by

$$\underline{\alpha}(\pi) = \inf_{x, y \in S} (\pi(x, \cdot) \wedge \pi(y, \cdot))(S). \tag{15}$$

(When (π^m) is a t.p.s., then $\underline{\alpha}(\pi^m) = \underline{\alpha}_m^{m+1}$ in our previous notation.) As a consequence of uniform coupling, it is easy to obtain the following fundamental inequality.

COROLLARY 7. *If π_0 and π_1 are transition functions, then*

$$1 - \underline{\alpha}(\pi_0 \circ \pi_1) \leqslant [1 - \underline{\alpha}(\pi_0)][1 - \underline{\alpha}(\pi_1)],$$

where $\pi_0 \circ \pi_1$ is the transition function defined by

$$(\pi_0 \circ \pi_1)(x, E) = \int_{y \in S} \pi_0(x, dy)\pi_1(y, E), \qquad x \in S, \quad E \in \mathscr{S}.$$

Proof. Consider a Markov process (ξ_n) on $(\Omega, \mathscr{B})$ with t.p.s. (π^m) such that $\pi^0 = \pi_0$ and $\pi^m = \pi_1$ for $m \geqslant 1$. Let (κ_n) be the trivariate process constructed in the proof of Theorem 3(a). Then for any $x, y \in S$,

$$1 - \alpha_0{}^2(x, y) \leqslant \hat{P}^0_{(x,y)}(\tilde{\xi}_2 \in \Delta) \leqslant \hat{P}^0_{(x,y)}(\eta_2 = 1) = [1 - \underline{\alpha}(\pi_0)][1 - \underline{\alpha}(\pi_1)].$$

The desired inequality is obtained by taking the supremum over x and y. ▮

Another corollary gives necessary and sufficient conditions for S- and ST-uniform ergodicity.

COROLLARY 8 (Hajnal–Iosifescu). *Let (ξ_n) be a Markov process on $(\Omega, \mathscr{B})$ with* t.p.s. (π^m). *Then (ξ_n) is S-uniformly ergodic if and only if there is a strictly increasing $(n_k) \subset \mathbb{N}$ such that*

$$\sum_{k=0}^{\infty} \underline{\alpha}_{n_k}^{n_k+1} = \infty,$$

and ST-uniformly ergodic if and only if $\underline{\underline{\alpha}}_l > 0$ for some l.

Proof. S-uniform ergodicity means that $\lim_{m\to\infty} \underline{\alpha}_m{}^n = 1$ for every $m \in \mathbb{N}$. Put $n_0 = 0$, choose n_1 so that $\underline{\alpha}_0^{n_1} \geqslant \frac{1}{2}$, and inductively choose n_{k+1} so that $\underline{\alpha}_{n_k}^{n_{k+1}} \geqslant \frac{1}{2}$. Then clearly the first condition of the corollary holds. ST-uniform ergodicity means that $\underline{\underline{\alpha}}_l \to 1$ as $l \to \infty$, and in this case the second condition is surely necessary. Conversely, suppose $\{n_k\}$ is an increasing sequence such that the sum of the $\underline{\alpha}_{n_k}^{n_k+1}$ diverges. Letting $(\zeta_n) = (\xi_{n_k})$ be the subsequential Markov process (with t.p.s. $(\bar{\pi}^m)$ such that $\bar{\pi}^m(\cdot, \cdot) = p_{n_m n_{m+1}}(\cdot, \cdot)$), it follows from Theorem 3 that (ζ_n) is S-uniformly ergodic, and so

$$\lim_{l\to\infty} \bar{\delta}_{n_j}^{n_{j+l}} = 0 \qquad \text{for any} \quad n_j \in (n_k). \tag{16}$$

Similarly, if $\underline{\underline{\alpha}}_l > 0$ for some l, then letting $(\zeta_n) = (\xi_{nl})$, Theorem 1 shows that

$$\lim_{n\to\infty} \sup_{m\in\mathbb{N}} \delta_{ml}^{(m+n)l} = 0. \tag{17}$$

Now by Corollary 7, and since $\bar{\delta} \leqslant 1$, we have $\bar{\delta}_m{}^n \leqslant \bar{\delta}_{m'}^{n'}$ whenever $m' \leqslant m \leqslant n \leqslant n'$. Using this, it is easy to see that properties (9) and (10) follow from (16) and (17), respectively. ∎

In the homogeneous case, S-uniform ergodicity coincides with geometric convergence to an invariant probability measure. Recall that μ is π-invariant for a homogeneous process (ξ_n) with t.f. π iff

$$\mu(E) = \int_S \mu(dx)\pi(x, E), \qquad E \in \mathscr{S}.$$

COROLLARY 9. *Let (ξ_n) be a homogeneous Markov process on $(\Omega, \mathscr{B})$ with transition function π. The following are equivalent:*

(i) *(ξ_n) is S-uniformly ergodic;*
(ii) *$\underline{\alpha}_0{}^l > 0$ for some $l \in \mathbb{N}$;*
(iii) *There is a unique π-invariant probability measure μ, and there are nonnegative constants K and $\epsilon < 1$ such that for every $\iota \in \mathscr{P}$,*

$$\|p_m(\iota, \cdot) - \mu\| \leqslant K\epsilon^m, \qquad m \in \mathbb{N}.$$

Proof. Since homogeneous S-uniformly ergodic processes are automatically ST-uniformly ergodic, (i) $\Leftrightarrow$ (ii) by Corollary 8. One easily shows that $\bar{\delta}_0{}^m = \sup_{\mu, \nu \in \mathscr{P}} \delta_0{}^m(\mu, \nu)$. Thus, assuming (i),

$$\lim_{m\to\infty} \sup_{n \in \mathbb{N}} \|p_m(\iota, \cdot) - p_{m+n}(\iota, \cdot)\| = \lim_{m\to\infty} \sup_{n \in \mathbb{N}} \delta_0{}^m(\iota, \iota^n) = 0,$$

and hence $\|p_m(\iota, \cdot) - \mu\| \to 0$ as $m \to \infty$ for some $\mu \in \mathscr{P}$. μ is invariant since

$$\mu(E) = \lim_{m\to\infty} p_{m+1}(\iota, E) = \lim_{m\to\infty} \int p_m(\iota, dx)\pi(x, E) = \int \mu(dx)\pi(x, E).$$

If $\mu' \in \mathscr{P}$ is invariant, then $\|\mu' - \mu\| = \delta_0{}^m(\mu', \mu) \to 0$ as $m \to \infty$, forcing $\mu' = \mu$. Finally, using Corollary 1 and the fact that $\delta_0{}^m$ decreases with m,

$$\delta_0{}^m(\iota, \mu) \leqslant \bar{\delta}_0{}^m \leqslant (\bar{\delta}_0{}^l)^{\left[\frac{m}{l}\right]} \leqslant (1 - \underline{\alpha}_0{}^l)^{\left(\frac{m}{l} - 1\right)} = K\epsilon^m, \qquad m \in \mathbb{N},$$

with $\epsilon < 1$ if $\underline{\alpha}_0{}^l > 0$. Thus (i) and (ii) imply (iii). (iii) $\Rightarrow$ (i) is trivial, from the triangle inequality. ∎

3. Applications to Homogeneous Processes

3.1. *Ergodic Theorems for Random Walks.*

Our purpose in this section is to prove some of the basic ergodic theorems for homogeneous Markov processes by means of coupling techniques. To start, let (ξ_n) be a random walk with transition function π. Write $s(\pi)$ for the support of π, and set $\mathscr{G}_\pi =$ the closed additive subgroup of $\mathbb{R}$ generated by $s(\pi)$. (ξ_n) is a *lattice random walk* if $\mathscr{G}_\pi = \{nu;\, n \in \mathbb{N}\}$ for some $u \in \mathbb{R}$; otherwise (ξ_n) is *nonlattice*. In the lattice case put $\mathscr{H}_\pi =$ the subgroup of $\mathscr{G}_\pi$ generated by $s(\pi) - s(\pi)$. The *period* of π is $\mathrm{per}(\pi) = |\mathscr{G}_\pi/\mathscr{H}_\pi| =$ the number of cosets of $\mathscr{H}_\pi$ in $\mathscr{G}_\pi$; we call (ξ_n) *aperiodic* if $\mathrm{per}(\pi) = 1$.

Ornstein's coupling serves to identify the P_ι-atomic decomposition of $\mathscr{T}$ for any lattice random walk such that $\iota(\mathscr{G}_\pi) = 1$, while a modified coupling shows that a nonlattice random walk has P_ι-trivial tail σ-algebra for every $\iota \in \mathscr{P}$ when some convolution power of π has an absolutely continuous part. The arguments make use of

PROPOSITION 1. *Let* (ξ_n) *be a random walk with transition function* π *satisfying* (i) $\int x\pi(dx) = 0$, *and* (ii) $\pi([-L, L]) = 1$ *for some* $L \in \mathbb{R}$. *Then whenever* $x - y \in \mathscr{G}_\pi$,

$$P_x(|\xi_n - y| \leqslant \epsilon \text{ i.o.}) = 1 \qquad \text{for all} \quad \epsilon > 0.$$

Proof. We may appeal to the Chung–Fuchs recurrence criterion since ξ_n is a sum of mean-0 i.i.d. random variables. In our case, though, the increments are bounded, so the proposition is a consequence of the most elementary weak law of large numbers. To illustrate this, we sketch an elegant proof due to Chung and Ornstein [6]: For any $n \in \mathbb{N}$ and $\epsilon > 0$, simple estimates yield

$$\sum_{k=0}^{\infty} P_0(|\xi_k| \leqslant 1) \geqslant \frac{1}{2n\epsilon} \sum_{k=1}^{\infty} P_0(|\xi_k| \leqslant n\epsilon) \geqslant \frac{1}{2n\epsilon} \sum_{k=1}^{n} P_0(|\xi_k| \leqslant k\epsilon).$$

By the weak law, $\lim_{k\to\infty} P_0(|\xi_k| \leqslant k\epsilon) = 1$, whence $\sum_{k=0}^{\infty} P_0(|\xi_k| \leqslant 1) \geqslant 1/(2\epsilon)$. Since ϵ is arbitrary, the usual recurrence condition holds. ∎

Proceeding to the coupling results, we consider first the lattice case.

LEMMA 1. *Let* (ξ_n) *be a lattice random walk. If* $(\tilde{P}_{(\mu,\nu)})$ *is the Ornstein coupling* (1.15) (*with a suitably large* K), *then*

$$\tilde{P}_{(x,y)}(\tau_\Delta < \infty) = 1 \qquad \textit{whenever} \quad x - y \in \mathscr{H}_\pi. \tag{1}$$

Proof. Until τ_Δ, the difference in the marginals, $(\xi_n^2 - \xi_n^1)$, of the Ornstein coupled process is a symmetric random walk with bounded increments. Such a random walk clearly satisfies the hypotheses of Proposition 1. To prove (1), it is therefore enough to show that $(x - y) - 0$ is in $\mathscr{G}_{\check{\pi}}$ whenever $x - y \in \mathscr{H}_\pi$, where $\check{\pi}$ is the t.f. for the difference process before

τ_Δ. The explicit form of $\breve{\pi}$ in the lattice case, as derived from (1.15), is

$$\breve{\pi}(z) = \left[\sum_{w\in s(\pi)} \pi(w)\pi(w+z)\right] 1_{\{|z|\leq K\}} + \left[\sum_{w\in s(\pi)} \pi(w) \sum_{\substack{v\in s(\pi):\\ |w-v|>K}} \pi(v)\right] 1_{\{z=0\}}. \quad (2)$$

We see from (2) that $s(\breve{\pi})$ contains any given point of $s(\pi) - s(\pi)$ if K is large enough. It follows that $\mathscr{G}_{\breve{\pi}}$ contains the generator u of $\mathscr{H}_\pi$ for sufficiently big K, and the lemma is proved. ∎

THEOREM 1. *Let (ξ_n) be a lattice random walk, with $\mathscr{G}_\pi = \{nu; n \in \mathbb{N}\}$, and let $E_r = \mathscr{H}_\pi + ru$; $0 \leq r < \mathrm{per}(\pi)$. If $\iota \in \mathscr{P}$ and $\iota(\mathscr{G}_\pi) = 1$, then the P_ι-decomposition of $\mathscr{T}$ is equivalent to the partition*

$$\Omega = \sum_{r:\, P_\iota(E_r)>0} \{\xi_0 \in E_r\} \qquad (\text{mod } P_\iota\text{-null sets}).$$

In particular, if (ξ_n) is aperiodic, then $\mathscr{T}$ is P_ι-trivial w.r.t. every ι such that $\iota(\mathscr{G}_\pi) = 1$.

Proof. Define $E_r^n = E_r + ns(\pi)$. Then the E_r^n; $0 \leq r < \mathrm{per}(\pi)$, $n \in \mathbb{N}$, satisfy (i)–(iii) of Theorem 1.2, so the conclusions of that result apply. ∎

Turning to the nonlattice case, a moment's thought reveals that when $\mathscr{G}_\pi = \mathbb{R}$, the Ornstein coupled process cannot in general hit the diagonal with positive probability. Being a truncation of the classical coupled process, $(\tilde{\xi}_n)$ will typically go to Δ with probability 0 at each step. It is therefore necessary to modify the Ornstein coupling by substituting the Vasershtein t.f. on a suitable set. We illustrate this technique, a key ingredient in any coupling argument on an uncountable state space, in the proof of the next theorem.

THEOREM 2. *Let (ξ_n) be a random walk on $\mathbb{R}$, and suppose that $\pi^{(m)} = p_m(0, \cdot)$ has a nonzero absolutely continuous part for some $m \geq 1$. Then $\delta_0(x, y) = 0$ for every $x, y \in S$, and so $\mathscr{T}$ is P_ι-trivial w.r.t. every $\iota \in \mathscr{P}$.*

Proof. With m as in the hypothesis, it suffices to show that $\delta_0^{2mn}(x, y) \to 0$ as $n \to \infty$, since δ_0^n decreases with n. To this end, we construct a coupling for $(\zeta_n) = (\xi_{2mn})$ such that $\tilde{P}_{(x,y)}(\tilde{\zeta}_n \in \Delta \text{ ev.}) = 1$ for all $x, y \in \mathbb{R}$. The t.f. for (ζ_n) is $\pi^{(2m)} = \pi^{(m)} * \pi^{(m)}$. Since $\pi^{(m)}$ has an absolutely continuous part, $\pi^{(2m)}(dx) = q(x)\,dx + \pi_0(dx)$, where q is a nonnegative continuous function such that $\int q(x)\,dx > 0$ and π_0 is a nonnegative measure. In particular we can find $x_0 \in \mathbb{R}$, $h_0, \epsilon > 0$ such that $q(x) \geq h_0$ on $[x_0 - 2\epsilon, x_0 + 2\epsilon]$. For the process (ζ_n), it follows that $\alpha_0{}^1(x, y) = (\pi^{(2m)}(\cdot) \wedge \pi^{(2m)}(\cdot + (y - x)))(\mathbb{R}) \geq h = 2\epsilon h_0 > 0$ whenever $|x - y| \leq \epsilon$. Consider a Markovian coupling $(\tilde{P}_{(\mu,\nu)})$ for (ζ_n) with t.f. $\tilde{\pi}((x, y), \cdot)$ satisfying (1.14), and for $x \neq y$, let $\tilde{\pi}((x, y), \cdot)$ be the Vasershtein coupling (1.16) when $(x, y) \in \Delta_\epsilon = \{(x, y): |x - y| \leqslant \epsilon\}$, the Ornstein coupling (1.15) otherwise. In all cases $\tilde{\pi}$ is defined in terms of the t.f. $\pi^{(2m)}$ for (ζ_n), and in the Ornstein coupling we choose $K > |x_0|$ to ensure that $\mathscr{G}_{\check{\pi}} = \mathbb{R}$ (with $\check{\pi}$ as in the previous theorem). Applying Proposition 1 to $\zeta_n{}^2 - \zeta_n{}^1$, we conclude that the hitting time for Δ_ϵ is finite from *any* (x, y). Moreover, from *any* point of Δ_ϵ, (ζ_n) goes to Δ in one step with at least probability $h > 0$. A routine argument now shows that $\tau_\Delta < \infty$ $\tilde{P}_{(x,y)}$-almost surely for every $x, y \in S$. ∎

Some condition on π is necessary to ensure an atomic tail σ-algebra in the nonlattice case, as is illustrated by a simple

EXAMPLE. (ξ_n) is a mean-0 random walk on $\mathbb{R}$ with $\pi(\sqrt{2}) = \sqrt{2} - 1$, $\pi(-1) = 2 - \sqrt{2}$. $\mathscr{G}_\pi = \mathbb{R}$, so by Proposition 1 (ξ_n) hits every interval infinitely often. But knowing the value of ξ_n we can recover the exact value of ξ_0 a.s., and it is not hard to see that the P_ι-decomposition of $\mathscr{T}$ has $F \cong \{\xi_0 \in$ the nonatomic part of $\iota\}$, and $A_r \cong \{\xi_0 = r\}$ for those $r \in \mathbb{R}$ such that $\iota(r) > 0$.

3.2. *Orey's Ergodic Theorem*

The remainder of the section will be devoted to proving ergodic theorems for Markov processes recurrent in the sense of Harris. In order to focus on coupling without becoming involved in the basic theory of such processes, we assume a few definitions and facts. The reader is referred to Orey [30] for a detailed exposition.

Let φ be a given (nontrivial) nonnegative σ-finite measure on $(S, \mathscr{S})$. For $E \in \mathscr{S}$, put $\tau_E(\omega) = \min\{n \geqslant 1 : \omega_n \in E\}$ ($= \infty$ if $\omega_n \notin E$ for all n). A homogeneous Markov process on $(\Omega, \mathscr{B})$ with t.f. π is called *φ-irreducible* (resp. *φ-recurrent*) iff $P_x(\tau_E < \infty) > 0$ (resp. $P_x(\xi_n \in E \text{ i.o.}) = 1$) for all $x \in S$ whenever $E \in \mathscr{S}$ and $\varphi(E) > 0$. In the Markov chain case $\varphi =$ counting measure yields the classical theory, but other measures are also of interest (e.g., $\varphi(S_0) = 0$ for some $S_0 \subset S$). For our purposes, any φ-irreducible process (ξ_n) satisfies conditions (i) and (ii) below.

(i) *Existence of a cycle*: There is a unique d, $1 \leqslant d < \infty$, and there are disjoint $E_r \in \mathscr{S}$, $0 \leqslant r \leqslant d - 1$, such that

(a) $\pi(x, E_{r+1}) = 1$ for all $x \in E_r$, $0 \leqslant r < d - 1$,
$\pi(x, E_0) = 1$ for all $x \in E_{d-1}$,

(b) $\varphi(S - \sum_{r=0}^{d-1} E_r) = 0$

(c) if $\{E_r'; 0 \leqslant r \leqslant d' - 1\}$ is any other collection of disjoint sets with property (a) of the E_r, then d' divides d.

d is called the *period* of π, per(π) in notation. The sets E_r are the *cyclically moving subclasses*, and are uniquely determined mod φ-null sets. Set $\bar{E} = \sum_{r=0}^{d-1} E_r$. If (ξ_n) is φ-recurrent, then

$$P_x(\tau_{\bar{E}} < \infty) = 1 \qquad \text{for every} \quad x \in S. \tag{3}$$

(ii) *Existence of "C-sets"*: Whenever $E \in \mathscr{S}$ with $\varphi(E) > 0$, there are $\epsilon > 0$, $m \geqslant 1$, and some $E_\varphi \subset E$ such that $E_\varphi \in \mathscr{S}$ and

$$\inf_{(x,y) \in E_\varphi \times E_\varphi} q_m(x, y) \geqslant \epsilon, \tag{4}$$

where $q_m(x, \cdot)$ is the Radon–Nikodym derivative of the absolutely continuous part of $p_m(x, \cdot)$ w.r.t. φ. Moreover,

$$\text{g.c.d.}\left\{m\colon \inf_{(x,y) \in E_\varphi \times E_\varphi} q_m(x, y) > 0\right\} = \text{per}(\pi)$$

for any such E_φ (g.c.d. = greatest common divisor).

Our main objective in this section is a coupling proof of Orey's ergodic theorem for φ-recurrent Markov processes. As a Corollary we will derive a description of the atomic decomposition of $\mathscr{T}$ for any φ-recurrent process. The argument presented here combines aspects of the classical, Vasershtein, and Ornstein couplings in a rather intricate non-Markovian construction. In this instance it is easier to give a sample path description of the coupling than a formal prescription of cylinder events.

THEOREM 3 (Orey). *Let (ξ_n) be an aperiodic φ-recurrent Markov process on $(\Omega, \mathscr{B})$. Then $\delta_0(x, y) = 0$ for all $x, y \in S$, and hence $\mathscr{T}$ is P_ι-trivial w.r.t. every $\iota \in \mathscr{P}$.*

Proof. Fix $\epsilon > 0$, $m \geqslant 1$, and E_φ satisfying (4). As in the proof of Theorem 2, we need only construct a coupling $(\tilde{P}_{(\mu,\nu)})$ for the process (ζ_n) with t.f. $\pi_m(\cdot, \cdot) = p_m(\cdot, \cdot)$, such that $\tilde{P}_{(x,y)}(\tau_\Delta < \infty) = 1$ for every $x, y \in S$. The coupled process $(\tilde{\zeta}_n)$ will be defined in terms of its marginals $({\zeta_n}^1)$ and $({\zeta_n}^2)$, which are allowed to evolve independently over certain time intervals, but dependently over others. The dependence is introduced at pairs of random times which differ for the two copies of (ζ_n); thus the non-Markovian nature of the coupling. In the construction, each time a r.v. (random variable) is introduced it should be taken independent of all quantities except those on which it depends explicitly.

Step 1: The object here is to force the two copies of (ζ_n) to the same state in E_φ at different random times ${T_1}^1$ and ${T_1}^2$ resp. This is accomplished as follows. Adjoin i.i.d. 0–1 valued r.v.'s Z_n, $n \in \mathbb{N}$, such that $\Pr(Z_n = 1) = \epsilon$. Starting at (x_0, y_0), an arbitrary point of S, let $({\zeta_n}^1)$ and $({\zeta_n}^2)$ run independently until the first times, $\tau_1{}^1$ and $\tau_1{}^2$ resp., when they are in E_φ ((ζ_n) inherits φ-recurrence from (ξ_n) so these are finite a.s.) Look at Z_0, and choose values for the marginals one step later so that

$$\begin{aligned}
&\Pr(\zeta^1_{\tau_1{}^1+1} \in E^1, \zeta^2_{\tau_1{}^2+1} \in E^2 \mid \zeta^1_{\tau_1{}^1} = x, \zeta^2_{\tau_1{}^2} = y, Z_0 = 1) \\
&\quad = \frac{\varphi(E^1 \cap E^2 \cap E_\varphi)}{\varphi(E_\varphi)}; \\
&\Pr(\zeta^1_{\tau_1{}^1+1} \in E^1, \zeta^2_{\tau_1{}^2+1} \in E^2 \mid \zeta^1_{\tau_1{}^1} = x, \zeta^2_{\tau_1{}^2} = y, Z_0 = 0) \\
&\quad = (1-\epsilon)^{-1}[(\pi_m(x,\cdot) \wedge \pi_m(y,\cdot))(E^1 \cap E^2) \\
&\qquad - \epsilon[\varphi(E_\varphi)]^{-1}\varphi(E^1 \cap E^2 \cap E_\varphi) + [1-\alpha_0{}^1(x,y)]^{-1}(\pi_m(x,\cdot) \\
&\qquad - \pi_m(y,\cdot)^+(E^1 \backslash E^2)(\pi_m(y,\cdot)) - \pi_m(x,\cdot)^+(E^2 \backslash E^1)]
\end{aligned} \tag{5}$$

(where $\alpha_0{}^1(x, y)$ is defined in terms of the t.f. π_m). Using (4), one checks that the marginal transition laws at the two times are well-defined and Markovian, each with t.f. π_m, and that $\zeta^1_{\tau_1{}^1+1} = \zeta^2_{\tau_1{}^2+1} \in E_\varphi$ a.s. *if* $Z_0 = 1$. If $Z_0 = 0$ we run the marginals independently until the times $\tau_2{}^1$, $\tau_2{}^2$ when they are in E_φ again, then determine the next step in analogy with (5) on the basis of Z_1. By repeating this procedure, eventually $\zeta^1_{\tau_n{}^1+1} = \zeta^2_{\tau_n{}^2+1} \in E_\varphi$ since $Z_n = 1$ for some n a.s. These are the desired times $T_1{}^1$ and $T_1{}^2$.

Step 2: Observe that if we were to repeat the procedure in Step 1 (with a new sequence of Z's), starting with $\zeta^1_{T_1{}^1} = x_1$, $\zeta^2_{T_1{}^2} = x_1$, then the marginals would both return to some $x_2 \in E_\varphi$ at a pair of random times $T_2{}^1$ and $T_2{}^2$ resp. For $x_1 \in E_\varphi$, let $Q_k(x_1) = \Pr(T_2{}^1 - T_1{}^1 = k \mid \zeta^1_{T_1{}^1} = x_1)$; $k \geqslant 1$. By construction,

$$\begin{aligned}
Q_k &:= \Pr(T_2{}^1 - T_1{}^1 = k) = \Pr((T_2{}^1 - 1) - (T_1{}^1 - 1) = k) \\
&= [\varphi(E_\varphi)]^{-1} \int_{x \in E_\varphi} Q_k(x)\varphi(dx).
\end{aligned}$$

We now apply a variation on Ornstein's trick. Fix a large integer K to be determined later. Take independent (Q_k)-distributed r.v.'s $Y_1{}^1$ and $Y_1{}^2$. If $|Y_1{}^2 - Y_1{}^1| \leqslant K$, follow Step 1 starting from $\zeta^1_{T_1{}^1} = \zeta^2_{T_1{}^2} = x_1$, and return to $\zeta^1_{T_2{}^1} = \zeta^2_{T_2{}^2} = x_2$ as above. If $|Y_1{}^2 - Y_1{}^1| > K$, however, we instead let $\zeta^1_{T_1{}^1+n} = \zeta^2_{T_1{}^2+n}$; $0 \leqslant n \leqslant Y_1{}^1$, with a common trajectory conditioned to return to E_φ after $Y_1{}^1$ steps, and set $T_2{}^1 = T_1{}^1 + Y_1{}^1$, $T_2{}^2 = T_1{}^2 + Y_1{}^1$.

Step 3 is to repeat Step 2 with new r.v.'s $Y_2{}^1$, $Y_2{}^2$, then $Y_3{}^1$, $Y_3{}^2$, etc., obtaining times $T_3{}^1$, $T_3{}^2$, then $T_4{}^1$, $T_4{}^2$, etc., when the copies occupy the same state of E_φ. By following this recipe, each marginal is Markovian with t.f. π_m. Moreover, the $T_n{}^2 - T_n{}^1$ are sums of i.i.d. bounded symmetric r.v.'s. The aperiodicity condition implies that $T_n{}^1 = T_n{}^2$ for some n with probability one if K is chosen sufficiently large (apply Proposition 1 just as in Lemma 1), i.e., $\zeta^1_{T_n{}^1} = \zeta^2_{T_n{}^2} = x_n$ for some n a.s. This completes the proof. ∎

COROLLARY (Blackwell–Freedman, Jamison–Orey). *Let (ξ_n) be a φ-recurrent Markov process with* t.f. π, $\operatorname{per}(\pi) = d \geqslant 1$. *For $0 \leqslant r < d$, define $A_r = \{\tau_{E_0} = r + nd$ for some $n \in \mathbb{N}\}$. If $\iota \in \mathscr{P}$, then the P_ι-decomposition of $\mathscr{T}$ is equivalent to*

$$\Omega = \sum_{r:\, P_\iota(A_r) > 0} A_r.$$

Proof. The process (ξ_{nd}) restricted to E_0 is aperiodic φ-recurrent, so by Orey's theorem $\lim_{n\to\infty} \delta_0{}^n(x, y) = \lim_{n\to\infty} \delta_0^{nd}(x, y) = 0$ for all $x, y \in E_0$. As in the proof of Theorem 1.2, we conclude that $P_x(B) = P_y(B) = 0$ *or* 1 whenever $B \in \mathscr{T}$ and $x, y \in E_0$. Since $x \in E_0$ implies $P_x(\xi_{nd} \in E_0) = 1$, we have

$$P_x(\theta^{l+nd}B) = \int_{y\in E_0} P_y(\theta^l B) P_x(\xi_{nd} \in dy) \qquad \text{for any} \quad l, n.$$

Hence $P_x(\theta^{l+nd}B) = P_y(\theta^l B)$ whenever $x, y \in E_0$, $0 \leqslant l < d$ and $n \in \mathbb{N}$. Also, from (3), $\tau_{E_0} < \tau_{\bar{E}} + d < \infty$ P_ι-a.s. for any $\iota \in \mathscr{P}$. Combining these facts,

$$P_\iota(B) = \sum_{k=0}^{\infty} P_\iota(\tau_{E_0} = k, B) = \sum_{\substack{0 \leqslant l < d:\\ P_x(\theta^{l+nd}B) = 1\\ \forall x \in E_0,\, n \in \mathbb{N}}} P_\iota(A_l).$$

Because $A_l \cong \{\xi_{l+nd} \in E_0 \text{ i.o.}\} \in \mathscr{T}$, the claim follows. ∎

3.3. *Strong Ergodic Theorems and Uniform Rates of Convergence*

Let (ξ_n) be a φ-recurrent Markov process. A set $E \in \mathscr{S}$ is called *φ-uniform* (resp. *φ-strongly uniform*) for (ξ_n) iff $\lim_{n\to\infty} \sup_{x\in E} P_x(\tau_{E'} > n) = 0$ (resp. $\sup_{x\in E} E_x(\tau_{E'}) < \infty$) for every $E' \in \mathscr{S}$ such that $\varphi(E') > 0$. A basic result in the theory states that any φ-recurrent process has an invariant σ-finite measure μ, unique up to a constant. The derivation of this fact, as presented by Orey [30], is based on the following proposition. The proof below is in [30]; we include it as an application of Corollary 2.9.

PROPOSITION 2. *If S is φ-uniform for (ξ_n), then (ξ_n) has a unique invariant probability measure μ, and there are nonnegative constants K and $\epsilon < 1$ such that*

$$\sup_{E\in\mathscr{S}} (p_m(\iota, E) - \mu(E)) \leqslant K\epsilon^m \qquad \textit{for all} \quad \iota \in \mathscr{P}, \quad m \in \mathbb{N}.$$

Proof. According to Corollary 2.9, and since $\bar{\delta}_0{}^n \lessapprox 2 \sup_{y\in S} \delta_0{}^n(x_0, y)$ for any fixed $x_0 \in S$, it suffices to show that

$$\sup_{y\in S} \delta_0{}^n(x_0, y) < \tfrac{1}{2} \qquad \text{for some} \quad n \in \mathbb{N}. \tag{6}$$

By Theorem 3 and Egorov's theorem, there is some $E_0 \in \mathscr{S}$ with $\varphi(E_0) > 0$ such that

$$\sup_{y \in E_0} \delta_0{}^n(x_0, y) \to 0 \qquad \text{as} \quad n \to \infty. \tag{7}$$

S is φ-uniform, so we can choose m such that $\sup_{y \in S} P_y(\tau_{E_0} > m) < \frac{1}{4}$. For $n > m$,

$$p_n(x_0, E) - p_n(y, E) \leqslant P_y(\tau_{E_0} > m) + p_n(x_0, E) P_y(\tau_{E_0} \leqslant m) - P_y(\tau_{E_0} \leqslant m, \xi_n \in E).$$

Using the first entry decomposition,

$$\sup_{y \in S} \delta_0{}^n(x_0, y) < \tfrac{1}{4} + \sup_{y \in S,\, E \in \mathscr{S}} \sum_{k=1}^{m} \int_{z \in E_0} P_y(\tau_{E_0} = k, \xi_k \in dz)$$
$$\cdot \left[(p_n(x_0, E) - p_{n-k}(x_0, E)) + (p_{n-k}(x_0, E) - p_{n-k}(z, E)) \right]$$
$$\leqslant \tfrac{1}{4} + \sum_{k=1}^{m} \delta_0^{n-k}(x_0{}^k, x_0) + \sup_{z \in E_0,\, E \in \mathscr{S}} (p_{n-k}(x_0, E) - p_{n-k}(z, E)).$$

Theorem 3 and (7) make the above sum small when n is large, so (6) holds for some such large n. ∎

Assuming the result mentioned above, let μ be an invariant measure for a φ-recurrent process (ξ_n). If μ is infinite, then a corollary of Theorem 3 asserts that $p_n(x, E) \to 0$ as $n \to \infty$ for every $x \in S$ and $E \in \mathscr{S}$ such that $\varphi(E) < \infty$. When μ is finite and (ξ_n) is aperiodic, Orey's theorem shows that

$$\lim_{m \to \infty} \sup_{E \in \mathscr{S}} (p_m(\iota, E) - \mu(E)) = 0, \qquad \iota \in \mathscr{P},$$

but one does not in general have the geometric convergence of Proposition 2. Our final application is based on the classical coupling, and makes use of certain observations for Markov chains due to Pitman [33]. We prove a uniform rate of convergence theorem for processes which possess a μ-strongly uniform set of positive μ-measure (in which case μ is necessarily finite).

First, though, we outline Pitman's treatment of classical coupling in the denumerable case. Suppose (ξ_n) is an irreducible aperiodic Markov chain with state space S and t.f. π. Let $(\bar{\xi}_n)$ denote the independent product process on $\tilde{S}$ with t.f. $\pi \times \pi$. Then $(\bar{\xi}_n)$ is also irreducible aperiodic, and (ξ_n) inherits positive recurrence or transience from $(\bar{\xi}_n)$, while if (ξ_n) is null recurrent, then $(\bar{\xi}_n)$ may be either null recurrent or transient. Consequently, the classical coupled process hits Δ in a finite expected time from any starting state (x, y) if (ξ_n) is positive recurrent, but otherwise examples show that the classical coupling may either succeed or fail.

These remarks yield

LEMMA 2. *Let Q_i^j; $i \geqslant 1$, $j = 1, 2$, be i.i.d. positive integer r.v.s such that $E(Q_i^j) < \infty$, and the group generated by $s(Q_i^j)$ is $\mathbb{Z}$. For $k \geqslant 1$, put*

$$\tilde{T}_k^{\,1} = \min\left\{m: \sum_{i=1}^{n_1} Q_i^{\,1} = k + \sum_{i=1}^{n_2} Q_i^{\,2} = m \text{ for some } n_1, n_2 \geqslant 1\right\},$$

$$\tilde{T}_k^{\,2} = \min\left\{m: k + \sum_{i=1}^{n_1} Q_i^{\,1} = \sum_{i=1}^{n_2} Q_i^{\,2} = m \text{ for some } n_1, n_2 \geqslant 1\right\}.$$

Then for $k \geqslant 1$, $E(T_k^{\,j}) \leqslant kE(T_1^{\,j}) < \infty$.

Proof. Let (ξ_n) be a homogeneous chain with state space $\mathbb{N}$ and t.f. $\pi(n, 0) = 1 - \pi(n, n+1) = \Pr(Q_i^j = n+1)$; $n \in \mathbb{N}$. Then (ξ_n) is irreducible aperiodic positive recurrent, with $P_0(\tau_0 = n) = \Pr(Q_i^j = n)$. If $(\bar{\xi}_n) = (\xi_n^{\,1}, \xi_n^{\,2})$ is the product process, then $E(\tilde{T}_1^{\,1})$ is the expected value of the first time m such that $\xi_m^{\,1} = 0$ and $\xi_{m-1}^2 = 0$. This is clearly majorized by $\bar{E}_{(0,0)}(\tau_{(0,1)})$, which is finite since $(\bar{\xi}_n)$ is positive recurrent. Symmetry shows that $E(\tilde{T}_k^{\,2}) = E(\tilde{T}_k^{\,1})$, and the remaining inequalities are proved by a straightforward induction on k. ∎

THEOREM 3. *Let (ξ_n) have invariant probability measure μ, and be aperiodic μ-recurrent. Suppose there is a strongly uniform set $E \in \mathscr{S}$ such that $\mu(E) > 0$. Then for $\mu \times \mu$-almost all $(x, y) \in \tilde{S}$,*

$$\lim_{n\to\infty} n\delta_0^{\,n}(x, y) = 0. \tag{8}$$

Proof. Since E is strongly uniform (w.r.t. μ), we can find $E_\mu \subset E$ satisfying (4) for some m, $\epsilon > 0$, and such that $\sup_{x\in E_\mu} E_x(\tau_{E_\mu}) < \infty$. Once again we pass to the subsequential process $(\zeta_n) = (\xi_{mn})$ with t.f. π_m. (ζ_n) inherits the invariant measure μ, aperiodicity, μ-recurrence, and E_μ is strongly uniform for (ζ_n). For the new process, let

$$\sup_{x\in E_\mu} E_x(\tau_{E_\mu}) = R < \infty.$$

To prove (8), we construct a coupling $(\tilde{P}_{(x,y)})$ for (ζ_n), and show that $\tilde{E}_{(x,y)}(\tau_\Delta) < \infty$ for $\mu \times \mu$-almost all $(x, y) \in \tilde{S}$. Together with monotonicity of δ and (1.10) (with γ replaced by δ), this gives

$$\limsup_{n\to\infty} n\delta_0^{\,n}(x, y) \leqslant \limsup_{n\to\infty} nP_{(x,y)}(\tau_\Delta > n) = 0.$$

Step 1: A simple check shows that $\tilde{\mu} = \mu \times \mu$ is an invariant probability measure for the product process $(\bar{\zeta}_n)$. If $\tilde{E} \in \tilde{\mathscr{S}}$, define $S_{x'} = \{y \in S:(x', y) \in \tilde{E}\}$. For any $(x, y) \in \tilde{S}$,

$$|\bar{P}_{(x,y)}(\bar{\zeta}_n \in \tilde{E}) - \mu(\tilde{E})| = \left| \int_{x' \in S} p_n(y, S_{x'})[p_n(x, dx') - \mu(dx')] \right.$$
$$\left. + \int_{x' \in S} [p_n(y, S_{x'}) - \mu(S_{x'})]\mu(dx') \right|$$
$$\leqslant \|p_n(x, \cdot) - \mu\| + \|p_n(y, \cdot) - \mu\| \to 0$$

as $n \to \infty$ by Orey's theorem on the marginals. From this one easily concludes that $(\bar{\zeta}_n)$ is $\mu \times \mu$-irreducible and aperiodic. By combining standard facts ([30, Theorem 8.1, p. 38; 7, Proposition 3.1]) it follows that for any $\tilde{E} \in \tilde{\mathscr{T}}$ such that $\mu \times \mu(\tilde{E}) > 0$, $\bar{E}_{(x,y)}(\tilde{\tau}_E) < \infty$ except on a $\mu \times \mu$-null set. In particular $\bar{E}_{(x,y)}(\tau_{E_\mu \times E_\mu}) < \infty$ for all (x, y) in some $\mu \times \mu$-full set $\tilde{S}_0 \subset \tilde{S}$. From now on, restrict x and y to $\tilde{S}_0$. The first step in our coupling is to let $(\zeta_n{}^1)$ and $(\zeta_n{}^2)$ run independently until they are simultaneously in $E_\mu \times E_\mu$. This time has finite expectation whenever $(x, y) \in \tilde{S}_0$, so for the remainder of the construction we may assume without loss of generality that $\bar{\zeta}_0 \in E_\mu \times E_\mu$.

Step 2: Now adjoin i.i.d. r.v.s Z_n such that $\Pr(Z_n = 1) = \epsilon$. Look at Z_0, and choose the values of the marginals one step later as in (5) (with $\tau_1{}^1 = \tau_1{}^2 = 0$). Just as we did there, if $Z_0 = 0$ let $(\zeta_n{}^1)$ and $(\zeta_n{}^2)$ run independently until times $\tau_2{}^1$ and $\tau_2{}^2$ when they are in E_μ again, then determine the next step on the basis of Z_1 and (5). Continuing in this manner, eventually $\zeta^1_{\tau_n{}^1+1} = \zeta^2_{\tau_n{}^2+1} \in E_\mu$ because $Z_n = 1$ eventually. Call these times $T_0{}^1$ and $T_0{}^2$ resp. By strong uniformity, $E(T_0{}^1) \leqslant 1 + (1 + R)E(\min\{n:Z_n = 1\}) < \infty$, and similarly $E(T_0{}^2) < \infty$.

Step 3 is to carry out Step 2 repeatedly, each time adjoining new Z's and using them to get times $T_i{}^1$, $T_i{}^2$ such that the marginals are at the same state of E_μ. As in the proof of Orey's theorem, the $Q_i{}^j = T_i{}^j - T_{i-1}^j$; $i \geqslant 1, j = 1, 2$, are all aperiodic i.i.d. r.v.s. Now if

$$T_0{}^1 + \sum_{i=1}^{n_1} Q_i{}^1 = T_0{}^2 + \sum_{i=1}^{n_2} Q_i{}^2$$

for some n_j, then our coupled process is on the diagonal at the time given by the common sum. To show that the expected time until τ_Δ is finite, it therefore suffices to show that

$$E(\tau_\Delta, T_0{}^1 = T_0{}^2) + E(\tilde{T}^1_{(T_0{}^2 - T_0{}^1)}, T_0{}^2 - T_0{}^1 > 0)$$
$$+ E(\tilde{T}^2_{(T_0{}^1 - T_0{}^2)}, T_0{}^1 - T_0{}^2 > 0)$$

is finite, with $\tilde{T}_i^{\,j}$ as in Lemma 2. Using that lemma, the above sum is dominated by

$$E(T_0^{\,1}) + E(\tilde{T}_1^{\,1})E(|T_0^{\,1} - T_0^{\,2}|) \leqslant E(T_0^{\,1}) + E(\tilde{T}_1^{\,1})(E(T_0^{\,1}) + E(T_0^{\,2})) < \infty.$$

This completes the proof. ∎

Under appropriate moment conditions, and using Theorem 3 as a starting point, one could undoubtedly develop a theory of uniform rates of convergence for processes with strongly uniform sets. The Markov chain case is presented in great detail by Pitman [33].

Appendix. Remarks on Continuous Time Processes

Corollary 2.2 shows that the structure of $\mathscr{T}$ is determined by any subsequential process. This fact leads to straightforward extension of most of the coupling theory to continuous time Markov processes. Such a process $(\xi_t)_{t \in \mathbb{R}^+}$ has tail σ-algebra $\mathscr{T} = \bigcap_{t \in \mathbb{R}^+} \mathscr{B}_t^{\,\infty}$. The quantities $\gamma_t^{\,u}(\mu, \nu)$ and $\delta_t^{\,u}(\mu, \nu)$ may be defined in strict analogy to the discrete case, and enjoy the same monotonicity properties. Markovian couplings are constructed in terms of the infinitesimal generator for the original process. For example, if (ξ_t) is a continuous time Markov chain uniquely determined by its infinitesimal transition functions $\pi^t(x, y)$ $(P_x^{\,t}(\xi_{dt} = y) \sim \pi^t(x, y)\,dt$ for $x \neq y)$, then the analog of the Vasershtein coupling is

$$\begin{aligned}
&\tilde{\pi}^t((x, y), (x, x)) = \pi^t(y, x); \qquad \tilde{\pi}^t((x, y), (y, y)) = \pi^t(x, y);\\
&\tilde{\pi}^t((x, y), (z, z)) = \pi^t(x, z) \wedge \pi^t(y, z);\\
&\tilde{\pi}^t((x, y), (x, z)) = (\pi^t(y, z) - \pi^t(x, z))^+;\\
&\tilde{\pi}^t((x, y), (z, y)) = (\pi^t(x, z) - \pi^t(y, z))^+;\\
&x, y, z \in S, \qquad x \neq y, \qquad z \neq x, y,
\end{aligned} \tag{$*$}$$

with

$$\tilde{\pi}^t((x, y), (x, y)) = -\sum_{(w,z) \neq (x,y)} \tilde{\pi}^t((x, y), (w, z)).$$

Construction of the maximal coupling is possible, but even more notationally tedious, since cylinders are indexed by arbitrary increasing sequences of times, and the induction arguments have to be modified.

Theorem 2.1 and its corollaries have obvious extensions, where limits are taken along subsequences. An analog of Theorem 2.3 may be proved using (∗). This was done in [16] for Markov chains, in which case the continuous time ergodic coefficients are

$$\underline{\beta}_t = \inf_{x \neq y} [\pi^t(x, y) + \pi^t(y, x) + \sum_{z \neq x,\, y} (\pi^t(x, z) \wedge \pi^t(y, z))]; \qquad \underline{\underline{\beta}} = \inf_{t \in \mathbb{R}^+} \underline{\beta}_t.$$

Various of the applications in Section 3, e.g., Orey's theorem, may also be obtained in the continuous setting without difficulty.

Bibliographical Notes

Introduction

Versions of Doeblin's proof are in [5, pp. 185–187] and [41]. The first general discussion of the coupling method of which I am aware is due to Vasershtein [37], though specific instances of the technique have been frequent since Doeblin. The papers [37] and [12] led to a proliferation of coupling arguments in the field of interacting particle systems (see [19] for references). A remark about the term "coupling": this is sometimes used to describe any bivariate process with prescribed marginals. In the more general sense coupling is commonly used to control one marginal by means of the other (cf. [20] and [28]). Such an idea enters occasionally into our discussion (e.g., in the proof of Theorem 2.3), but the term "coupling" will be reserved for joint measures on two copies of the same process such that the diagonal Δ is absorbing.

Section 1

§1.1. The technical assumption that S be polish is made to ensure (i) measurability of Δ in $S \times S$, (ii) validity of the Kolmogorov extension theorem, and (most importantly) (iii) existence of regular conditional probabilities. This also guarantees measurability of $\gamma_m(\xi_m, m)$ and validity of Egorov's theorem in Proposition 3.2. More notation used in the text: $|S|$ and 1_S are, respectively, the cardinality and the indicator function of a set S. $\triangle$ denotes symmetric difference, $\otimes$ means product σ-algebra, and $\sum$ is used for a disjoint union of sets. $\equiv$ means identically equal, $:=$ is occasionally used for a defining notational equality. w.r.t. abbreviates "with respect to." ▌ signals the end of a proof.

§1.2. A version of Proposition 1 with reference to Blackwell's paper [2] may be found in [10]. Alternate proofs of Corollary 1 are in [9], [25], and [34]. Cohn first proved that $\mathcal{T}$ is finite when S is finite in [8], while the fact that $|I| \leqslant |S|$ is due to Senchenko [34]. *Homogeneous* chains with full and fully nonatomic $\mathcal{T}$ may be found in [14, p. 46] and [30, p. 21]; it is worth noting how relatively simple Example 3 is in comparison.

§1.3. 0–λ laws were discussed in [1], and more systematically in [23]. The notions of λ-mixing and λ-ergodicity are also given in [23], while the property of λ-constant space–time harmonics generalizes the $\lambda = 1$ case exploited by Jamison and Orey [24] in the homogeneous setting.

§1.4–1.6 extend the presentation of [15] from homogeneous chains to nonhomogeneous processes. Ornstein's coupling is based on a somewhat different construction in [31]; the Vasershtein coupling is derived from [37]. Greater generality makes the proof of Theorem 3 slightly simpler than the original proof in [15]; let us note in passing that the equation in [15] corresponding to (28) in the present chapter contained unfortunate typographical errors. See [23] for a proof of Corollary 2 which does not use coupling.

Section 2

§2.1–2.3 generalize the material of [17]. Theorem 1 has an analog based on reverse martingales, and due to Cohn [9], which states that

$$\lim_{m\to\infty}\lim_{n\to\infty}\sup_{B\in\mathscr{B}_0^m}\{P_\iota(B)-P_\iota(B|\mathscr{B}_n^\infty)(\omega)\} \qquad (*)$$

takes on the same values as $1-\alpha(\omega)$ P_ι-a.s. Corollaries 1 and 2 are also consequences of Cohn's structure theorem (∗).

Theorem 2 is the most general version of a series of equivalence theorems for homogeneous and nonhomogeneous processes developed in [3], [24], [30], and [23]. The sufficient conditions (i)–(iv) are new. Condition (i′) for chains appeared in [15], and is related to the original proof of Orey's ergodic theorem (cf. [29]). A strong version of (i) in the homogeneous case says roughly that the process (ξ_n) is totally forgetful w.r.t. every $\iota\in\mathscr{P}$ iff starting from x the process spends a lot of time in $H_n^-(x, y)$ *or* starting from y it visits $H_n^+(x, y)$ repeatedly. In other words, (ξ_n) often visits states which are more likely from another starting position. The result (ii′) ⇔ (1.2) is due to Iosifescu [23].

Corollary 3 is based on a remark in [1], while Proposition 3 generalizes a result for chains from [23]. Corollaries 5 and 6 extend slightly results due to Cohn which may also be proved using (∗).

§2.4 is taken from [16], where S is countable. I learned the method of proof formalized in Proposition 5 from Harris [19]. References for Theorem 3 are [27], [11], and [21]. Corollary 7 is due to Dobrushin [11]; the results of Corollary 8 were first proved in [18] and [21]. Some interesting remarks relating to this material are in [25]. Corollary 9 is well known. The literature on ergodic coefficients is immense; for a sampling the reader is referred to [22], [32], [35], and [36]. A colorful problem for *finite* homogeneous chains is the following. Suppose $|S| = N < \infty$. By Corollary 9, the distribution of ξ_n converges geometrically to a limiting probability measure iff $\underline{\alpha}_0{}^l > 0$ for some $l > 0$. Can we find some $m = m(N)$ so that for any N-state chain $\underline{\alpha}_0{}^m = 0 \Rightarrow \underline{\alpha}_0{}^n = 0$ for all $n\in\mathbb{N}$, but such that there is an N-state chain with $\underline{\alpha}_0^{m-1} = 0$, $\underline{\alpha}_0{}^m > 0$? The recent solution by Madsen [26] gives $m(N) = (N^2/2) - \{(N/2) + [N/2]\} + 1$.

Section 3

§3.1. The program here is Ornstein's [31]. A more detailed version of the Chung–Ornstein proof of Proposition 1 appears in [4, pp. 56–58]. Theorem 1 can also be proved à la Blackwell and Freedman [3]; their proof using the Hewitt–Savage law generalizes to $S =$ a countable abelian group, whereas ours depends on the structure of $\mathbb{R}$ (i.e., on Proposition 1). The proof we give of Theorem 2 is different from Ornstein's, though similar in spirit.

§3.2–3.3. Orey's theorem was proved in [29] for Markov chains. The Corollary appeared in [3] (with Theorem 3 as a corollary), again in the denumerable case. The extension to φ-recurrent processes is due to Jamison and Orey [24]; their proof used space–time martingales. The coupling argument given here is new; H. Kesten showed me the key step. The terminology of uniform sets in Section 3.3 is taken from Cogburn [7]. Positive recurrence of $(\tilde{\xi}_n)$ yields the Markov chain case of Theorem 3, which was shown by Pitman [33]. Our generalization is new; again H. Kesten was instrumental in the proof. I would like to thank G. Letac for pointing out an erroneous remark in [15], where it is asserted that the classical coupling is never successful when (ξ_n) is transient. The coupled process for an aperiodic transient random walk on $\mathbb{Z} =$ the integers, with $S(\pi)$ finite, has $\tau_\Delta < \infty$ a.s. from any (x, y) as a consequence of Proposition 1.

Appendix

Similar general remarks are made in [23] and [9].

Acknowledgments

I would like to thank Frank Spitzer and Richard Holley, who introduced me to coupling methods, and Harry Kesten for invaluable help with Section 3 of this paper. My gratitude also extends to the many mathematicians whose ideas formed the basis for my work, especially R. Dobrushin, S. Orey, H. Cohn, and J. Pitman. Finally, thanks to H. O. Georgii for pointing out a number of misprints in the original manuscript.

References

1. P. Bartfay and P. Révész, On a zero-one law, *Z. Wahrscheinlichkeitstheorie und Verw. Gebiete* **7** (1967), 43–47.
2. D. Blackwell, Finite nonhomogeneous Markov chains, *Ann. of Math.* **46** (1945), 594–599.
3. D. Blackwell and D. Freedman, The tail σ-field of a Markov chain and a theorem of Orey, *Ann. Math. Statist.* **35** (1964), 1291–1295.
4. L. Breiman, "Probability," Addison-Wesley, Reading, Mass., 1968.
5. L. Breiman, "Probability and Stochastic Processes with a View Toward Applications," Houghton, Boston, 1969.

6. K. L. Chung and D. Ornstein, On the recurrence of sums of random variables, *Bull. Amer. Math. Soc.* **68** (1962), 30–32.
7. R. Cogburn, A uniform theory for sums of Markov chain transition probabilities, *Ann. Probability* **3** (1975), 191–214.
8. H. Cohn, On the tail σ-algebra of the finite inhomogeneous Markov chains, *Ann. Math. Statist.* **41** (1970), 2175–2176.
9. H. Cohn, On the tail events of a Markov chain, *Z. Wahrscheinlichkeitstheorie und Verw. Gebiete* **29** (1974), 65–72.
10. H. Cohn, A ratio limit theorem for the finite nonhomogeneous Markov chains, *Israel J. Math.* **19** (1975), 329–334.
11. R. L. Dobrushin, Central limit theorem for nonstationary Markov chains, *Theor. Probability Appl.* **1** (1956), 65–80; 329–383.
12. R. L. Dobrushin, Markov processes with a large number of locally interacting components, *Problemy Peredači Informacii* **7** (1971), 70–87.
13. W. Doeblin, Exposé de la théorie des chaînes simples constantes de Markov à un nombre fini d'états, *Rev. Math. Union Interbalkanique* **2** (1938), 77–105.
14. D. Freedman, "Markov Chains," Holden-Day, San Francisco, 1971.
15. D. Griffeath, A maximal coupling for Markov chains, *Z. Wahrscheinlichkeitstheorie und Verw. Gebiete* **31** (1975), 95–106.
16. D. Griffeath, Uniform coupling of nonhomogeneous Markov chains, *J. Appl. Probability* **12** (1975), 753–762.
17. D. Griffeath, Partial coupling and loss of memory for Markov chains, *Ann. Probability* **4** (1976), 850–858.
18. J. Hajnal, Weak ergodicity in nonhomogeneous Markov chains, *Proc. Cambridge Philos. Soc.* **54** (1958), 233–246.
19. T. E. Harris, Contact interactions on a lattice, *Ann. Probability* **2** (1974), 969–988.
20. R. Holley, Recent results on the stochastic Ising model, *Rocky Mt. J. Math.* **4** (1974), 479–496.
21. M. Iosifescu, Conditions nécessaires et suffisantes pour l'ergodicité uniforme des chaînes de Markoff variables et multiples, *Rev. Roumaine Math. Pures Appl.* **11** (1966), 325–330.
22. M. Iosifescu, On two recent papers on ergodicity in nonhomogeneous Markov chains, *Ann. Math. Statist.* **43** (1972), 1732–1736.
23. M. Iosifescu, On finite tail σ-algebras, *Z. Wahrscheinlichkeitstheorie und Verw. Gebiete* **24** (1972), 159–166.
24. B. Jamison and S. Orey, Markov chains recurrent in the sense of Harris, *Z. Wahrscheinlichkeitstheorie und Verw. Gebiete* **8** (1967), 41–48.
25. J. F. C. Kingman, Geometrical aspects of the theory of nonhomogeneous Markov chains, *Math. Proc. Cambridge Philos. Soc.* **77** (1975), 171–183.
26. R. Madsen, Decidability of $\alpha(P^k) > 0$ for some k, *J. Appl. Probability* **12** (1975), 333–340.
27. A. A. Markov, Investigation of an important case of dependent trials, *Izv. Akad. Nauk SPB (6)* **1** (1907), 61–80.
28. G. L. O'Brien, The comparison method for stochastic processes, *Ann. Probability* **3** (1975), 80–88.
29. S. Orey, An ergodic theorem for Markov chains, *Z. Wahrscheinlichkeitstheorie und Verw. Gebiete* **1** (1962), 174–176.
30. S. Orey, "Limit Theorems for Markov Chains," Van Nostrand, New York, 1971.
31. D. Ornstein, Random walk I, *Trans. Amer. Math. Soc.* **138** (1969), 1–43.
32. A. Paz, Ergodic theorems for infinite probabilistic tables, *Ann. Math. Statist.* **41** (1970), 539–550.
33. J. W. Pitman, Uniform rates of convergence for Markov chain transition probabilities,

Z. Wahrscheinlichkeitstheorie und Verw. Gebiete **29** (1974), 193–227.
34. D. V. SENCHENKO, The final σ-algebra of an inhomogeneous Markov chain with a finite number of states, *Math. Notes* **12** (1972), 610–613.
35. E. SENETA, On the historical development of the theory of finite inhomogeneous Markov chains. *Proc. Cambridge Philos. Soc.* **74** (1973), 507–513.
36. E. SENETA, "Non-negative Matrices," Wiley, New York, 1973.
37. L. N. VASERSHTEIN, Markov processes on countable product spaces describing large systems of automata, *Problemy Peredači Informacii* **3** (1969), 64–72.

SUPPLEMENTARY REFERENCES

After completing the present paper, the author learned of the following references:

38. J. W. PITMAN, Stopping time identities and limit theorems for Markov chains, Ph.D. Thesis, Univ. of Sheffield, 1974
39. J. W. PITMAN, On coupling of Markov chains, *Z. Wahrscheinlichkeitstheorie und Verw. Gebiete* **35** (1976), 315–322.
40. P. HOEL, S. PORT, AND C. STONE, "Introduction to Stochastic Processes," Houghton, New York, 1972.
41. T. LINDVALL, A probabilistic proof of Blackwell's renewal theorem, *Ann. Probability* **5** (1977), 482–485.

Ams (MOS) 1970 subject classification: 60J05.

STUDIES IN PROBABILITY AND ERGODIC THEORY
ADVANCES IN MATHEMATICS SUPPLEMENTARY STUDIES, VOL. 2

On Fluctuations of Sums of Random Variables

LAJOS TAKÁCS

Department of Mathematics and Statistics, Case Western Reserve University, Cleveland, Ohio

The main object of this paper is to give a method for determining the stochastic laws of the fluctuations of the partial sums for a sequence of independent and identically distributed real random variables, and for a semi-Markov sequence of real random variables. The indicated method is based on the solutions of various recurrence equations in a commutative Banach algebra $\mathbf{R}_1$ and in a noncommutative Banach algebra $\mathbf{R}_2$.

1. INTRODUCTION

Let $\zeta_0, \xi_1, \xi_2, \ldots, \xi_n, \ldots$ be real random variables. Define ζ_n for $n = 1, 2, \ldots$ by the recurrence formula

$$\zeta_n = \zeta_{n-1} + \xi_n \tag{1}$$

and η_n for $n = 1, 2, \ldots$ by the recurrence formula

$$\eta_n = [\eta_{n-1} + \xi_n]^+ \tag{2}$$

where $\eta_0 = \zeta_0$ and $[x]^+ = \max(0, x)$.

In the theory of probability there are many problems which require the determination of the stochastic laws of the fluctuations of the sequences $\{\zeta_n\}$ and $\{\eta_n\}$ for a wide class of random variables $\{\xi_n\}$. In this paper we assume either that $\{\xi_n\}$ is a sequence of mutually independent and identically distributed real random variables, or, more generally, that $\{\xi_n\}$ is a semi-Markov sequence of real random variables and determine the distributions of several random variables depending on (1) and (2). In addition to ζ_n and η_n we shall find the distributions and various joint distributions of the random variables η_n^*, ν, ν^*, Δ_n, Δ_n^*, ω_n, ω_n^*, and $\eta_{n,k}$ defined below.

We define

$$\eta_n^* = \max(\zeta_0, \zeta_1, \ldots, \zeta_n) \tag{3}$$

for $n \geqslant 0$,

$$\nu = \inf\{n : \zeta_n \leqslant 0 \text{ and } n \geqslant 1\} \tag{4}$$

ISBN 0-12-599102-9

($\nu = \infty$ if $\zeta_n > 0$ for all $n \geqslant 1$), and

$$\nu^* = \inf\{n:\zeta_n < 0 \text{ and } n \geqslant 1\}, \tag{5}$$

($\nu^* = \infty$ if $\zeta_n \geqslant 0$ for all $n \geqslant 1$).

We define Δ_n as the number of positive elements in the sequence $\zeta_1, \zeta_2, \ldots, \zeta_n$ ($\Delta_0 = 0$), Δ_n^* as the number of nonnegative elements in the sequence $\zeta_1, \zeta_2, \ldots, \zeta_n$ ($\Delta_0^* = 0$), ω_n as the number of subscripts $k = 1, 2, \ldots, n$ for which either $\zeta_k > 0$ and $\zeta_{k-1} \leqslant 0$ or $\zeta_k \leqslant 0$ and $\zeta_{k-1} > 0$ ($\omega_0 = 0$), and ω_n^* as the number of subscripts $k = 1, 2, \ldots, n$ for which either $\zeta_k \geqslant 0$ and $\zeta_{k-1} < 0$ or $\zeta_k < 0$ and $\zeta_{k-1} \geqslant 0$ ($\omega_0^* = 0$).

If $x_0, x_1, \ldots, x_n$ are real numbers, then let $R_{n,k}(x_0, x_1, \ldots, x_n)$ ($0 \leqslant k \leqslant n$) be the numbers $x_0, x_1, \ldots, x_n$ arranged in nondecreasing order of magnitude. We assume that in the ordered sequence x_i precedes x_j if either $x_i < x_j$ or $x_i = x_j$ and $i < j$. Let us define

$$\eta_{n,k} = R_{n,k}(\zeta_0, \zeta_1, \ldots, \zeta_n) \tag{6}$$

for $0 \leqslant k \leqslant n$.

In this paper we consider two Banach algebras $\mathbf{R}_1$ and $\mathbf{R}_2$ and give the solutions of various recurrence equations in $\mathbf{R}_1$ and $\mathbf{R}_2$. By making use of these results we shall be able to determine the distributions of the indicated random variables.

2. A Space $\mathbf{R}_1$

Let us define $\mathbf{R}_1$ as the space of functions $\Phi(s)$ defined for $\text{Re}(s) = 0$ on the complex plane which can be represented in the form

$$\Phi(s) = \mathbf{E}\{\zeta e^{-s\eta}\} \tag{7}$$

where η is a real random variable and ζ is a complex (or real) random variable for which $\mathbf{E}\{|\zeta|\} < \infty$. Let us define in $\mathbf{R}_1$ the operations of addition, multiplication, and multiplication by a complex (or real) constant according to the rules of algebra. We define the norm of $\Phi(s) \in \mathbf{R}_1$ by

$$\|\Phi\| = \inf_{\zeta} \mathbf{E}\{|\zeta|\} \tag{8}$$

where the infimum is taken for all admissible ζ in the representation (7). We have $|\Phi(s)| \leqslant \|\Phi\|$ for $\text{Re}(s) = 0$.

If $\Phi(s) \in \mathbf{R}_1$, $\Phi_1(s) \in \mathbf{R}_1$, $\Phi_2(s) \in \mathbf{R}_1$, and c is a complex (or real) constant, then $c\Phi(s) \in \mathbf{R}_1$, $\Phi_1(s) + \Phi_2(s) \in \mathbf{R}_1$, $\Phi_1(s)\Phi_2(s) \in \mathbf{R}_1$, and $\|c\Phi\| = |c|\ \|\Phi\|$, $\|\Phi_1 + \Phi_2\| \leqslant \|\Phi_1\| + \|\Phi_2\|$, $\|\Phi_1\Phi_2\| \leqslant \|\Phi_1\|\ \|\Phi_2\|$. The space $\mathbf{R}_1$ is a linear normed space.

LEMMA 1. *If* $\Phi_k(s) \in \mathbf{R}_1$ *for* $k = 0, 1, 2, \ldots$, *and if*

$$\sum_{k=0}^{\infty} \|\Phi_k\| < \infty, \tag{9}$$

then

$$\Phi(s) = \sum_{k=0}^{\infty} \Phi_k(s) \tag{10}$$

belongs to $\mathbf{R}_1$ *and*

$$\|\Phi\| \leqslant \sum_{k=0}^{\infty} \|\Phi_k\|. \tag{11}$$

Proof. Let ϵ be any positive number and for each $k = 0, 1, 2, \ldots$ let us represent $\Phi_k(s)$ in the form of $\Phi_k(s) = \mathbf{E}\{\zeta_k e^{-s\eta_k}\}$, where $\mathbf{E}\{|\zeta_k|\} \leqslant (1 + \epsilon)\|\Phi_k\|$. Let us define a discrete random variable ω such that it takes on nonnegative integers only, it is independent of the sequence (ζ_k, η_k) $(k = 0, 1, 2, \ldots)$ and $\mathbf{P}\{\omega = k\} = 1/(k + 1)(k + 2)$ for $k = 0, 1, 2, \ldots$. Define $\zeta = (\omega + 1)(\omega + 2)\zeta_\omega$ and $\eta = \eta_\omega$. Then by the theorem of total expectation we obtain

$$\mathbf{E}\{|\zeta|\} = \sum_{k=0}^{\infty} \mathbf{E}\{|\zeta_k|\} \leqslant (1 + \epsilon) \sum_{k=0}^{\infty} \|\Phi_k\| < \infty \tag{12}$$

and therefore

$$\mathbf{E}\{\zeta e^{-s\eta}\} = \sum_{k=0}^{\infty} \mathbf{E}\{\zeta_k e^{-s\eta_k}\} = \sum_{k=0}^{\infty} \Phi_k(s) \tag{13}$$

belongs to $\mathbf{R}_1$. Accordingly, if $\Phi(s) = \mathbf{E}\{\zeta e^{-s\eta}\}$, then $\Phi(s) \in \mathbf{R}_1$, $\Phi(s)$ is represented by (10), and

$$\|\Phi\| \leqslant \mathbf{E}\{|\zeta|\} \leqslant (1 + \epsilon) \sum_{k=0}^{\infty} \|\Phi_k\|. \tag{14}$$

Since ϵ is an arbitrary positive number, this last inequality implies (11).

By Lemma 1 we can prove the following property of $\mathbf{R}_1$.

LEMMA 2. *If* $\Phi_k(s) \in \mathbf{R}_1$ *for* $k = 0\ 1, 2, \ldots$, *and if* (9) *holds, then there exists a* $\Phi(s) \in \mathbf{R}_1$ *such that*

$$\lim_{n\to\infty} \left\| \Phi - \sum_{k=0}^{n} \Phi_k \right\| = 0. \tag{15}$$

The function $\Phi(s)$ *is given by* (10) *and it satisfies* (11).

Proof. It is sufficient to prove that (10) implies (15). If $\Phi(s)$ is defined by (10), then $\Phi(s) \in \mathbf{R}_1$ and by Lemma 1

$$\left\| \Phi - \sum_{k=0}^{n} \Phi_k \right\| = \left\| \sum_{k=n+1}^{\infty} \Phi_k \right\| \leqslant \sum_{k=n+1}^{\infty} \|\Phi_k\|. \tag{16}$$

If $n \to \infty$, then the extreme right member in (16) tends to 0 and this implies (15).

Conversely, (15) implies (10) too. If (15) holds and if we use the inequality

$$\left| \Phi(s) - \sum_{k=0}^{n} \Phi_k(s) \right| \leqslant \left\| \Phi - \sum_{k=0}^{n} \Phi_k \right\|, \tag{17}$$

then by (15) it follows that

$$\lim_{n \to \infty} \sum_{k=0}^{n} \Phi_k(s) = \Phi(s) \tag{18}$$

for $\mathrm{Re}(s) = 0$. This proves (10).

By Lemma 2 we can conclude that the space $\mathbf{R}_1$ is complete, that is, $\mathbf{R}_1$ is a Banach space. More precisely, $\mathbf{R}_1$ is a commutative Banach algebra. However, we are not making use of the theory of Banach spaces, we are using only the aforementioned properties of the space $\mathbf{R}_1$.

In particular, it follows from Lemma 1 that if $\Psi(s) \in \mathbf{R}_1$, then for any ρ

$$\exp(\rho\Psi(s)) = \sum_{k=0}^{\infty} \frac{\rho^k}{k!} [\Psi(s)]^k \tag{19}$$

belongs to $\mathbf{R}_1$ and

$$\|\exp(\rho\Psi)\| \leqslant \exp(|\rho|\ \|\Psi\|). \tag{20}$$

Furthermore, if $\Psi(s) \in \mathbf{R}_1$, and if $|\rho|\ \|\Psi\| < 1$, then

$$\log[1 - \rho\Psi(s)] = -\sum_{k=1}^{\infty} \frac{\rho^k}{k} [\Psi(s)]^k \tag{21}$$

belongs to $\mathbf{R}_1$ and

$$\|\log[1 - \rho\Psi]\| \leqslant -\log(1 - |\rho|\ \|\Psi\|). \tag{22}$$

We shall consider transformations $\mathbf{T}$ in $\mathbf{R}_1$ which satisfy the following conditions:

(i) The transformation $\mathbf{T}$ is a bounded linear transformation of $\mathbf{R}_1$ into itself.

$$\mathbf{T}^2\{\Phi(s)\} = \mathbf{T}\{\Phi(s)\} \tag{23}$$

for all $\Phi(s) \in \mathbf{R}_1$.

(iii) If $\Phi_1(s) \in \mathbf{R}_1$ and $\Phi_2(s) \in \mathbf{R}_1$, then

$$\begin{aligned}\mathbf{T}\{\Phi_1(s)\Phi_2(s)\} = \mathbf{T}\{\mathbf{T}\{\Phi_1(s)\}\Phi_2(s)\} + \mathbf{T}\{\Phi_1(s)\mathbf{T}\{\Phi_2(s)\}\}\\ - \mathbf{T}\{\Phi_1(s)\}\mathbf{T}\{\Phi_2(s)\}.\end{aligned} \tag{24}$$

By (i) we mean that if $\Phi(s) \in \mathbf{R}_1$, $\Phi_1(s) \in \mathbf{R}_1$, $\Phi_2(s) \in \mathbf{R}_1$ and c is a complex (or real) constant, then $\mathbf{T}\{\Phi(s)\} \in \mathbf{R}_1$, $\mathbf{T}\{c\Phi(s)\} = c\mathbf{T}\{\Phi(s)\}$, and $\mathbf{T}\{\Phi_1(s) + \Phi_2(s)\} = \mathbf{T}\{\Phi_1(s)\} + \mathbf{T}\{\Phi_2(s)\}$. Furthermore, $\|\mathbf{T}\{\Phi\}\| \leqslant K\|\Phi\|$, where K is a finite nonnegative constant. The smallest possible value of K is denoted by $\|\mathbf{T}\|$ and is called the norm of $\mathbf{T}$.

By (ii) it follows that if $\mathbf{T}$ is not the zero transformation, then $\|\mathbf{T}\| \geqslant 1$.

By (iii) it follows that if $\Psi_i(s) \in \mathbf{R}_1$ $(i = 1, 2)$ and $\mathbf{T}\{\Psi_i(s)\} = 0$ $(i = 1, 2)$, then

$$\mathbf{T}\{\Psi_1(s)\Psi_2(s)\} = 0, \tag{25}$$

and furthermore, if $\Psi_i(s) \in \mathbf{R}_1$ $(i = 1, 2)$ and $\mathbf{T}\{\Psi_i(s)\} = \Psi_i(s)$ $(i = 1, 2)$, then

$$\mathbf{T}\{\Psi_1(s)\Psi_2(s)\} = \Psi_1(s)\Psi_2(s). \tag{26}$$

Conversely, (25) and (26) imply (24). If we apply (25) to the functions $\Psi_i(s) = \Phi_i(s) - \mathbf{T}\{\Phi_i(s)\}$ $(i = 1, 2)$ and (26) to the functions $\Psi_i(s) = \mathbf{T}\{\Phi_i(s)\}$ $(i = 1, 2)$ then we get (24).

Transformations of type $\mathbf{T}$ have extensively been studied in the literature. See Baxter [5], Rota [26], and others.

We note that if $\mathbf{T}$ satisfies (i), if $\Phi_k(s) \in \mathbf{R}_1$ $(k = 0, 1, 2, \ldots)$ satisfy (9), and if $\Phi(s)$ is defined by (10), then we have

$$\mathbf{T}\{\Phi(s)\} = \sum_{k=0}^{\infty} \mathbf{T}\{\Phi_k(s)\}. \tag{27}$$

To prove (27) let us write

$$\mathbf{T}\{\Phi(s)\} - \sum_{k=0}^{n} \mathbf{T}\{\Phi_k(s)\} = \mathbf{T}\left\{\sum_{k=n+1}^{\infty} \Phi_k(s)\right\} \tag{28}$$

for $n = 0, 1, 2, \ldots$. By (11) we have

$$\begin{aligned}\left|\mathbf{T}\left\{\sum_{k=n+1}^{\infty} \Phi_k(s)\right\}\right| \leqslant \left\|\mathbf{T}\left\{\sum_{k=n+1}^{\infty} \Phi_k\right\}\right\| \leqslant \|\mathbf{T}\| \left\|\sum_{k=n+1}^{\infty} \Phi_k\right\|\\ \leqslant \|\mathbf{T}\| \sum_{k=n+1}^{\infty} \|\Phi_k\|\end{aligned} \tag{29}$$

for $\mathrm{Re}(s) = 0$. Since the extreme right member of (29) tends to 0 as $n \to \infty$, it follows that the right-hand side of (28) tends to 0 as $n \to \infty$. This proves (27).

In particular, it follows from (19) that

$$\mathbf{T}\{\exp\{\rho\Psi(s)\}\} = \sum_{k=0}^{\infty} \frac{\rho^k}{k!} \mathbf{T}\{[\Psi(s)]^k\} \tag{30}$$

for any $\Psi(s) \in \mathbf{R}_1$ and ρ. Furthermore, by (21) we have

$$\mathbf{T}\{\log[1 - \rho\Psi(s)]\} = -\sum_{k=1}^{\infty} \frac{\rho^k}{k} \mathbf{T}\{[\Psi(s)]^k\} \tag{31}$$

for any $\Psi(s) \in \mathbf{R}_1$ and $|\rho|\,\|\Psi\| < 1$.

For any transformation $\mathbf{T}$ in $\mathbf{R}_1$ we define $\mathbf{T}^*$ as the transformation for which

$$\mathbf{T}^*\{\Phi(s)\} = \Phi(s) - \mathbf{T}\{\Phi(s)\} \tag{32}$$

for any $\Phi(s) \in \mathbf{R}_1$. If $\mathbf{T}$ satisfies (i), (ii), (iii), then $\mathbf{T}^*$ too satisfies (i), (ii), and (iii).

We shall also consider transformations $\mathbf{L}$ which satisfy the conditions (i), (ii), (iii) and can be represented in the form

$$\mathbf{L}\{\Phi(s)\} = \mathbf{T}\{\Phi(s)\} - \alpha(\Phi) \tag{33}$$

where $\mathbf{T}$ is a given transformation satisfying (i), (ii), (iii), and $\alpha(\Phi)$ is a complex (or real) functional on $\mathbf{R}_1$. We define $\mathbf{L}^*$ by

$$\mathbf{L}^*\{\Phi(s)\} = \Phi(s) - \mathbf{L}\{\Phi(s)\} \tag{34}$$

for any $\Phi(s) \in \mathbf{R}_1$.

3. Some Recurrence Equations in $\mathbf{R}_1$

Our aim is to give a method for solving the recurrence equation

$$U_n(s) = w\mathbf{L}\{U_{n-1}(s)\psi(s)\} + z\mathbf{L}^*\{U_{n-1}(s)\psi(s)\} \tag{35}$$

for $n = 1, 2, \ldots$, where $U_0(s) \in \mathbf{R}_1$, $\psi(s) \in \mathbf{R}_1$, w and z are complex (or real) numbers, and $\mathbf{L}$ satisfies the conditions (i), (ii), (iii) and can be represented in the form of (33).

Obviously, $U_n(s) \in \mathbf{R}_1$ for $n = 1, 2, \ldots$, and we shall see that if

$$|\rho| \max(|w|, |z|)\|\psi\| < 1, \tag{36}$$

then

$$U(s, \rho) = \sum_{n=0}^{\infty} U_n(s)\rho^n \tag{37}$$

belongs to $\mathbf{R}_1$. The solution of (35) is determined by the generating function (37).

If we multiply (35) by ρ^n and add for $n = 1, 2, \ldots$, then we obtain

$$\mathbf{L}\{U(s, \rho)[1 - \rho w\psi(s)]\} + \mathbf{L}^*\{U(s, \rho)[1 - \rho z\psi(s)]\} = U_0(s) \tag{38}$$

whenever (36) is satisfied.

Conversely, if

$$U(s, \rho) = \sum_{n=0}^{\infty} U_n^*(s)\rho^n \tag{39}$$

belongs to $\mathbf{R}_1$ for $|\rho| < r$ where r is some positive number, and if (39) satisfies (38), then by forming the coefficient of ρ^n for $n = 0, 1, 2, \ldots$, we obtain that $U_0^*(s) = U_0(s)$ and $U_n^*(s)$ $(n = 1, 2, \ldots)$ satisfies the same recurrence formula as $U_n(s)$ $(n = 1, 2, \ldots)$. Thus necessarily $U_n^*(s) = U_n(s)$ for $n \geqslant 0$. Hence we can draw the conclusion that if $U(s, \rho)$ is any function which can be represented in the form of (39) and if $U(s, \rho)$ satisfies (38), then $U(s, \rho)$ is necessarily equal to the right-hand side of (37).

First let us consider a particular case of (35).

THEOREM 1. *Let* $A_0(s) \equiv 1$, $\psi(s) \in \mathbf{R}_1$ *and define* $A_n(s)$ *for* $n = 1, 2, \ldots$ *and* $\mathrm{Re}(s) = 0$ *by the recurrence formula*

$$A_n(s) = \mathbf{T}\{A_{n-1}(s)\psi(s)\} \tag{40}$$

where the transformation $\mathbf{T}$ *satisfies* (i), (ii), (iii). *If* $|\rho|\ \|\psi\| < 1$, *then*

$$A(s, \rho) = \sum_{n=0}^{\infty} A_n(s)\rho^n \tag{41}$$

belongs to $\mathbf{R}_1$ *and we have*

$$A(s, \rho) = \exp\{-\mathbf{T}\{\log[1 - p\psi(s)]\}\}. \tag{42}$$

Proof. We can prove by mathematical induction that

$$nA_n(s) = \sum_{k=1}^{n} A_{n-k}(s)\mathbf{T}\{[\psi(s)]^k\} \tag{43}$$

for $n = 1, 2, \ldots$. If $n = 1$, then (43) is obviously true. Let us suppose that (43) is true for n where $n \geqslant 1$. We shall prove that (43) remains true if n is replaced by $n + 1$. Consequently, (43) is true for all $n \geqslant 1$. If (43) is true for $n \geqslant 1$, then by (40) we have

$$nA_{n+1}(s) = n\mathbf{T}\{A_n(s)\psi(s)\} = \sum_{k=1}^{n} \mathbf{T}\{A_{n-k}(s)\psi(s)\mathbf{T}\{[\psi(s)]^k\}\} \tag{44}$$

for $n \geqslant 1$. Now by (24) and (40) it follows that

$$\begin{aligned}\mathbf{T}\{A_{n-k}(s)\psi(s)\mathbf{T}\{[\psi(s)]^k\}\} \\ = \mathbf{T}\{A_{n-k}(s)[\psi(s)]^{k+1}\} - \mathbf{T}\{A_{n-k+1}(s)[\psi(s)]^k\} \\ + A_{n-k+1}(s)\mathbf{T}\{[\psi(s)]^k\}\end{aligned} \tag{45}$$

for $1 \leqslant k \leqslant n$. If we put (45) into (44) and if we use (40) again, then we get

$$(n+1)A_{n+1}(s) = \sum_{k=1}^{n+1} A_{n+1-k}(s)\mathbf{T}\{[\psi(s)]^k\}. \tag{46}$$

Accordingly, (43) is true if n is replaced by $n+1$. This proves that (43) is true for all $n \geqslant 1$.

From (43) it follows by induction that

$$\|A_n\| \leqslant \binom{\|\mathbf{T}\| + n - 1}{n}(\|\psi\|)^n \tag{47}$$

for $n \geqslant 0$. This implies that if $|\rho|\ \|\psi\| < 1$, then (41) belongs to $\mathbf{R}_1$. If we multiply (43) by ρ^n and add for $n = 1, 2, \ldots$, then we obtain that

$$\frac{\partial A(s, \rho)}{\partial \rho} = A(s, \rho) \sum_{k=1}^{\infty} \rho^{k-1}\mathbf{T}\{[\psi(s)]^k\} \tag{48}$$

for $|\rho|\ \|\psi\| < 1$. Since $A(s, 0) = 1$, therefore

$$\log A(s, \rho) = \sum_{k=1}^{\infty} \frac{\rho^k}{k} \mathbf{T}\{[\psi(s)]^k\} \tag{49}$$

for $|\rho|\ \|\psi\| < 1$. By (31) and (49) we get (42). This completes the proof of the theorem.

If in Theorem 1 we replace $\mathbf{T}$ by $\mathbf{T}^*$, then we obtain the following result.

Let $B_0(s) \equiv 1$, $\psi(s) \in \mathbf{R}_1$ and define $B_n(s)$ for $n = 1, 2, \ldots$ and $\mathrm{Re}(s) = 0$ by the recurrence formula

$$B_n(s) = \mathbf{T}^*\{B_{n-1}(s)\psi(s)\} \tag{50}$$

where the transformation $\mathbf{T}^*$ satisfies (i), (ii), (iii). If $|\rho|\ \|\psi\| < 1$, then

$$B(s, \rho) = \sum_{n=0}^{\infty} B_n(s)\rho^n \tag{51}$$

belongs to $\mathbf{R}_1$ and we have

$$B(s, \rho) = \exp\{-\mathbf{T}^*\{\log[1 - \rho\psi(s)]\}\}. \tag{52}$$

We shall prove later that the generating function (37) can be expressed by the functions $A(s, \rho)$ and $B(s, \rho)$. However, first let us observe that $A(s, \rho)$ and $B(s, \rho)$ satisfy the following properties for $|\rho|\,\|\psi\| < 1$.

PROPERTY (a_1). $A(s, \rho) \in \mathbf{R}_1, [A(s, \rho)]^{-1} \in \mathbf{R}_1, \mathbf{T}\{A(s, \rho) - 1\} = A(s, \rho) - 1, \mathbf{T}\{[A(s, \rho)]^{-1} - 1\} = [A(s, \rho)]^{-1} - 1$, $A(s, 0) = 1$, $A(s, \rho)$ and $[A(s, \rho)]^{-1}$ can be expanded into a Taylor series about $\rho = 0$.

PROPERTY (b_1). $B(s, \rho) \in \mathbf{R}_1, [B(s, \rho)]^{-1} \in \mathbf{R}_1, \mathbf{T}^*\{B(s, \rho) - 1\} = B(s, \rho) - 1, \mathbf{T}^*\{[B(s, \rho)]^{-1} - 1\} = [B(s, \rho)]^{-1} - 1$, $B(s, 0) = 1$, $B(s, \rho)$ and $[B(s, \rho)]^{-1}$ can be expanded into a Taylor series about $\rho = 0$.

Furthermore, we have

$$1 - \rho\psi(s) = [A(s, \rho)]^{-1}[B(s, \rho)]^{-1} \tag{53}$$

for $\mathrm{Re}(s) = 0$ and $|\rho|\,\|\psi\| < 1$.

We shall demonstrate after the proof of the next theorem that conditions (a_1) and (b_1) uniquely determine $A(s, \rho)$ and $B(s, \rho)$ in (53).

Now let us consider the solution of (35).

THEOREM 2. *If $U_n(s)$ $(n = 1, 2, \ldots)$ satisfies* (35) *and if* (36) *holds, then the generating function* (37) *is given by*

$$\begin{aligned} U(s, \rho) = [\mathbf{L}\{U_0(s)B(s, \rho w)[B(s, \rho z)]^{-1}\} \\ + \mathbf{L}^*\{U_0(s)A(s, \rho z)[A(s, \rho w)]^{-1}\}]A(s, \rho w)B(s, \rho z) \end{aligned} \tag{54}$$

where $A(s, \rho)$ is defined by (42) *and $B(s, \rho)$ by* (52).

Proof. If we assume that $A(s, \rho)$ and $B(s, \rho)$ satisfy the properties (a_1) and (b_1) and (53), then by (24) we can prove that (54) satisfies (38). This proves that (37) is indeed equal to (54).

If, in particular, $\mathbf{L} = \mathbf{T}$, $U_0(s) \equiv 1$, $w = 1$, and $z = 0$, then (54) reduces to $U(s, \rho) = A(s, \rho)$. As we have already demonstrated, in this case, $U(s, \rho)$ is given by (42). This proves that if (a_1), (b_1), and (54) are satisfied, then $A(s, \rho)$ is necessarily equal to the right-hand side of (42). If, in particular, $\mathbf{L} = \mathbf{T}$, $U_0(s) \equiv 1$, $w = 0$, and $z = 1$, then (54) reduces to $U(s, \rho) = B(s, \rho)$. As we have already demonstrated, in this case, $U(s, \rho)$ is given by (52). This proves that if (a_1), (b_1), and (54) are satisfied, then $B(s, \rho)$ is necessarily equal to the right-hand side of (52).

If, in particular, $\mathbf{T}\{U_0(s)\} = U_0(s)$, $w = 1$, and $z = 0$, then (54) reduces to

$$U(s, \rho) = \{\mathbf{L}\{U_0(s)B(s, \rho)\} + \alpha(U_0)[\mathbf{T}\{1\} + \alpha([A(s, \rho)]^{-1})]\}A(s, \rho) \tag{55}$$

for $\mathrm{Re}(s) = 0$ and $|\rho|\,\|\psi\| < 1$.

4. Some Particular Transformations in $\mathbf{R}_1$

In this section we shall consider some transformations in $\mathbf{R}_1$ which satisfy the conditions (i), (ii), and (iii).

If $\Phi(s) \in \mathbf{R}_1$ is given by (7), then let us define

$$\Phi^+(s) = \mathbf{E}\{\zeta e^{-s\eta^+}\} \tag{56}$$

for $\mathrm{Re}(s) \geqslant 0$ and

$$\Phi^-(s) = \mathbf{E}\{\zeta(e^{-s\eta} - e^{-s\eta^+})\} \tag{57}$$

for $\mathrm{Re}(s) \leqslant 0$ where $\eta^+ = \max(0, \eta)$. We have $\Phi^+(s) \in \mathbf{R}_1$, $\Phi^-(s) \in \mathbf{R}_1$, $|\Phi^+(s)| \leqslant \|\Phi\|$ for $\mathrm{Re}(s) \geqslant 0$, $|\Phi^-(s)| \leqslant 2\|\Phi\|$ for $\mathrm{Re}(s) \leqslant 0$, and

$$\Phi(s) = \Phi^+(s) + \Phi^-(s) \tag{58}$$

for $\mathrm{Re}(s) = 0$.

The function $\Phi^+(s)$ is regular for $\mathrm{Re}(s) > 0$, continuous and bounded for $\mathrm{Re}(s) \geqslant 0$ and $\Phi^+(0) = \Phi(0)$.

The function $\Phi^-(s)$ is regular for $\mathrm{Re}(s) < 0$, continuous and bounded for $\mathrm{Re}(s) \leqslant 0$ and $\Phi^-(0) = 0$.

Now we shall prove that the above properties uniquely determine $\Phi^+(s)$ and $\Phi^-(s)$ in the representation (58). This shows that (56) and (57) are uniquely determined by $\Phi(s)$, that is, (56) and (57) do not depend on the particular representation (7) of $\Phi(s)$.

To prove the above statement, let us suppose that $\Phi(s) = \Psi^+(s) + \Psi^-(s)$ for $\mathrm{Re}(s) = 0$, where $\Psi^+(s)$ and $\Psi^-(s)$ satisfy the same conditions as $\Phi^+(s)$ and $\Phi^-(s)$. If we define

$$G(s) = \begin{cases} \Phi^+(s) - \Psi^+(s) & \text{for} \quad \mathrm{Re}(s) \geqslant 0, \\ \Psi^-(s) - \Phi^-(s) & \text{for} \quad \mathrm{Re}(s) \leqslant 0, \end{cases} \tag{59}$$

then $G(s)$ is bounded and continuous on the whole complex plane and regular for $\mathrm{Re}(s) > 0$ and $\mathrm{Re}(s) < 0$. By a theorem of Morera [19] (see also Osgood [20, p. 122]) it follows that $G(s)$ is necessarily a regular function of s on the whole complex plane. By Liouville's theorem it follows that $G(s)$ is constant, that is, $G(s) = G(0) = 0$ for all s. Consequently, $\Psi^+(s) = \Phi^+(s)$ for $\mathrm{Re}(s) \geqslant 0$ and $\Psi^-(s) = \Phi^-(s)$ for $\mathrm{Re}(s) \leqslant 0$. This proves the statement.

If $\Phi(s) \in \mathbf{R}_1$, then for $\mathrm{Re}(s) > 0$ we have

$$\Phi^+(s) = \tfrac{1}{2}\Phi(0) + \lim_{\epsilon \to 0} \frac{s}{2\pi i} \int_{L_\epsilon} \frac{\Phi(z)}{z(s - z)} dz \tag{60}$$

where $L_\epsilon = \{z : z = iy, \ -\infty < y \leqslant -\epsilon < \epsilon \leqslant y < \infty\}$. See [28].

We note that if $\Phi(s+u) \in \mathbf{R}_1$ for $0 \leqslant u \leqslant \epsilon$, then

$$\Phi^+(s) = \frac{s}{2\pi i} \int_{\epsilon - i\infty}^{\epsilon + i\infty} \frac{\Phi(z)}{z(s-z)} dz \tag{61}$$

for $\text{Re}(s) > \epsilon$, and if $\Phi(s-u) \in \mathbf{R}_1$ for $0 \leqslant u \leqslant \epsilon$, then

$$\Phi^+(s) = \Phi(0) + \frac{s}{2\pi i} \int_{-\epsilon - i\infty}^{-\epsilon + \infty} \frac{\Phi(z)}{z(s-z)} dz \tag{62}$$

for $\text{Re}(s) > -\epsilon$.

For any event A let us define $\delta(A)$ as the indicator variable of A, that is, $\delta(A) = 1$ if A occurs and $\delta(A) = 0$ if A does not occur.

Now we define three transformations $\mathbf{T}$, $\mathbf{T}_0$, $\mathbf{T}_1$, in $\mathbf{R}_1$ which satisfy the conditions (i), (ii), (iii). If $\Phi(s) \in \mathbf{R}_1$ is given by (7), then let

$$\mathbf{T}\{\Phi(s)\} = \Phi^+(s) = \mathbf{E}\{\zeta e^{-s\eta^+}\}, \tag{63}$$

$$\mathbf{T}_0\{\Phi(s)\} = \Phi^+(s) - \Phi^+(\infty) = \mathbf{E}\{\zeta e^{-s\eta}\delta(\eta > 0)\} \tag{64}$$

and

$$\mathbf{T}_1\{\Phi(s)\} = \Phi^+(s) + \Phi^-(-\infty) = \mathbf{E}\{\zeta e^{-s\eta}\delta(\eta \geqslant 0)\}. \tag{65}$$

We define $\mathbf{T}^*$, $\mathbf{T}_0{}^*$, and $\mathbf{T}_1{}^*$ by (32). The transformations (63), (64), and (65) satisfy (i), (ii), (iii) and $\|\mathbf{T}\| = \|\mathbf{T}_0\| = \|\mathbf{T}_1\| = \|\mathbf{T}_0{}^*\| = \|\mathbf{T}_1{}^*\| = 1$, $\|\mathbf{T}^*\| = 2$. Only (iii) requires proof. To prove that each transformation satisfies (24) let us define $\Phi_1(s) = \mathbf{E}\{\zeta_1 e^{-s\eta_1}\}$ and $\Phi_2(s) = \mathbf{E}\{\zeta_2 e^{-s\eta_2}\}$, where (ζ_1, η_1) and (ζ_2, η_2) are independent vector variables. If in the identity

$$e^{-s(x+y)^+} = e^{-s(x^+ + y)^+} + e^{-s(x+y^+)^+} - e^{-s(x^+ + y^+)} \tag{66}$$

we put $x = \eta_1$ and $y = \eta_2$, multiply it by $\zeta_1\zeta_2$ and form its expectation, then we can conclude that (24) is valid for $\mathbf{T}$. If in the identity

$$\delta(x+y>0) = \delta(x>0)\delta(x+y>0) + \delta(y>0)\delta(x+y>0) - \delta(x>0)\delta(y>0) \tag{67}$$

we put $x = \eta_1$ and $y = \eta_2$, multiply it by $\zeta_1\zeta_2 e^{-s(\eta_1+\eta_2)}$ and form its expectation, then we can conclude that (24) is valid for $\mathbf{T}_0$. In a similar way we can prove that (24) is valid for $\mathbf{T}_1$ too.

If $\mathbf{L}$ is any one of the transformations $\mathbf{T}$, $\mathbf{T}_0$, $\mathbf{T}_1$ defined by (63), (64), and (65), respectively, then $\mathbf{L}\{\Phi(s)\}$ can be represented in the form of (33) where $\mathbf{T}\{\Phi(s)\}$ is defined by (63), and $\alpha(\Phi) = 0$ for $\mathbf{L} = \mathbf{T}$, $\alpha(\Phi) = \Phi^+(\infty)$ for $\mathbf{L} = \mathbf{T}_0$, and $\alpha(\Phi) = -\Phi^-(-\infty)$ for $\mathbf{L} = \mathbf{T}_1$.

If we assume that $\mathbf{T}$ is defined by (63), then in Theorem 2 we can also determine $U(s, \rho)$ by the following factorization of $1 - \rho\psi(s)$.

Let $|\rho| \, \|\psi\| < 1$ and

$$1 - \rho\psi(s) = \psi^+(s, \rho)\psi^-(s, \rho) \tag{68}$$

for $\operatorname{Re}(s) = 0$ where

(α_1) the function $\psi^+(s, \rho)$ is regular for $\operatorname{Re}(s) > 0$, continuous, bounded and free from zeros for $\operatorname{Re}(s) \geqslant 0$,

(β_1) the function $\psi^-(s, \rho)$ is regular for $\operatorname{Re}(s) < 0$, continuous, bounded, and free from zeros for $\operatorname{Re}(s) \leqslant 0$.

We say that (68) is a canonical factorization of $1 - \rho\psi(s)$. Such a factorization always exists. For example, if

$$\psi^+(s, \rho) = C_1(\rho)[A(s, \rho)]^{-1} \tag{69}$$

for $\operatorname{Re}(s) \geqslant 0$ and $|\rho| \, \|\psi\| < 1$ and if

$$\psi(s, \rho) = C_2(\rho)[B(s, \rho)]^{-1} \tag{70}$$

for $\operatorname{Re}(s) \leqslant 0$ and $|\rho| \, \|\psi\| < 1$ where $A(s, \rho)$ is defined by (42), $B(s, \rho)$ is defined by (52) and $C_1(\rho)C_2(\rho) = 1$, then (α_1), (β_1), and (68) are satisfied. Conversely, it follows from Liouville's theorem that conditions (α_1), (β_1), and (68) determine $\psi^+(s, \rho)$ and $\psi^-(s, \rho)$ up to a nonvanishing factor depending only on ρ. Thus (69) and (70) are the general forms of $\psi^+(s, \rho)$ and $\psi^-(s, \rho)$, respectively.

If we use the canonical factorization (68), then in Theorem 2 we can express (54) in the following way. If $|\rho| \, \|\psi\| < 1$, then

$$\begin{aligned} U(s, \rho) = [&\mathbf{L}\{U_0(s)\psi^-(s, \rho z)[\psi^-(s, \rho w)]^{-1}\} \\ &+ \mathbf{L}^*\{U_0(s)\psi^+(s, \rho w)[\psi^+(s, \rho z)]^{-1}\}] \cdot [\psi^+(s, \rho w)]^{-1}[\psi^-(s, \rho z)]^{-1}. \end{aligned} \tag{71}$$

If we put (69) and (70) into (71) and take into consideration that $C_1(\rho)C_2(\rho) = 1$, then (71) reduces to (54).

We note that in (α_1) and (β_1) the requirement of boundedness can be replaced by the weaker conditions $\lim_{|s| \to \infty} [\log \psi^+(s, \rho)]/s = 0$ ($\operatorname{Re}(s) \geqslant 0$) and $\lim_{|s| \to \infty} [\log \psi^-(s, \rho)]/s = 0$ ($\operatorname{Re}(s) \leqslant 0$).

The solution of the equation

$$U_n(s) = \mathbf{L}\{U_{n-1}(s)\psi(s)\} \tag{72}$$

for $n = 1, 2, \ldots$ is given by (71) where now $w = 1$ and $z = 0$.

If, in particular, $\mathbf{T}\{U_0(s)\} = U_0(s)$, $w = 1$ and $z = 0$, then (71) reduces to

$$U(s, \rho) = [\mathbf{L}\{U_0(s)[\psi^-(s, \rho)]^{-1}\} + \epsilon(\mathbf{L})U_0(\infty)\psi^+(\infty, \rho)][\psi^+(s, \rho)]^{-1} \tag{73}$$

for $\operatorname{Re}(s) \geqslant 0$ and $|\rho| \, \|\psi\| < 1$ where $\epsilon(\mathbf{T}) = \epsilon(\mathbf{T}_1) = 0$ and $\epsilon(\mathbf{T}_0) = 1$. This follows from (55).

If $U_0(s) \equiv 1$, $w = 1$, $z = 0$, $\mathrm{Re}(s) \geqslant 0$ and $|\rho| \, \|\psi\| < 1$, then by (71) or by (73) we obtain that

$$U(s, \rho) = [\psi^{-}(0, \rho)]^{-1}[\psi^{+}(s, \rho)]^{-1} \tag{74}$$

for $\mathbf{L} = \mathbf{T}$,

$$U(s, \rho) = \psi^{+}(\infty, \rho)[\psi^{+}(s, \rho)]^{-1} \tag{75}$$

for $\mathbf{L} = \mathbf{T}_0$, and

$$U(s, \rho) = [\psi^{-}(-\infty, \rho)]^{-1}[\psi^{+}(s, \rho)]^{-1} \tag{76}$$

for $\mathbf{L} = \mathbf{T}_1$.

5. A System of Recurrence Equations in $\mathbf{R}_1$

Now let us suppose that $U_0 \in \mathbf{R}_1$, $V_0 \in \mathbf{R}_1$, $\psi \in \mathbf{R}_1$, w and z are complex (or real) numbers, and that $\mathbf{L}$ satisfies the conditions (i), (ii), (iii) and can be represented in the form of (33). We are interested in finding the solution of the system of recurrence equations

$$U_n(s) = \mathbf{L}\{[wU_{n-1}(s) + zV_{n-1}(s)]\psi(s)\} \tag{77}$$

and

$$V_n(s) = \mathbf{L}^*\{[zU_{n-1}(s) + wV_{n-1}(s)]\psi(s)\}. \tag{78}$$

To solve this system let us introduce a new Banach algebra $\mathbf{S}_1$ with elements

$$\boldsymbol{\Phi}(s) = \begin{bmatrix} \Phi_{11}(s) & \Phi_{12}(s) \\ \Phi_{21}(s) & \Phi_{22}(s) \end{bmatrix} \tag{79}$$

where $\Phi_{ij}(s) \in \mathbf{R}_1$ for $i = 1, 2$, $j = 1, 2$. In $\mathbf{S}_1$ let us define the operations of addition, multiplication, and multiplication by a complex (or real) constant according to the rules of matrix algebra. Define the norm of (79) either by

$$\|\boldsymbol{\Phi}\|_{\mathbf{S}_1} = \max(\|\Phi_{11}\|_{\mathbf{R}_1} + \|\Phi_{12}\|_{\mathbf{R}_1}, \|\Phi_{21}\|_{\mathbf{R}_1} + \|\Phi_{22}\|_{\mathbf{R}_1}) \tag{80}$$

or by

$$\|\boldsymbol{\Phi}\|_{\mathbf{S}_1} = \max(\|\Phi_{11}\|_{\mathbf{R}_1} + \|\Phi_{21}\|_{\mathbf{R}_1}, \|\Phi_{12}\|_{\mathbf{R}_1} + \|\Phi_{22}\|_{\mathbf{R}_1}). \tag{81}$$

We can easily see that $\mathbf{S}_1$ is a noncommutative Banach algebra with zero element

$$\begin{bmatrix} 0 & 0 \\ 0 & 0 \end{bmatrix}$$

and identity element

$$\begin{bmatrix} 1 & 0 \\ 0 & 1 \end{bmatrix}.$$

If we extend the definition of the transformation $\mathbf{L}$ from the space $\mathbf{R}_1$ to the space $\mathbf{S}_1$ in such a way that we form the transformation element by element for a matrix function $\boldsymbol{\Phi}(s)$ given by (79), then $\mathbf{L}$ satisfies the same properties in $\mathbf{S}_1$ as in $\mathbf{R}_1$.

The system of recurrence equations (77), (78) can be replaced by the following recurrence equation in $\mathbf{S}_1$:

$$\begin{bmatrix} U_n(s) & V_n(s) \\ 0 & 0 \end{bmatrix} = \mathbf{L}\left\{\begin{bmatrix} U_{n-1}(s) & V_{n-1}(s) \\ 0 & 0 \end{bmatrix}\begin{bmatrix} w\psi(s) & 0 \\ z\psi(s) & 0 \end{bmatrix}\right\} + \mathbf{L}^*\left\{\begin{bmatrix} U_{n-1}(s) & V_{n-1}(s) \\ 0 & 0 \end{bmatrix}\begin{bmatrix} 0 & z\psi(s) \\ 0 & w\psi(s) \end{bmatrix}\right\} \tag{82}$$

for $n = 1, 2, \ldots$.

The solution of this recurrence equation will be discussed in the next section where we consider a more general space $\mathbf{R}_2$ which contains $\mathbf{S}_1$ as a particular case.

6. A Space $\mathbf{R}_2$

Let I be a fixed finite or countably infinite set. We consider complex (or real) matrices $\mathbf{A} = [a_{ij}]$, $i \in I$, $j \in I$, for which

$$\mathbf{M}\{\mathbf{A}\} = \sup_{i \in I} \sum_{j \in I} |a_{ij}| < \infty. \tag{83}$$

We shall denote by $\mathbf{0}$ the zero matrix all of whose elements are zeros, and by $\mathbf{I}$ the identity matrix. ($\mathbf{I} = [\delta_{ij}]$, $i \in I$, $j \in I$, where $\delta_{ij} = 1$ for $i = j$ and $\delta_{ij} = 0$ for $i \neq j$.) If $\mathbf{M}\{\mathbf{A}\} < \infty$, $\mathbf{M}\{\mathbf{B}\} < \infty$ and $\mathbf{AB} = \mathbf{BA} = \mathbf{I}$, then we say that $\mathbf{A}$ and $\mathbf{B}$ are inverse matrices and write $\mathbf{B} = \mathbf{A}^{-1}$.

We say that a matrix function $\mathbf{A}(s) = [a_{ij}(s)]$, $i \in I$, $j \in I$, is continuous, or regular, or bounded on a set D according to whether every $a_{ij}(s)$ is continuous on D, or every $a_{ij}(s)$ is regular on D, or $\mathbf{M}\{\mathbf{A}(s)\} < K$ for $s \in D$ where K is a positive constant.

Let $\mathbf{R}_2$ be the space of all matrix functions

$$\boldsymbol{\Phi}(s) = [\Phi_{ij}(s)]_{i,\, j \in I} \tag{84}$$

defined for $\mathrm{Re}(s) = 0$ on the complex plane such that I is a fixed countable set, $\Phi_{ij}(s) \in \mathbf{R}_1$ and

$$\|\boldsymbol{\Phi}\| = \sup_{i \in I} \sum_{j \in I} \|\Phi_{ij}\|_{\mathbf{R}_1} < \infty. \tag{85}$$

We define the norm of $\boldsymbol{\Phi}(s) \in \mathbf{R}_2$ by (85). We have $\mathbf{M}\{\boldsymbol{\Phi}(s)\} \leqslant \|\boldsymbol{\Phi}\|$ for $\mathrm{Re}(s) = 0$. We define the operations of addition, multiplication, and multiplication by a scalar in $\mathbf{R}_2$ according to the rules of matrix algebra.

We can easily see that if $\boldsymbol{\Phi}(s) \in \mathbf{R}_2$, $\boldsymbol{\Phi}_1(s) \in \mathbf{R}_2$, $\boldsymbol{\Phi}_2(s) \in \mathbf{R}_2$ and c is a complex (or real) number, then $c\boldsymbol{\Phi}(s) \in \mathbf{R}_2$, $\boldsymbol{\Phi}_1(s) + \boldsymbol{\Phi}_2(s) \in \mathbf{R}_2$, $\boldsymbol{\Phi}_1(s)\boldsymbol{\Phi}_2(s) \in \mathbf{R}_2$ and $\|c\boldsymbol{\Phi}\| = |c| \, \|\boldsymbol{\Phi}\|$, $\|\boldsymbol{\Phi}_1 + \boldsymbol{\Phi}_2\| \leqslant \|\boldsymbol{\Phi}_1\| + \|\boldsymbol{\Phi}_2\|$, $\|\boldsymbol{\Phi}_1\boldsymbol{\Phi}_2\| \leqslant \|\boldsymbol{\Phi}_1\| \, \|\boldsymbol{\Phi}_2\|$. From Lemma 1 and Lemma 2 it follows immediately the following property of $\mathbf{R}_2$.

LEMMA 3. *If $\boldsymbol{\Phi}_k(s) \in \mathbf{R}_2$ for $k = 0, 1, 2, \ldots$, and if*

$$\sum_{k=0}^{\infty} \|\boldsymbol{\Phi}_k\| < \infty, \tag{86}$$

then

$$\boldsymbol{\Phi}(s) = \sum_{k=0}^{\infty} \boldsymbol{\Phi}_k(s) \tag{87}$$

belongs to $\mathbf{R}_2$,

$$\lim_{n \to \infty} \|\boldsymbol{\Phi} - \sum_{k=0}^{n} \boldsymbol{\Phi}_k\| = 0 \tag{88}$$

and

$$\|\boldsymbol{\Phi}\| \leqslant \sum_{k=0}^{\infty} \|\boldsymbol{\Phi}_k\|. \tag{89}$$

The space $\mathbf{R}_2$ is a normed linear space. Lemma 3 implies that $\mathbf{R}_2$ is complete, that is, $\mathbf{R}_2$ is a Banach space. More precisely, $\mathbf{R}_2$ is a noncommutative Banach algebra.

In $\mathbf{R}_1$ we considered a transformation $\mathbf{T}$ which satisfies (i), (ii) and (iii). Now let us extend the definition of $\mathbf{T}$ to the space $\mathbf{R}_2$ in such a way that we form $\mathbf{T}$ element by element for $\boldsymbol{\Phi}(s) \in \mathbf{R}_2$, that is, if $\boldsymbol{\Phi}(s)$ is given by (79), then

$$\mathbf{T}\{\boldsymbol{\Phi}(s)\} = [\mathbf{T}\{\Phi_{ij}(s)\}]_{i,\, j \in I}. \tag{90}$$

We can easily see that $\mathbf{T}$ satisfies the conditions (i), (ii), and (iii) in the space $\mathbf{R}_2$ too. In particular we have

$$\begin{aligned}\mathbf{T}\{\boldsymbol{\Phi}_1(s)\boldsymbol{\Phi}_2(s)\} &= \mathbf{T}\{\mathbf{T}\{\boldsymbol{\Phi}_1(s)\}\boldsymbol{\Phi}_2(s)\} \\ &\quad + \mathbf{T}\{\boldsymbol{\Phi}_1(s)\mathbf{T}\{\boldsymbol{\Phi}_2(s)\}\} - \mathbf{T}\{\boldsymbol{\Phi}_1(s)\}\mathbf{T}\{\boldsymbol{\Phi}_2(s)\}\end{aligned} \tag{91}$$

for any $\boldsymbol{\Phi}_1(s) \in \mathbf{R}_2$ and $\boldsymbol{\Phi}_2(s) \in \mathbf{R}_2$. Furthermore, the norm of $\mathbf{T}$ is the same in the space $\mathbf{R}_2$ as in the space $\mathbf{R}_1$. Formulas (27), (30), and (31) remain valid unchangeably in the space $\mathbf{R}_2$ too.

For any transformation $\mathbf{T}$ in $\mathbf{R}_2$ we define $\mathbf{T}^*$ as the transformation for which

$$\mathbf{T}^*\{\boldsymbol{\Phi}(s)\} = \boldsymbol{\Phi}(s) - \mathbf{T}\{\boldsymbol{\Phi}(s)\} \tag{92}$$

for any $\boldsymbol{\Phi}(s) \in \mathbf{R}_2$. If $\mathbf{T}$ satisfies (i), (ii), (iii), then $\mathbf{T}^*$ too satisfies (i), (ii), and (iii).

We shall also consider transformations $\mathbf{L}$ in $\mathbf{R}_2$ which satisfy the conditions (i), (ii), (iii) and can be represented in the form

$$\mathbf{L}\{\boldsymbol{\Phi}(s)\} = \mathbf{T}\{\boldsymbol{\Phi}(s)\} - \boldsymbol{\alpha}(\boldsymbol{\Phi}) \tag{93}$$

where $\mathbf{T}$ is a given transformation in $\mathbf{R}_2$ satisfying (i), (ii), (iii), and $\boldsymbol{\alpha}(\boldsymbol{\Phi})$ is a matrix functional on $\mathbf{R}_2$, that is, $\boldsymbol{\alpha}(\boldsymbol{\Phi})$ is matrix which does not depend on s. We define $\mathbf{L}^*$ by

$$\mathbf{L}^*\{\boldsymbol{\Phi}(s)\} = \boldsymbol{\Phi}(s) - \mathbf{L}\{\boldsymbol{\Phi}(s)\} \tag{94}$$

for any $\boldsymbol{\Phi}(s) \in \mathbf{R}_2$.

7. Some Recurrence Equations in $\mathbf{R}_2$

In the space $\mathbf{R}_2$ there are four possible versions of the recurrence equation (35). Since $\mathbf{R}_2$ is noncommutative we can form the product in two ways in each term of (35). First, let us consider the recurrence equations

$$\mathbf{U}_n(s) = w\mathbf{L}\{\mathbf{U}_{n-1}(s)\boldsymbol{\Psi}(s)\} + z\mathbf{L}\{\boldsymbol{\Psi}(s)\mathbf{U}_{n-1}(s)\} \tag{95}$$

and

$$\mathbf{U}_n(s) = w\mathbf{L}\{\boldsymbol{\Psi}(s)\mathbf{U}_{n-1}(s)\} + z\mathbf{L}^*\{\mathbf{U}_{n-1}(s)\boldsymbol{\Psi}(s)\} \tag{96}$$

for $n = 1, 2, \ldots$ where $\mathbf{U}_0(s) \in \mathbf{R}_2$, $\boldsymbol{\Psi}(s) \in \mathbf{R}_2$, w and z are complex (or real) numbers, and $\mathbf{L}$ satisfies the conditions (i), (ii), (iii) and can be represented in the form of (93).

In both cases $\mathbf{U}_n(s) \in \mathbf{R}_2$ for $n = 1, 2, \ldots$ and

$$\mathbf{U}(s, \rho) = \sum_{n=0}^{\infty} \mathbf{U}_n(s)\rho^n \tag{97}$$

belongs to $\mathbf{R}_2$ if

$$|\rho| \max(|w|, |z|)\|\boldsymbol{\Psi}\| < \gamma(\mathbf{L}) \tag{98}$$

where $\min(\|\mathbf{L}\|^{-1}, \|\mathbf{L}^*\|^{-1}) \leqslant \gamma(\mathbf{L}) \leqslant 1$ and $\gamma(\mathbf{L}) = \gamma(\mathbf{T})$.

If we multiply (95) by ρ^n and add for $n = 1, 2, \ldots$, then we obtain that

$$\mathbf{L}\{\mathbf{U}(s, \rho)[\mathbf{I} - \rho w \boldsymbol{\Psi}(s)]\} + \mathbf{L}^*\{[\mathbf{I} - \rho z \boldsymbol{\Psi}(s)]\mathbf{U}(s, \rho)\} = \mathbf{U}_0(s) \tag{99}$$

whenever (98) is satisfied.

If we multiply (96) by ρ^n and add for $n = 1, 2, \ldots$, then we obtain that

$$\mathbf{L}\{[\mathbf{I} - \rho w \boldsymbol{\Psi}(s)]\mathbf{U}(s, \rho)\} + \mathbf{L}^*\{\mathbf{U}(s, \rho)[\mathbf{I} - \rho z \boldsymbol{\Psi}(s)]\} = \mathbf{U}_0(s) \tag{100}$$

whenever (98) is satisfied.

Conversely, if

$$\mathbf{U}(s, \rho) = \sum_{n=0}^{\infty} \mathbf{U}_n^*(s)\rho^n \tag{101}$$

belongs to $\mathbf{R}_2$ for $|\rho| < r$, where r is some positive number and if (101) satisfies (99) then $\mathbf{U}_n^*(s) = \mathbf{U}_n(s)$ for $n = 0, 1, 2, \ldots$, where $\mathbf{U}_n(s)$ $(n = 1, 2, \ldots)$ is defined by (95). If (101) satisfies (100), then $\mathbf{U}_n^*(s) = \mathbf{U}_n(s)$ for $n = 0, 1, 2, \ldots$, where $\mathbf{U}_n(s)$ $(n = 1, 2, \ldots)$ is defined by (96).

The recurrence equation (95) can be solved by using the following factorization. Let us suppose that

$$\mathbf{I} - \rho \boldsymbol{\Psi}(s) = [\mathbf{A}(s, \rho)]^{-1}[\mathbf{B}(s, \rho)]^{-1} \tag{102}$$

for $|\rho|\,\|\boldsymbol{\Psi}\| < \gamma(\mathbf{T})$ and $\mathrm{Re}(s) = 0$ where

(a_2) $\mathbf{A}(s, \rho) \in \mathbf{R}_2$, $[\mathbf{A}(s, \rho)]^{-1} \in \mathbf{R}_2$, $\mathbf{T}\{\mathbf{A}(s, \rho) - \mathbf{I}\} = \mathbf{A}(s, \rho) - \mathbf{I}$, $\mathbf{T}\{[\mathbf{A}(s, \rho)]^{-1} - \mathbf{I}\} = [\mathbf{A}(s, \rho)]^{-1} - \mathbf{I}$, $\mathbf{A}(s, 0) = \mathbf{I}$, $\mathbf{A}(s, \rho)$ and $[\mathbf{A}(s, \rho)]^{-1}$ can be expanded into a Taylor series about $\rho = 0$, and

(b_2) $\mathbf{B}(s, \rho) \in \mathbf{R}_2$, $[\mathbf{B}(s, \rho)]^{-1} \in \mathbf{R}_2$, $\mathbf{T}^*\{\mathbf{B}(s, \rho) - \mathbf{I}\} = \mathbf{B}(s, \rho) - \mathbf{I}$, $\mathbf{T}^*\{[\mathbf{B}(s, \rho)]^{-1} - \mathbf{I}\} = [\mathbf{B}(s, \rho)]^{-1} - \mathbf{I}$, $\mathbf{B}(s, 0) = \mathbf{I}$, $\mathbf{B}(s, \rho)$ and $[\mathbf{B}(s, \rho)]^{-1}$ can be expanded into a Taylor series about $\rho = 0$.

The factorization (102) always exists and actually $\mathbf{A}(s, \rho)$ and $\mathbf{B}(s, \rho)$ are uniquely determined by the above conditions. See [30].

If we define $\mathbf{A}_n(s)$ $(n = 0, 1, 2, \ldots)$ and $\mathbf{B}_n(s)$ $(n = 0, 1, 2, \ldots)$ by the recurrence formulas $\mathbf{A}_n(s) = \mathbf{T}\{\mathbf{A}_{n-1}(s)\boldsymbol{\Psi}(s)\}$ $(n = 1, 2, \ldots)$ and $\mathbf{B}_n(s) = \mathbf{T}^*\{\boldsymbol{\Psi}(s)\mathbf{B}_{n-1}(s)\}$ $(n = 1, 2, \ldots)$, where $\mathbf{A}_0(s) = \mathbf{B}_0(s) = \mathbf{I}$, then

$$\mathbf{A}(s, \rho) = \sum_{n=0}^{\infty} \mathbf{A}_n(s)\rho^n, \tag{103}$$

$$\mathbf{B}(s, \rho) = \sum_{n=0}^{\infty} \mathbf{B}_n(s)\rho^n \tag{104}$$

belong to $\mathbf{R}_2$ whenever $|\rho|\,\|\boldsymbol{\Psi}\| < \gamma(\mathbf{T})$. By (91) we can prove that (103) and (104) satisfy (a_2), (b_2), and (102). From the next theorem we can draw the conclusion that $\mathbf{A}(s, \rho)$ and $\mathbf{B}(s, \rho)$ are uniquely determined by (a_2), (b_2), and (102).

THEOREM 3. *If* $\mathbf{U}_n(s)$ $(n = 1, 2, \ldots)$ *satisfies* (95) *and if* (98) *holds, then the generating function* (97) *is given by*

$$\mathbf{U}(s, \rho) = \mathbf{B}(s, \rho z)[\mathbf{L}\{[\mathbf{B}(s, \rho z)]^{-1}\mathbf{U}_0(s)\mathbf{B}(s, \rho w)\} + \mathbf{L}^*\{\mathbf{A}(s, \rho z)\mathbf{U}_0(s)[\mathbf{A}(s, \rho w)]^{-1}\}]\mathbf{A}(s, \rho w) \tag{105}$$

where $\mathbf{A}(s, \rho)$ *and* $\mathbf{B}(s, \rho)$ *satisfy* (a_2), (b_2), *and* (102).

Proof. If we assume that $\mathbf{A}(s, \rho)$ and $\mathbf{B}(s, \rho)$ satisfy (a_2), (b_2), and (102), then by (91) we can prove that (105) satisfies (99). This proves that (97) is indeed equal to (105).

If, in particular, $\mathbf{L} = \mathbf{T}$, $\mathbf{U}_0(s) = \mathbf{I}$, $w = 1$ and $z = 0$, then (105) reduces to $\mathbf{U}(s, \rho) = \mathbf{A}(s, \rho)$. Since in this case $\mathbf{U}(s, \rho)$ is given by (103), it follows that if (a_2), (b_2), and (102) are satisfied, then $\mathbf{A}(s, \rho)$ is necessarily equal to the right-hand side of (103). If, in particular, $\mathbf{L} = \mathbf{T}$, $\mathbf{U}_0(s) = \mathbf{I}$, $w = 0$ and $z = 1$, then (105) reduces to $\mathbf{U}(s, \rho) = \mathbf{B}(s, \rho)$. Since in this case $\mathbf{U}(s, \rho)$ is given by (104), it follows that if (a_2), (b_2), and (102) are satisfied, then $\mathbf{B}(s, \rho)$ is necessarily equal to the right-hand side of (104).

The recurrence equation (96) can be solved by using the following factorization. Let us suppose that

$$\mathbf{I} - \rho\boldsymbol{\Psi}(s) = [\mathbf{D}(s, \rho)]^{-1}[\mathbf{C}(s, \rho)]^{-1} \tag{106}$$

for $|\rho|\,\|\boldsymbol{\Psi}\| < \gamma(\mathbf{T})$ and $\mathrm{Re}(s) = 0$ where $\mathbf{C}(s, \rho)$ satisfies (a_2) and $\mathbf{D}(s, \rho)$ satisfies (b_2). The factorization (106) always exists and $\mathbf{C}(s, \rho)$ and $\mathbf{D}(s, \rho)$ are uniquely determined by the above conditions.

If we define $\mathbf{C}_n(s)$ $(n = 0, 1, 2, \ldots)$ and $\mathbf{D}_n(s)$ $(n = 0, 1, 2, \ldots)$ by the recurrence formulas $\mathbf{C}_n(s) = \mathbf{T}\{\boldsymbol{\Psi}(s)\mathbf{C}_{n-1}(s)\}$ $(n = 1, 2, \ldots)$ and $\mathbf{D}_n(s) = \mathbf{T}^*\{\mathbf{D}_{n-1}(s)\boldsymbol{\Psi}(s)\}$ $(n = 1, 2, \ldots)$, where $\mathbf{C}_0(s) = \mathbf{D}_0(s) = \mathbf{I}$, then

$$\mathbf{C}(s, \rho) = \sum_{n=0}^{\infty} \mathbf{C}_n(s)\rho^n \tag{107}$$

and

$$\mathbf{D}(s, \rho) = \sum_{n=0}^{\infty} \mathbf{D}_n(s)\rho^n \tag{108}$$

satisfy (a_2), (b_2), and (106).

THEOREM 4. *If* $\mathbf{U}_n(s)$ $(n = 1, 2, \ldots)$ *satisfies* (96) *and if* (98) *holds, then the generating function* (97) *is given by*

$$\mathbf{U}(s, \rho) = \mathbf{C}(s, \rho w)[\mathbf{L}\{\mathbf{D}(s, \rho w)\mathbf{U}_0(s)[\mathbf{D}(s, \rho z)]^{-1}\} + \mathbf{L}^*\{[\mathbf{C}(s, \rho w)]^{-1}\mathbf{U}_0(s)\mathbf{C}(s, \rho z)\}]\mathbf{D}(s, \rho z) \tag{109}$$

where $\mathbf{C}(s, \rho)$ *and* $\mathbf{D}(s, \rho)$ *satisfy* (a_2), (b_2), *and* (106).

Proof. If we assume that $\mathbf{C}(s, \rho)$ and $\mathbf{D}(s, \rho)$ satisfy (a_2), (b_2), and (106), then by (91) we can prove that (109) satisfies (100). This proves that (97) is indeed equal to (109).

From Theorem 4 we can conclude that $\mathbf{C}(s, \rho)$ and $\mathbf{D}(s, \rho)$ are uniquely determined by (a_2), (b_2), and (106).

Second, let us consider the recurrence equations

$$\mathbf{U}_n(s) = w\mathbf{L}\{\mathbf{U}_{n-1}(s)\boldsymbol{\Psi}(s)\} + z\mathbf{L}^*\{\mathbf{U}_{n-1}(s)\boldsymbol{\Psi}(s)\} \tag{110}$$

and

$$\mathbf{U}_n(s) = w\mathbf{L}\{\boldsymbol{\Psi}(s)\mathbf{U}_{n-1}(s)\} + z\mathbf{L}^*\{\boldsymbol{\Psi}(s)\mathbf{U}_{n-1}(s)\} \tag{111}$$

for $n = 1, 2, \ldots$ where $\mathbf{U}_0(s) \in \mathbf{R}_2$, $\boldsymbol{\Psi}(s) \in \mathbf{R}_2$, w and z are complex (or real) numbers and $\mathbf{L}$ satisfies the conditions (i), (ii), (iii), and can be represented in the form of (93).

In both cases $\mathbf{U}_n(s) \in \mathbf{R}_2$ for $n = 1, 2, \ldots$ and if

$$|\rho|[\min(|w|, |z|) + |w - z|]\ \|\boldsymbol{\Psi}\| < \gamma(\mathbf{L}) \tag{112}$$

where $\gamma(\mathbf{L})$ satisfies the properties stated after (98), then

$$\mathbf{U}(s, \rho) = \sum_{n=0}^{\infty} \mathbf{U}_n(s)\rho^n \tag{113}$$

belongs to $\mathbf{R}_2$.

If we multiply (110) by ρ^n and add for $n = 1, 2, \ldots$, then we obtain that

$$\mathbf{L}\{\mathbf{U}(s, \rho)[\mathbf{I} - \rho w\boldsymbol{\Psi}(s)]\} + \mathbf{L}^*\{\mathbf{U}(s, \rho)[\mathbf{I} - \rho z\boldsymbol{\Psi}(s)]\} = \mathbf{U}_0(s) \tag{114}$$

whenever (112) is satisfied.

If we multiply (111) by ρ^n and add for $n = 1, 2, \ldots$, then we obtain that

$$\mathbf{L}\{[\mathbf{I} - \rho w\boldsymbol{\Psi}(s)]\mathbf{U}(s, \rho)\} + \mathbf{L}^*\{[\mathbf{I} - \rho z\boldsymbol{\Psi}(s)]\mathbf{U}(s, \rho)\} = \mathbf{U}_0(s) \tag{115}$$

whenever (112) is satisfied.

Conversely, if

$$\mathbf{U}(s, \rho) = \sum_{n=0}^{\infty} \mathbf{U}_n^*(s)\rho^n \tag{116}$$

belongs to $\mathbf{R}_2$ for $|\rho| < r$, where r is some positive number, and if (116) satisfies (114), then $\mathbf{U}_n^*(s) = \mathbf{U}_n(s)$ for $n = 0, 1, 2, \ldots$, where $\mathbf{U}_n(s)$ $(n = 1, 2, \ldots)$ is defined by (110). If (116) satisfies (115), then $\mathbf{U}_n^*(s) = \mathbf{U}_n(s)$ for $n = 0, 1, 2, \ldots$, where $\mathbf{U}_n(s)$ $(n = 1, 2, \ldots)$ is defined by (111).

The recurrence equation (110) can be solved by using the following factorization. Let us suppose that

$$[\mathbf{I} - \rho z \boldsymbol{\Psi}(s)]^{-1}[\mathbf{I} - \rho w \boldsymbol{\Psi}(s)] = [\mathbf{A}(s, \rho w, \rho z)]^{-1}[\mathbf{B}(s, \rho w, \rho z)]^{-1} \tag{117}$$

for $\text{Re}(s) = 0$ whenever (112) holds and that $\mathbf{A}(s, \rho w, \rho z)$ satisfies (a_2) and $\mathbf{B}(s, \rho w, \rho z)$ satisfies (b_2). By (102) we can conclude that such a factorization exists and $\mathbf{A}(s, \rho w, \rho z)$ and $\mathbf{B}(s, \rho w, \rho z)$ are uniquely determined by the stated conditions.

THEOREM 5. *If* $\mathbf{U}_n(s)$ $(n = 1, 2, \ldots)$ *satisfies* (110) *and if* (112) *holds, then the generating function* (113) *is given by*

$$\begin{aligned}\mathbf{U}(s, \rho) = [\mathbf{L}\{\mathbf{U}_0(s)\mathbf{B}(s, \rho w, \rho z)\} \\ + \mathbf{L}^*\{\mathbf{U}_0(s)[\mathbf{A}(s, \rho w, \rho z)]^{-1}\}]\mathbf{A}(s, \rho w, \rho z)[\mathbf{I} - \rho z \boldsymbol{\Psi}(s)]^{-1}\end{aligned} \tag{118}$$

where $\mathbf{A}(s, \rho w, \rho z)$ *and* $\mathbf{B}(s, \rho w, \rho z)$ *satisfy* (a_2), (b_2), *and* (117).

Proof. If we assume that $\mathbf{A}(s, \rho w, \rho z)$ and $\mathbf{B}(s, \rho w, \rho z)$ satisfy (a_2), (b_2), and (117), then by (91) we can prove that (118) satisfies (114). This proves that (113) is indeed equal to (118).

The recurrence equation (111) can be solved by using the following factorization. Let us suppose that

$$[\mathbf{I} - \rho z \boldsymbol{\Psi}(s)]^{-1}[\mathbf{I} - \rho w \boldsymbol{\Psi}(s)] = [\mathbf{D}(s, \rho w, \rho z)]^{-1}[\mathbf{C}(s, \rho w, \rho z)]^{-1} \tag{119}$$

for $\text{Re}(s) = 0$ whenever (112) holds and that $\mathbf{C}(s, \rho w, \rho z)$ satisfies (a_2) and $\mathbf{D}(s, \rho w, \rho z)$ satisfies (b_2). By (106) we can conclude that such a factorization exists and $\mathbf{C}(s, \rho w, \rho z)$ and $\mathbf{D}(s, \rho w, \rho z)$ are uniquely determined by the stated conditions.

THEOREM 6. *If* $\mathbf{U}_n(s)$ $(n = 1, 2, \ldots)$ *satisfies* (111) *and if* (112) *holds, then the generating function* (113) *is given by*

$$\begin{aligned}\mathbf{U}(s, \rho) = [\mathbf{I} - \rho z \boldsymbol{\Psi}(s)]^{-1}\mathbf{C}(s, \rho w, \rho z)[\mathbf{L}\{\mathbf{D}(s, \rho w, \rho z)\mathbf{U}_0(s)\} \\ + \mathbf{L}^*\{[\mathbf{C}(s, \rho w, \rho z)]^{-1}\mathbf{U}_0(s)\}]\end{aligned} \tag{120}$$

where $\mathbf{C}(s, \rho w, \rho z)$ *and* $\mathbf{D}(s, \rho w, \rho z)$ *satisfy* (a_2), (b_2), *and* (119).

Proof. If we assume that $\mathbf{C}(s, \rho w, \rho z)$ and $\mathbf{D}(s, \rho w, \rho z)$ satisfy (a_2), (b_2), and (119), then by (91) we can prove that (120) satisfies (115). This proves that (113) is indeed equal to (120).

8. Some Particular Transformations in $\mathbf{R}_2$

In this section we shall consider some transformations in $\mathbf{R}_2$ which satisfy the conditions (i), (ii), and (iii).

If $\boldsymbol{\Phi}(s) \in \mathbf{R}_2$ is given by (84), then let

$$\boldsymbol{\Phi}^+(s) = [\Phi_{ij}^+(s)]_{i,\, j \in I} \tag{121}$$

for Re(s) $\geqslant 0$ and

$$\boldsymbol{\Phi}^-(s) = [\Phi_{ij}^-(s)]_{i,\, j \in I} \tag{122}$$

for Re(s) $\leqslant 0$ where $\Phi_{ij}^-(s)$ is defined by (56) and $\Phi_{ij}^-(s)$ by (57).

Obviously, $\boldsymbol{\Phi}^+(s) \in \mathbf{R}_2$, $\boldsymbol{\Phi}^-(s) \in \mathbf{R}_2$ and

$$\boldsymbol{\Phi}(s) = \boldsymbol{\Phi}^+(s) + \boldsymbol{\Phi}^-(s) \tag{123}$$

for Re$(s) = 0$. We have $\mathbf{M}\{\boldsymbol{\Phi}^+(s)\} \leqslant \|\boldsymbol{\Phi}\|$ for Re$(s) \geqslant 0$ and $\mathbf{M}\{\boldsymbol{\Phi}^-(s)\} \leqslant 2\|\boldsymbol{\Phi}\|$ for Re$(s) \leqslant 0$.

The matrix function $\boldsymbol{\Phi}^+(s)$ is regular for Re$(s) > 0$, continuous and bounded for Re$(s) \geqslant 0$ and $\boldsymbol{\Phi}^+(0) = \boldsymbol{\Phi}(0)$.

The matrix function $\boldsymbol{\Phi}^-(s)$ is regular for Re$(s) < 0$, continuous and bounded for Re$(s) \leqslant 0$ and $\boldsymbol{\Phi}^-(0) = 0$.

By Liouville's theorem it follows that the above properties uniquely determine $\boldsymbol{\Phi}^+(s)$ and $\boldsymbol{\Phi}^-(s)$ in the representation (123).

Now let us extend the definition of the transformations (63), (64), (65) from the space $\mathbf{R}_1$ to the space $\mathbf{R}_2$ in such a way that we form these transformations element by element for $\boldsymbol{\Phi}(s) \in \mathbf{R}_2$, that is,

$$\mathbf{T}\{\boldsymbol{\Phi}(s)\} = \boldsymbol{\Phi}^+(s), \tag{124}$$

$$\mathbf{T}_0\{\boldsymbol{\Phi}(s)\} = \boldsymbol{\Phi}^+(s) - \boldsymbol{\Phi}^+(\infty), \tag{125}$$

and

$$\mathbf{T}_1\{\boldsymbol{\Phi}(s)\} = \boldsymbol{\Phi}^+(s) + \boldsymbol{\Phi}^-(-\infty). \tag{126}$$

We define $\mathbf{T}^*$, $\mathbf{T}_0^*$, $\mathbf{T}_1^*$ by (92). We can easily see that these transformations satisfy (i), (ii), (iii), $\|\mathbf{T}\| = \|\mathbf{T}_0\| = \|\mathbf{T}_1\| = \|\mathbf{T}_0^*\| = \|\mathbf{T}_1^*\| = 1$ and $\|\mathbf{T}^*\| = 2$.

If $\mathbf{L}$ is any one of the transformations (124), (125), (126), and if $\boldsymbol{\Phi}(s) \in \mathbf{R}_2$, then $\mathbf{L}\{\mathbf{C}\boldsymbol{\Phi}(s)\} = \mathbf{C}\mathbf{L}\{\boldsymbol{\Phi}(s)\}$ and $\mathbf{L}\{\boldsymbol{\Phi}(s)\mathbf{C}\} = \mathbf{L}\{\boldsymbol{\Phi}(s)\}\mathbf{C}$ for any constant matrix $\mathbf{C}$ for which $\mathbf{M}\{\mathbf{C}\} < \infty$. Furthermore, $\mathbf{L}\{\boldsymbol{\Phi}(s)\}$ can be represented in the form of (93) where $\mathbf{T}\{\boldsymbol{\Phi}(s)\}$ is defined by (124), $\boldsymbol{\alpha}(\boldsymbol{\Phi}) = \mathbf{0}$ for $\mathbf{L} = \mathbf{T}$, $\boldsymbol{\alpha}(\boldsymbol{\Phi}) = \boldsymbol{\Phi}^{+}(\infty)$ for $\mathbf{L} = \mathbf{T}_0$, and $\boldsymbol{\alpha}(\boldsymbol{\Phi}) = -\boldsymbol{\Phi}^{-}(-\infty)$ for $\mathbf{L} = \mathbf{T}_1$. This implies that $\gamma(\mathbf{L}) = 1$ necessarily holds.

If we assume that $\mathbf{T}$ is defined by (124), then in Theorem 3 we can determine $\mathbf{U}(s, \rho)$ by the following factorization too.

Let $|\rho| \, \|\boldsymbol{\Psi}\| < 1$ and

$$\mathbf{I} - \rho\boldsymbol{\Psi}(s) = \boldsymbol{\Psi}^{+}(s, \rho)\boldsymbol{\Psi}^{-}(s, \rho) \tag{127}$$

for $\mathrm{Re}(s) = 0$ where

(α_2) the matrix function $\boldsymbol{\Psi}^{+}(s, \rho)$ has an inverse $[\boldsymbol{\Psi}^{+}(s, \rho)]^{-1}$ for $\mathrm{Re}(s) \geqslant 0$, and $\boldsymbol{\Psi}^{+}(s, \rho)$ and $[\boldsymbol{\Psi}^{+}(s, \rho)]^{-1}$ are bounded and continuous for $\mathrm{Re}(s) \geqslant 0$ and regular for $\mathrm{Re}(s) > 0$,

(β_2) the matrix function $\boldsymbol{\Psi}^{-}(s, \rho)$ has an inverse $[\boldsymbol{\Psi}^{-}(s, \rho)]^{-1}$ for $\mathrm{Re}(s) \leqslant 0$, and $\boldsymbol{\Psi}^{-}(s, \rho)$ and $[\boldsymbol{\Psi}^{-}(s, \rho)]^{-1}$ are bounded and continuous for $\mathrm{Re}(s) \leqslant 0$ and regular for $\mathrm{Re}(s) < 0$.

The factorization (127) satisfying (α_2) and (β_2) always exists. For example, if

$$\boldsymbol{\Psi}^{+}(s, \rho) = [\mathbf{A}(s, \rho)]^{-1}\mathbf{C}_1(\rho) \tag{128}$$

for $\mathrm{Re}(s) \geqslant 0$ and

$$\boldsymbol{\Psi}^{-}(s, \rho) = \mathbf{C}_2(\rho)[\mathbf{B}(s, \rho)]^{-1} \tag{129}$$

for $\mathrm{Re}(s) \leqslant 0$, where $\mathbf{A}(s, \rho)$ and $\mathbf{B}(s, \rho)$ satisfy (a_2), (b_2), and (102), $\mathbf{M}\{\mathbf{C}_1(\rho)\} < \infty$, $\mathbf{M}\{\mathbf{C}_2(\rho)\} < \infty$ and $\mathbf{C}_1(\rho)\mathbf{C}_2(\rho) = \mathbf{I}$, then ($\alpha_2$), ($\beta_2$), and (127) are satisfied. Conversely, it follows from Liouville's theorem that conditions (α_2), (β_2), and (127) determine $\boldsymbol{\Psi}^{+}(s, \rho)$ and $\boldsymbol{\Psi}^{-}(s, \rho)$ up to a matrix factor independent of s. This implies that (128) and (129) are the general forms of $\boldsymbol{\Psi}^{+}(s, \rho)$ and $\boldsymbol{\Psi}^{-}(s, \rho)$, respectively.

If $\mathbf{T}$ is defined by (124) and if $\mathbf{L}$ is any one of the transformations (124), (125), (126), then in Theorem 3 we can write that

$$\begin{aligned}\mathbf{U}(s, \rho) = [\boldsymbol{\Psi}^{-}(s, \rho z)]^{-1}[\mathbf{L}\{&\boldsymbol{\Psi}^{-}(s, \rho z)\mathbf{U}_0(s)[\boldsymbol{\Psi}^{-}(s, \rho w)]^{-1}\} \\ &+ \mathbf{L}^{*}\{\boldsymbol{\Psi}^{+}(s, \rho z)\mathbf{U}_0(s)\boldsymbol{\Psi}^{+}(s, \rho w\}][\boldsymbol{\Psi}^{+}(s, \rho w)]^{-1}\end{aligned} \tag{130}$$

where $\boldsymbol{\Psi}^{+}(s, \rho)$ and $\boldsymbol{\Psi}^{-}(s, \rho)$ satisfy (α_2), (β_2), and (127). For if we put (128) and (129) into (130) and take into consideration that $\mathbf{C}_1(\rho)\mathbf{C}_2(\rho) = \mathbf{I}$, then (130) reduces to (105).

If $\mathbf{T}$ is defined by (124), then in Theorem 4 we can determine $\mathbf{U}(s, \rho)$ by the following factorization too.

Let $|\rho|\,\|\boldsymbol{\Psi}\| < 1$ and

$$\mathbf{I} - \rho\boldsymbol{\Psi}(s) = \boldsymbol{\Phi}^{-}(s, \rho)\boldsymbol{\Phi}^{+}(s, \rho) \tag{131}$$

for $\operatorname{Re}(s) = 0$, where $\boldsymbol{\Phi}^{+}(s, \rho)$ satisfies (α_2) and $\boldsymbol{\Phi}^{-}(s, \rho)$ satisfies (β_2). In a similar way as above we can prove that the factorization (131) always exists and in Theorem 4 we can write that

$$\begin{aligned}\mathbf{U}(s, \rho) = [\boldsymbol{\Phi}^{+}(s, \rho w)]^{-1}[\mathbf{L}\{[\boldsymbol{\Phi}^{-}(s, \rho w)]^{-1}\mathbf{U}_0(s)\boldsymbol{\Phi}^{-}(s, \rho z)\}\\ + \mathbf{L}^*\{\boldsymbol{\Phi}^{+}(s, \rho w)\mathbf{U}_0(s)[\boldsymbol{\Phi}^{+}(s, \rho z)]^{-1}\}][\boldsymbol{\Phi}^{-}(s, \rho z)]^{-1}\end{aligned} \tag{132}$$

where $\boldsymbol{\Phi}^{+}(s, \rho)$ and $\boldsymbol{\Phi}^{-}(s, \rho)$ satisfy (α_2), (β_2), and (131).

If $\mathbf{T}$ is defined by (124) and if $\mathbf{L}$ is any one of the transformations (124), (125), (126), then Theorem 5 and Theorem 6 can also be expressed in the following forms.

In Theorem 5 we can write that

$$\begin{aligned}\mathbf{U}(s, \rho) = [\mathbf{L}\{\mathbf{U}_0(s)[\boldsymbol{\Psi}^{-}(s, \rho w, \rho z)]^{-1}\}\\ + \mathbf{L}^*\{\mathbf{U}_0(s)\boldsymbol{\Psi}^{+}(s, \rho w, \rho z)\}][\boldsymbol{\Psi}^{+}(s, \rho w, \rho z)]^{-1}[\mathbf{I} - \rho z\boldsymbol{\Psi}^{+}(s)]^{-1}\end{aligned} \tag{133}$$

where

$$[\mathbf{I} - \rho w\boldsymbol{\Psi}(s)]^{-1}[\mathbf{I} - \rho z\boldsymbol{\Psi}(s)] = \boldsymbol{\Psi}^{+}(s, \rho w, \rho z)\boldsymbol{\Psi}^{-}(s, \rho w, \rho z) \tag{134}$$

for $\operatorname{Re}(s) = 0$ and $\boldsymbol{\Psi}^{+}(s, \rho w, \rho z)$ satisfies property (α_2) and $\boldsymbol{\Psi}^{-}(s, \rho w, \rho z)$ satisfies property (β_2).

In Theorem 6 we can write that

$$\begin{aligned}\mathbf{U}(s, \rho) = [\mathbf{I} - \rho z\boldsymbol{\Psi}(s)]^{-1}[\boldsymbol{\Phi}^{+}(s, \rho w, \rho z)]^{-1}[\mathbf{L}\{[\boldsymbol{\Phi}^{-}(s, \rho w, \rho z)]^{-1}\mathbf{U}_0(s)\}\\ + \mathbf{L}^*\{\boldsymbol{\Phi}^{+}(s, \rho w, \rho z)\mathbf{U}_0(s)\}]\end{aligned} \tag{135}$$

where

$$[\mathbf{I} - \rho w\boldsymbol{\Psi}(s)]^{-1}[\mathbf{I} - \rho z\boldsymbol{\Psi}(s)] = \boldsymbol{\Phi}^{-}(s, \rho w, \rho z)\boldsymbol{\Phi}^{+}(s, \rho w, \rho z) \tag{136}$$

for $\operatorname{Re}(s) = 0$ and $\boldsymbol{\Phi}^{+}(s, \rho w, \rho z)$ satisfies property (α_2) and $\boldsymbol{\Phi}^{-}(s, \rho w, \rho z)$ satisfies property (β_2).

The solutions of the equations (95) and (96) have a special interest in the case where $\mathbf{L}$ is any one of the transformations (124), (125), (126), $w = 1$ and $z = 0$.

If

$$\mathbf{U}_n(s) = \mathbf{L}\{\mathbf{U}_{n-1}(s)\boldsymbol{\Psi}(s)\} \tag{137}$$

for $n = 1, 2, \ldots$, then by (130) we get

$$\mathbf{U}(s, \rho) = [\mathbf{L}\{\mathbf{U}_0(s)[\boldsymbol{\Psi}^{-}(s, \rho)]^{-1}\} + \mathbf{L}^*\{\mathbf{U}_0(s)\boldsymbol{\Psi}^{+}(s, \rho)\}][\boldsymbol{\Psi}^{+}(s, \rho)]^{-1} \tag{138}$$

where $\boldsymbol{\Psi}^{+}(s, \rho)$ and $\boldsymbol{\Psi}^{-}(s, \rho)$ are defined by (127).

If, in particular, $\mathbf{T}\{\mathbf{U}_0(s)\} = \mathbf{U}_0(s)$, then (138) reduces to

$$\mathbf{U}(s, \rho) = [\mathbf{L}\{\mathbf{U}_0(s)[\boldsymbol{\Psi}^-(s, \rho)]^{-1}\} + \epsilon(\mathbf{L})\mathbf{U}_0(\infty)\boldsymbol{\Psi}^+(\infty, \rho)][\boldsymbol{\Psi}^+(s, \rho)]^{-1} \tag{139}$$

where $\epsilon(\mathbf{T}) = \epsilon(\mathbf{T}_1) = 0$ and $\epsilon(\mathbf{T}_0) = 1$.

If, in particular, $\mathbf{U}_0(s) = \mathbf{I}$, then (138) reduces to

$$\mathbf{U}(s, \rho) = [\boldsymbol{\Psi}^-(0, \rho)]^{-1}[\boldsymbol{\Psi}^+(s, \rho)]^{-1} \tag{140}$$

for $\mathbf{L} = \mathbf{T}$,

$$\mathbf{U}(s, \rho) = [\boldsymbol{\Psi}^+(\infty, \rho)][\boldsymbol{\Psi}^+(s, \rho)]^{-1} \tag{141}$$

for $\mathbf{L} = \mathbf{T}_0$, and

$$\mathbf{U}(s, \rho) = [\boldsymbol{\Psi}^-(-\infty, \rho)]^{-1}[\boldsymbol{\Psi}^+(s, \rho)]^{-1} \tag{142}$$

for $\mathbf{L} = \mathbf{T}_1$.

If

$$\mathbf{U}_n(s) = \mathbf{L}\{\boldsymbol{\Psi}(s)\mathbf{U}_{n-1}(s)\} \tag{143}$$

for $n = 1, 2, \ldots$, then by (132) we get

$$\mathbf{U}(s, \rho) = [\boldsymbol{\Phi}^+(s, \rho)]^{-1}[\mathbf{L}\{[\boldsymbol{\Phi}^-(s, \rho)]^{-1}\mathbf{U}_0(s)\} + \mathbf{L}^*[\boldsymbol{\Phi}^+(s, \rho)\mathbf{U}_0(s)\}] \tag{144}$$

where $\boldsymbol{\Phi}^+(s, \rho)$ and $\boldsymbol{\Phi}^-(s, \rho)$ are defined by (131).

If, in particular, $\mathbf{T}\{\mathbf{U}_0(s)\} = \mathbf{U}_0(s)$, then (144) reduces to

$$\mathbf{U}(s, \rho) = [\boldsymbol{\Phi}^+(s, \rho)]^{-1}[\mathbf{L}\{[\boldsymbol{\Phi}^-(s, \rho)]^{-1}\mathbf{U}_0(s)\} + \epsilon(\mathbf{L})\boldsymbol{\Phi}^+(\infty, \rho)\mathbf{U}_0(\infty)] \tag{145}$$

where $\epsilon(\mathbf{T}) = \epsilon(\mathbf{T}_1) = 0$ and $\epsilon(\mathbf{T}_0) = 1$.

If, in particular, $\mathbf{U}_0(s) = \mathbf{I}$, then (145) reduces to

$$\mathbf{U}(s, \rho) = [\boldsymbol{\Phi}^+(s, \rho)]^{-1}[\boldsymbol{\Phi}^-(0, \rho)]^{-1} \tag{146}$$

for $\mathbf{L} = \mathbf{T}$,

$$\mathbf{U}(s, \rho) = [\boldsymbol{\Phi}^+(s, \rho)]^{-1}[\boldsymbol{\Phi}^+(\infty, \rho)] \tag{147}$$

for $\mathbf{L} = \mathbf{T}_0$, and

$$\mathbf{U}(s, \rho) = [\boldsymbol{\Phi}^+(s, \rho)]^{-1}[\boldsymbol{\Phi}^-(-\infty, \rho)]^{-1} \tag{148}$$

for $\mathbf{L} = \mathbf{T}_1$.

In conclusion, we remark that the method of matrix factorization has already been used in several fields of mathematics, namely, in the theory of systems of integral equations, in the theory of linear prediction of multivariate stationary stochastic processes, and in the theory of Markov chains. We refer to the works of Masani [16], Gohberg and Krein [11], Miller [17, 18], and Presman [25].

9. A System of Recurrence Equations in $\mathbf{R}_2$

In a similar way as in $\mathbf{R}_1$ we can consider various systems of recurrence equations in the space $\mathbf{R}_2$. For example, let us suppose that

$$\mathbf{U}_n(s) = \mathbf{L}\{[w\mathbf{U}_{n-1}(s) + z\mathbf{V}_{n-1}(s)]\boldsymbol{\Psi}(s)\} \tag{149}$$

and

$$\mathbf{V}_n(s) = \mathbf{L}^*\{[z\mathbf{U}_{n-1}(s) + w\mathbf{V}_{n-1}(s)]\boldsymbol{\Psi}(s)\} \tag{150}$$

for $n = 1, 2, \ldots,$ where $\mathbf{U}_0(s) \in \mathbf{R}_2$, $\mathbf{V}_0(s) \in \mathbf{R}_2$, $\boldsymbol{\Psi}(s) \in \mathbf{R}_2$, w and z are complex (or real) numbers and $\mathbf{L}$ satisfies the conditions (i), (ii), (iii) and can be represented in the form of (93). To solve this system we introduce a new Banach algebra $\mathbf{S}_2$ with elements

$$\tilde{\boldsymbol{\Phi}}(s) = \begin{bmatrix} \boldsymbol{\Phi}_{11}(s) & \boldsymbol{\Phi}_{12}(s) \\ \boldsymbol{\Phi}_{21}(s) & \boldsymbol{\Phi}_{22}(s) \end{bmatrix} \tag{151}$$

where $\boldsymbol{\Phi}_{ij}(s) \in \mathbf{R}_2$ for $i = 1, 2$, $j = 1, 2$. In $\mathbf{S}_2$ let us define the operations of addition, multiplication, and multiplication by a complex (or real) constant according to the rules of matrix algebra and according to the rules established in $\mathbf{R}_2$. Define the norm of (151) either by

$$\|\tilde{\boldsymbol{\Phi}}\|_{\mathbf{S}_2} = \max(\|\boldsymbol{\Phi}_{11}\|_{\mathbf{R}_2} + \|\boldsymbol{\Phi}_{12}\|_{\mathbf{R}_2}, \|\boldsymbol{\Phi}_{21}\|_{\mathbf{R}_2} + \|\boldsymbol{\Phi}_{22}\|_{\mathbf{R}_2}) \tag{152}$$

or by

$$\|\tilde{\boldsymbol{\Phi}}\|_{\mathbf{S}_2} = \max(\|\boldsymbol{\Phi}_{11}\|_{\mathbf{R}_2} + \|\boldsymbol{\Phi}_{21}\|_{\mathbf{R}_2}, \|\boldsymbol{\Phi}_{12}\|_{\mathbf{R}_2} + \|\boldsymbol{\Phi}_{22}\|_{\mathbf{R}_2}). \tag{153}$$

We can easily see that $\mathbf{S}_2$ is a noncommutative Banach algebra with zero element

$$\begin{bmatrix} \mathbf{0} & \mathbf{0} \\ \mathbf{0} & \mathbf{0} \end{bmatrix}$$

and identity element

$$\begin{bmatrix} \mathbf{I} & \mathbf{0} \\ \mathbf{0} & \mathbf{I} \end{bmatrix}.$$

If we extend the definition of the transformation $\mathbf{L}$ from the space $\mathbf{R}_2$ to the space $\mathbf{S}_2$ in such a way that we form the transformation element by element for the partitioned matrix (151), then $\mathbf{L}$ satisfies the same properties in $\mathbf{S}_2$ as in $\mathbf{R}_2$.

The system of recurrence equations (149), (150) can be replaced by the following recurrence equation in $\mathbf{S}_2$:

$$\begin{bmatrix} \mathbf{U}_n(s) & \mathbf{V}_n(s) \\ \mathbf{0} & \mathbf{0} \end{bmatrix} = \mathbf{L}\left\{\begin{bmatrix} \mathbf{U}_{n-1}(s) & \mathbf{V}_{n-1}(s) \\ \mathbf{0} & \mathbf{0} \end{bmatrix}\begin{bmatrix} w\boldsymbol{\Psi}(s) & \mathbf{0} \\ z\boldsymbol{\Psi}(s) & \mathbf{0} \end{bmatrix}\right\} + \mathbf{L}^*\left\{\begin{bmatrix} \mathbf{U}_{n-1}(s) & \mathbf{V}_{n-1}(s) \\ \mathbf{0} & \mathbf{0} \end{bmatrix}\begin{bmatrix} \mathbf{0} & z\boldsymbol{\Psi}(s) \\ \mathbf{0} & w\boldsymbol{\Psi}(s) \end{bmatrix}\right\} \tag{154}$$

for $n = 1, 2, \ldots$.

This equation is exactly the same type as equation (110) in $\mathbf{R}_2$ and can be solved in exactly the same way as (110) in $\mathbf{R}_2$.

10. Independent and Identically Distributed Random Variables

Now let us suppose that $\xi_1, \xi_2, \ldots, \xi_n, \ldots$ is a sequence of mutually independent and identically distributed real random variables. Let

$$\psi(s) = \mathbf{E}\{e^{-s\xi_n}\} \tag{155}$$

for $\mathrm{Re}(s) = 0$ and $n \geqslant 1$. Obviously, $\psi(s) \in \mathbf{R}_1$ and $\|\psi\| = 1$. Let us suppose that ζ_0 is a real random variable which is independent of the sequence $\xi_1, \xi_2, \ldots, \xi_n, \ldots$. We use the notations introduced in Section 1.

In this case the solutions of many fluctuation problems can be reduced to the solution of the recurrence equation (35) where $\mathbf{L}$ is one of the transformations $\mathbf{T}, \mathbf{T}_0, \mathbf{T}_1$ defined in Section 4. The general solution of (35) is given by (71), and particular solutions by (73), (74), (75), and (76).

Now we shall consider the random variables introduced in Section 1 and describe the methods of finding their distributions.

The Maximal Partial Sum

The problem of finding the distribution of η_n^* defined by (3) can be reduced to the problem of finding the distribution of η_n defined by (2) in the particular case when $\zeta_0 = 0$. If $\mathbf{P}\{\zeta_0 = 0\} = 1$, then η_n^* and η_n have the same distribution.

By the repeated applications of (2) we obtain that

$$\eta_n = \max(\zeta_n - \zeta_n, \zeta_n - \zeta_{n-1}, \ldots, \zeta_n - \zeta_1, \zeta_n) \tag{156}$$

for $n \geqslant 1$. This random variable has the same distribution as

$$\bar{\eta}_n = \max(\zeta_0 - \zeta_0, \zeta_1 - \zeta_0, \ldots, \zeta_{n-1} - \zeta_0, \zeta_n). \tag{157}$$

If $\zeta_0 = 0$ in (157), then by (3) we can write that

$$\eta_n{}^* = \zeta_0 + \bar{\eta}_n. \tag{158}$$

In this representation ζ_0 and $\bar{\eta}_n$ are independent and $\bar{\eta}_n$ has the same distribution as η_n in the particular case when $\zeta_0 = 0$.

Let

$$U_n(s) = \mathbf{E}\{e^{-s\eta_n}\} \tag{159}$$

for $\mathrm{Re}(s) = 0$ and $n = 0, 1, 2, \ldots$. Since

$$\mathbf{E}\{e^{-s\eta_n}\} = \mathbf{T}\{\mathbf{E}\{e^{-s(\eta_{n-1}+\xi_n)}\}\}, \tag{160}$$

it follows that

$$U_n(s) = \mathbf{T}\{U_{n-1}(s)\psi(s)\} \tag{161}$$

for $n = 1, 2, \ldots$ and $\mathrm{Re}(s) = 0$. If $|\rho| < 1$, then the generating function

$$U(s, \rho) = \sum_{n=0}^{\infty} U_n(s)\rho^n \tag{162}$$

belongs to $\mathbf{R}_1$ and is given by (71).

We note that if

$$U_n(s, v) = \mathbf{E}\{e^{-s\eta_n - v\zeta_n}\} \tag{163}$$

for $\mathrm{Re}(s) = 0$, $\mathrm{Re}(v) = 0$ and $n = 0, 1, 2, \ldots$, then

$$U_n(s, v) = \mathbf{T}\{U_{n-1}(s, v)\psi(s + v)\} \tag{164}$$

for $n = 1, 2, \ldots$ and $\mathrm{Re}(s) = \mathrm{Re}(v) = 0$. In (164) $\mathbf{T}$ operates on the variable s and v is merely a parameter. If $|\rho| < 1$, then the generating function

$$U(s, v, \rho) = \sum_{n=0}^{\infty} U_n(s, v)\rho^n \tag{165}$$

belongs to $\mathbf{R}_1$ and is given by (71) where now $\psi(s)$ should be replaced by $\psi(s + v)$.

The generating function $U(s, \rho)$ was found in 1952 by Pollaczek [21] in the case where $U_0(s) = e^{-cs}$ and c is a nonnegative constant and $\psi(s + \epsilon) \in \mathbf{R}_1$ for some real $\epsilon \neq 0$. In 1956 Spitzer [27] demonstrated that if $U_0(s) = 1$ and $\psi(s) \in \mathbf{R}_1$, then

$$U(s, \rho) = \exp\left\{\sum_{k=1}^{\infty} \frac{\rho^k}{k} \mathbf{T}\{[\psi(s)]^k\}\right\} \tag{166}$$

for $\mathrm{Re}(s) \geqslant 0$ and $|\rho| < 1$. See also Wendel [31] and Kingman [14, 15].

First Passage Times

Now let

$$\Phi_n(s) = \mathbf{E}\{e^{-s\zeta_n}\delta(\nu = n)\} \tag{167}$$

for $n \geqslant 1$ and $\mathrm{Re}(s) \leqslant 0$, where ν is defined by (4). Our aim is to find

$$\Phi(s, \rho) = \mathbf{E}\{e^{-s\zeta_\nu}\rho^\nu\} = \sum_{n=1}^{\infty} \Phi_n(s)\rho^n \tag{168}$$

for $\mathrm{Re}(s) \leqslant 0$ and $|\rho| < 1$.

Let

$$U_n(s) = \mathbf{E}\{e^{-s\zeta_n}\delta(\nu > n)\} \tag{169}$$

for $n \geqslant 0$ and $\mathrm{Re}(s) = 0$, and

$$U(s, \rho) = \sum_{n=0}^{\infty} U_n(s)\rho^n \tag{170}$$

for $\mathrm{Re}(s) = 0$ and $|\rho| < 1$.

Obviously, we have

$$\Phi_n(s) = U_{n-1}(s)\psi(s) - U_n(s) \tag{171}$$

for $n \geqslant 1$ and $\mathrm{Re}(s) = 0$, and thus

$$\Phi(s, \rho) = U_0(s) - U(s, \rho)[1 - \rho\psi(s)] \tag{172}$$

for $\mathrm{Re}(s) = 0$ and $|\rho| < 1$.

Since

$$\mathbf{E}\{e^{-s\zeta_n}\delta(\nu > n)\} = \mathbf{T}_0\{\mathbf{E}\{e^{-s\zeta_{n-1}-s\xi_n}\delta(\nu > n - 1)\}\} \tag{173}$$

for $n \geqslant 1$, it follows that

$$U_n(s) = \mathbf{T}_0\{U_{n-1}(s)\psi(s)\} \tag{174}$$

for $n \geqslant 1$ and $\mathrm{Re}(s) \geqslant 0$. Accordingly, $U(s, \rho)$ can be obtained by (71) and $\Phi(s, \rho)$ by (172).

If, in particular, $\mathbf{P}\{\zeta_0 = 0\} = 1$, then by (75) we obtain that

$$\Phi(s, \rho) = 1 - \exp\{\mathbf{T}_0{}^*\{\log[1 - \rho\psi(s)]\}\} \tag{175}$$

for $\mathrm{Re}(s) \leqslant 0$ and $|\rho| < 1$. If we let $\rho \to 1$ in (175), then we get

$$\mathbf{E}\{e^{-s\zeta_\nu}\} = 1 - \exp\left\{-\sum_{k=1}^{\infty} \frac{1}{k}\mathbf{E}\{e^{-s\zeta_k}\delta(\zeta_k \leqslant 0)\}\right\} \tag{176}$$

for $\mathrm{Re}(s) < 0$.

Define

$$\Phi_n^*(s) = \mathbf{E}\{e^{-s\zeta_n}\delta(v^* = n)\} \tag{177}$$

for $n \geqslant 1$ and $\mathrm{Re}(s) \leqslant 0$, where v^* is defined by (5). In a similar way as we found (168) we can determine

$$\Phi^*(s, \rho) = \mathbf{E}\{e^{-s\zeta_{v^*}}\rho^{v^*}\} = \sum_{n=0}^{\infty} \Phi_n^*(s)\rho^n \tag{178}$$

for $\mathrm{Re}(s) \leqslant 0$ and $|\rho| < 1$.

Let

$$U_n^*(s) = \mathbf{E}\{e^{-s\zeta_n}\delta(v^* > n)\} \tag{179}$$

for $n \geqslant 0$ and $\mathrm{Re}(s) = 0$, and

$$U^*(s, \rho) = \sum_{n=0}^{\infty} U_n^*(s)\rho^n \tag{180}$$

for $\mathrm{Re}(s) = 0$ and $|\rho| < 1$.

Obviously, we have

$$\Phi_n^*(s) = U_{n-1}^*(s)\psi(s) - U_n^*(s) \tag{181}$$

for $n \geqslant 1$ and $\mathrm{Re}(s) = 0$, and thus

$$\Phi^*(s, \rho) = U_0^*(s) - U^*(s, \rho)[1 - \rho\psi(s)] \tag{182}$$

for $\mathrm{Re}(s) = 0$ and $|\rho| < 1$.

Since

$$U_n^*(s) = \mathbf{T}_1\{U_{n-1}^*(s)\psi(s)\} \tag{183}$$

for $n \geqslant 1$ and $\mathrm{Re}(s) \geqslant 0$, we can obtain $U^*(s, \rho)$ by (71) and $\Phi^*(s, \rho)$ by (182).

If, in particular, $\mathbf{P}\{\zeta_0 = 0\} = 1$, then by (76) we obtain that

$$\Phi^*(s, \rho) = 1 - \exp\{\mathbf{T}_1^*\{\log[1 - \rho\psi(s)]\}\} \tag{184}$$

for $\mathrm{Re}(s) \leqslant 0$ and $|\rho| < 1$. If we let $\rho \to 1$ in (184), then we get

$$\mathbf{E}\{e^{-s\zeta_{v^*}}\} = 1 - \exp\left\{-\sum_{k=1}^{\infty} \frac{1}{k}\mathbf{E}\{e^{-s\zeta_k}\delta(\zeta_k < 0)\}\right\} \tag{185}$$

for $\mathrm{Re}(s) < 0$.

Formula (175) was found in 1952 by Pollaczek [22] in the context of queuing theory. In 1954 Andersen [3] proved (175) and (184) for $s = 0$. In 1958 Baxter [5] proved (184). See also Feller [10] for a combinatorial proof of (184), and Kemperman [12].

Positive Partial Sums

Let

$$U_n(s, w) = \mathbf{E}\{e^{-s\zeta_n} w^{\Delta_n}\} \tag{186}$$

for $n \geqslant 0$ and $\mathrm{Re}(s) = 0$, where Δ_n is defined in Section 1. The generating function

$$U(s, w, \rho) = \sum_{n=0}^{\infty} U_n(s, w)\rho^n \tag{187}$$

can be determined again by (71) for $\mathrm{Re}(s) = 0$, $|\rho| < 1$ and $|\rho w| < 1$.
Since

$$w^{\Delta_n} = w^{\Delta_{n-1}}[w\delta(\zeta_n > 0) + \delta(\zeta_n \leqslant 0)], \tag{188}$$

if we multiply (188) by $e^{-s\zeta_n}$ and form its expectation, then we obtain that

$$U_n(s, w) = w\mathbf{T}_0\{U_{n-1}(s, w)\psi(s)\} + \mathbf{T}_0^*\{U_{n-1}(s, w)\psi(s)\} \tag{189}$$

for $n = 1, 2, \ldots$. This equation is of type (35) and $U(s, w, \rho)$ is given by (71) for $|\rho| < 1$, $|\rho w| < 1$ and $\mathrm{Re}(s) = 0$.
Define

$$U_n^*(s, w) = \mathbf{E}\{e^{-s\zeta_n} w^{\Delta_n^*}\} \tag{190}$$

for $n \geqslant 0$ and $\mathrm{Re}(s) = 0$, where Δ_n^* is defined in Section 1. Now we have

$$U_n^*(s, w) = w\mathbf{T}_1\{U_{n-1}^*(s, w)\psi(s)\} + \mathbf{T}_1^*\{U_{n-1}^*(s, w)\psi(s)\} \tag{191}$$

for $n = 1, 2, \ldots$. Thus the generating function

$$U^*(s, w, \rho) = \sum_{n=0}^{\infty} U_n^*(s, w)\rho^n \tag{192}$$

is given by (71) for $\mathrm{Re}(s) = 0$, $|\rho| < 1$ and $|\rho w| < 1$.

In the particular case when $\mathbf{P}\{\zeta_0 = 0\} = 1$, the generating functions (187) and (192) were found in 1953 by Andersen [1, 2], in 1961 by Baxter [7], and in 1962 by Darling [8].

Ordered Partial Sums

In this section we assume that $\mathbf{P}\{\zeta_0 = 0\} = 1$. The general case can immediately be reduced to this particular case. First, we shall prove a general identity for the random variables $\eta_{n,k}(0 \leqslant k \leqslant n)$ defined by (6). Let us define $\eta_{n,k}^{(1)}(0 \leqslant k \leqslant n)$ for the random variables $\xi_2, \xi_3, \ldots, \xi_{n+1}$ in exactly the same way as we defined $\eta_{n,k}$ for the random variables $\xi_1, \xi_2, \ldots, \xi_n$. Then we have the identity

$$\sum_{k=0}^{n+1} e^{-s\eta_{n+1,k}}w^k = w\sum_{k=0}^{n} e^{-s[\xi_1+\eta_{n,k}^{(1)}]^+}w^k + \sum_{k=0}^{n}(e^{-s[\xi_1+\eta_{n,k}^{(1)}]^-}-1)w^k + 1 \quad (193)$$

for any s, w and $n \geqslant 0$, where $[x]^+ = \max(0, x)$ and $[x]^- = \min(0, x)$. The proof of (193) is based on the identity

$$e^{-sx} = e^{-sx^+} + e^{-sx^-} - 1 \quad (194)$$

and the relation

$$\eta_{n+1,k} = [\xi_1 + \eta_{n,k-1}^{(1)}]^+ + [\xi_1 + \eta_{n,k}^{(1)}]^- \quad (195)$$

for $0 \leqslant k \leqslant n+1$ where on the right-hand side of (195) the first term is 0 for $k = 0$ and the second term is 0 for $k = n+1$. If we put (195) into (194), multiply it by w^k and add for $k = 0, 1, \ldots, n+1$, then we get (193). The identity (193) is valid for any sequence of real random variables (or real numbers) $\xi_1, \xi_2, \ldots, \xi_{n+1}$.

Now let us suppose that $\xi_1, \xi_2, \ldots, \xi_n, \ldots$ are mutually independent and identically distributed real random variables each having the Laplace–Stieltjes transform (155). Define

$$\Phi_{n,k}(s, v) = \mathbf{E}\{e^{-s\eta_{n,k}-v\zeta_n}\} \quad (196)$$

for $\mathrm{Re}(s) = \mathrm{Re}(v) = 0$ and $0 \leqslant k \leqslant n$,

$$\Phi_n(s, v, w) = \sum_{k=0}^{n} \Phi_{n,k}(s, v)w^k \quad (197)$$

for $\mathrm{Re}(s) = \mathrm{Re}(v) = 0$ and $n \geqslant 0$, and

$$\Phi(s, v, w, \rho) = \sum_{n=0}^{\infty} \Phi_n(s, v, w)\rho^n \quad (198)$$

for $\mathrm{Re}(s) = \mathrm{Re}(v) = 0$, $|\rho| < 1$ and $|\rho w| < 1$.

If we multiply (193) by $e^{-v\zeta_{n+1}}$ and form its expectation, then we obtain that

$$\Phi_{n+1}(s, v, w) = w\mathbf{T}\{\psi(s+v)\Phi_n(s, v, w)\} + \mathbf{T}^*\{\psi(s+v)\Phi_n(s, v, w)\} + [\psi(v)]^{n+1} \quad (199)$$

for $n \geqslant 0$ and $\Phi_0(s, v, w) \equiv 1$.

Let

$$U_n(s, v, w) = \Phi_n(s, v, w) - \psi(v)\Phi_{n-1}(s, v, w) \quad (200)$$

for $n \geqslant 1$ and $U_0(s, v, w) = \Phi_0(s, v, w) \equiv 1$. Then by (199) we get

$$U_{n+1}(s, v, w) = w\mathbf{T}\{\psi(s+v)U_n(s, v, w)\} + \mathbf{T}^*\{\psi(s+v)U_n(s, v, w)\} \quad (201)$$

for $n \geqslant 1$. This equation is of type (35) and

$$U(s, v, w, \rho) = \sum_{n=0}^{\infty} U_n(s, v, w)\rho^n \tag{202}$$

is given by (71) for $|\rho| < 1$, $|\rho w| < 1$ and $\mathrm{Re}(s) = \mathrm{Re}(v) = 0$. Finally, by (200) it follows that

$$\Phi(s, v, w, \rho) = U(s, v, w, \rho)[1 - \rho\psi(v)]^{-1} \tag{203}$$

for $|\rho| < 1$, $|\rho w| < 1$ and $\mathrm{Re}(s) = \mathrm{Re}(v) = 0$.

In 1952 Pollaczek [22] gave a method of finding $U(s, v, w, \rho)$. See also Wendel [32], Baxter [7], Darling [8], Port [24], Andersen [4], De Smit [9], and Pollaczek [23].

Changes of Sign

Define ω_n and $\omega_n{}^*$ as in the Introduction. Let

$$U_n(s, z) = \mathbf{E}\{e^{-s\zeta_n} z^{\omega_n} \delta(\zeta_n > 0)\} \tag{204}$$

and

$$V_n(s, z) = \mathbf{E}\{e^{-s\zeta_n} z^{\omega_n} \delta(\zeta_n \leqslant 0)\} \tag{205}$$

for $n \geqslant 0$ and $\mathrm{Re}(s) = 0$.

If we take into consideration that

$$z^{\omega_n}\delta(\zeta_n > 0) = [z^{\omega_{n-1}}\delta(\zeta_{n-1} > 0) + z z^{\omega_{n-1}}\delta(\zeta_{n-1} \leqslant 0)]\delta(\zeta_n > 0), \tag{206}$$

multiply (206) by $e^{-s\zeta_n}$ and form its expectation, then we obtain that

$$U_n(s, z) = \mathbf{T}_0\{[U_{n-1}(s, z) + zV_{n-1}(s, z)]\psi(s)\} \tag{207}$$

for $n \geqslant 1$ and $\mathrm{Re}(s) = 0$. In a similar way we get

$$V_n(s, z) = \mathbf{T}_0{}^*\{[zU_{n-1}(s, z) + V_{n-1}(s, z)]\psi(s)\} \tag{208}$$

for $n \geqslant 1$ and $\mathrm{Re}(s) = 0$.

If we extend the definition of the transformation $\mathbf{T}_0$ and $\mathbf{T}_0{}^*$ to the space $\mathbf{R}_2$ where $I = \{1, 2\}$, then by (207) and (208) we can write that

$$\begin{bmatrix} U_n(s, z) & V_n(s, z) \\ 0 & 0 \end{bmatrix} = \mathbf{T}_0\left\{\begin{bmatrix} U_{n-1}(s, z) & V_{n-1}(s, z) \\ 0 & 0 \end{bmatrix}\begin{bmatrix} \psi(s) & 0 \\ z\psi(s) & 0 \end{bmatrix}\right\} + \mathbf{T}_0\left\{\begin{bmatrix} U_{n-1}(s, z) & V_{n-1}(s, z) \\ 0 & 0 \end{bmatrix}\begin{bmatrix} 0 & z\psi(s) \\ 0 & \psi(s) \end{bmatrix}\right\} \tag{209}$$

for $n \geq 1$ and $\mathrm{Re}(s) = 0$. This equation is of similar type as (110), and we can solve it in the same way as (110). Let us suppose that

$$\left\{\begin{bmatrix} 1 & 0 \\ 0 & 1 \end{bmatrix} - \rho \begin{bmatrix} 0 & z\psi(s) \\ 0 & \psi(s) \end{bmatrix}\right\}^{-1} \left\{\begin{bmatrix} 1 & 0 \\ 0 & 1 \end{bmatrix} - \rho \begin{bmatrix} \psi(s) & 0 \\ z\psi(s) & 0 \end{bmatrix}\right\}$$
$$= [1 - \rho\psi(s)]^{-1} \begin{bmatrix} 1 - \rho\psi(s) & \rho z\psi(s) \\ 0 & 1 \end{bmatrix} \begin{bmatrix} 1 - \rho\psi(s) & 0 \\ -\rho z\psi(s) & 1 \end{bmatrix}$$
$$= \boldsymbol{\Psi}^{+}(s, \rho, \rho z)\boldsymbol{\Psi}^{-}(s, \rho, \rho z) \tag{210}$$

for $\mathrm{Re}(s) = 0$, $|\rho| < 1$ and $|\rho z| < 1$, where the matrices $\boldsymbol{\Psi}^{+}(s, \rho, \rho z)$ and $\boldsymbol{\Psi}^{-}(s, \rho, \rho z)$ satisfy conditions (α_2) and (β_2), respectively, stated after (127).

If we write

$$\mathbf{U}_n(s, z) = \begin{bmatrix} U_n(s, z) & V_n(s, z) \\ 0 & 0 \end{bmatrix}, \tag{211}$$

then

$$\mathbf{U}(s, z, \rho) = \sum_{n=0}^{\infty} \mathbf{U}_n(s, z)\rho^n \tag{212}$$

can be expressed as follows:

$$\mathbf{U}(s, z, \rho) = [\mathbf{T}_0\{\mathbf{U}_0(s, z)[\boldsymbol{\Psi}^{-}(s, \rho, \rho z)]^{-1}\}$$
$$+ \mathbf{T}_0{}^*\{\mathbf{U}_0(s, z)\boldsymbol{\Psi}^{+}(s, \rho, \rho z)\}][\boldsymbol{\Psi}^{+}(s, \rho, \rho z)]^{-1} \begin{bmatrix} 1 & -\rho z\psi(s) \\ 0 & 1 - \rho\psi(s) \end{bmatrix}^{-1} \tag{213}$$

for $\mathrm{Re}(s) = 0$, $|\rho| < 1$ and $|\rho z| < 1$.

Furthermore, let

$$U_n{}^*(s, z) = \mathbf{E}\{e^{-s\zeta_n} z^{\omega_n{}^*} \delta(\zeta_n \geq 0)\} \tag{214}$$

and

$$V_n{}^*(s, z) = \mathbf{E}\{e^{-s\zeta_n} z^{\omega_n{}^*} \delta(\zeta_n < 0)\} \tag{215}$$

for $n \geq 0$.

If we take into consideration that

$$z^{\omega_n}\delta(\zeta_n \geq 0) = [z^{\omega_{n-1}}\delta(\zeta_{n-1} \geq 0) + z z^{\omega_{n-1}}\delta(\zeta_{n-1} < 0)]\delta(\zeta_n \geq 0), \tag{216}$$

multiply (216) by $e^{-s\zeta_n}$ and form its expectation, then we obtain that

$$U_n{}^*(s, z) = \mathbf{T}_1\{[U_{n-1}^*(s, z) + zV_{n-1}^*(s, z)]\psi(s)\} \tag{217}$$

for $n \geqslant 1$ and $\mathrm{Re}(s) = 0$. In a similar way we get

$$V_n^*(s, z) = \mathbf{T}_1^*\{[zU_{n-1}^*(s, z) + V_{n-1}^*(s, z)]\psi(s)\} \tag{218}$$

for $n \geqslant 1$ and $\mathrm{Re}(s) = 0$.

If we extend the definition of the transformations $\mathbf{T}_1$ and $\mathbf{T}_1^*$ to the space $\mathbf{R}_2$ where $I = \{1, 2\}$, then by (217) and (218) we can write that

$$\begin{aligned}\begin{bmatrix} U_n^*(s, z) & V_n^*(s, z) \\ 0 & 0 \end{bmatrix} &= \mathbf{T}_1\left\{\begin{bmatrix} U_{n-1}^*(s, z) & V_{n-1}^*(s, z) \\ 0 & 0 \end{bmatrix}\begin{bmatrix} \psi(s) & 0 \\ z\psi(s) & 0 \end{bmatrix}\right\} \\ &+ \mathbf{T}_1^*\left\{\begin{bmatrix} U_{n-1}^*(s, z) & V_{n-1}^*(s, z) \\ 0 & 0 \end{bmatrix}\begin{bmatrix} 0 & z\psi(s) \\ 0 & \psi(s) \end{bmatrix}\right\}\end{aligned} \tag{219}$$

for $n \geqslant 1$ and $\mathrm{Re}(s) = 0$. We can solve this recurrence equation in the same way as (209) except that now $\mathbf{T}_0$ should be replaced by $\mathbf{T}_1$.

In the particular case when $\mathbf{P}\{\zeta_0 = 0\} = 1$, the problem of finding (212) was investigated in 1960 by Baxter [6]. See also Kemperman [13].

11. Compound Recurrent Processes

Let $\chi_1, \chi_2, \ldots, \chi_n, \ldots$ and $\tau_1, \tau_2 - \tau_1, \ldots, \tau_n - \tau_{n-1}, \ldots$ be independent sequences of mutually independent and identically distributed real random variables. We assume that $\tau_n - \tau_{n-1}$ $(n = 1, 2, \ldots; \tau_0 = 0)$ are positive random variables, whereas the random variables χ_n $(n = 1, 2, \ldots)$ may take positive and negative values too. Define

$$\chi(u) = \sum_{0 < \tau_n \leqslant u} \chi_n \tag{220}$$

for $u \geqslant 0$. We say that the family of random variables $\{\chi(u), 0 \leqslant u < \infty\}$ forms a compound recurrent process.

Let

$$\varphi(s) = \mathbf{E}\{e^{-s(\tau_n - \tau_{n-1})}\} \tag{221}$$

for $\mathrm{Re}(s) \geqslant 0$ and $n \geqslant 1$ and

$$\psi(s) = \mathbf{E}\{e^{-s\chi_n}\} \tag{222}$$

for $\mathrm{Re}(s) = 0$ and $n \geqslant 1$. We write

$$\zeta_n = \chi_1 + \chi_2 + \cdots + \chi_n \tag{223}$$

for $n \geqslant 1$ and $\zeta_0 = 0$.

If $\operatorname{Re}(s) = 0$ and $\operatorname{Re}(q) > 0$, then we have

$$q \int_0^\infty e^{-qt} \mathbf{E}\{e^{-s\chi(t)}\}\, dt = \frac{1 - \varphi(q)}{1 - \varphi(q)\psi(s)}. \tag{224}$$

This can be proved as follows: For almost all realizations of the process $\{\chi(t), 0 \leqslant t < \infty\}$ we have

$$\begin{aligned} q \int_0^\infty e^{-qt} e^{-s\chi(t)}\, dt &= q \sum_{n=0}^{\infty} e^{-s\zeta_n} \int_{\tau_n}^{\tau_{n+1}} e^{-qt}\, dt \\ &= \sum_{n=0}^{\infty} e^{-s\zeta_n}(e^{-q\tau_n} - e^{-q\tau_{n+1}}) \end{aligned} \tag{225}$$

if $\operatorname{Re}(s) = 0$ and $\operatorname{Re}(q) > 0$. If we form the expectation of (225), then we get (224).

Let

$$\xi(u) = \chi(u) - cu + \eta_0 \tag{226}$$

for $u \geqslant 0$, where c is a real constant and η_0 is a real random variable independent of the process $\{\chi(u), 0 \leqslant u < \infty\}$.

In the theories of queues, dams, storage, and insurance risk there are many problems which require the determination of the distribution of the random variable

$$\eta(t) = \sup_{0 \leqslant u \leqslant t} \xi(u) \tag{227}$$

for $t \geqslant 0$. See [29].

Let us introduce the notations

$$\Phi(t, s, v) = \mathbf{E}\{e^{-s\eta(t) - (v-s)\xi(t)}\} \tag{228}$$

for $\operatorname{Re}(s) = \operatorname{Re}(v) = 0$ and $t \geqslant 0$ and

$$\delta(t, q, s, v) = e^{-qt - s[\eta(t) - \xi(t)] - v\xi(t)} \tag{229}$$

for $t \geqslant 0$, where q, s, v are complex (or real) numbers.

If $c \geqslant 0$, $\operatorname{Re}(q) > 0$ and $\operatorname{Re}(s) = \operatorname{Re}(v) = 0$, then for almost all realizations of the process $\{\xi(u), 0 \leqslant u < \infty\}$ we have

$$\begin{aligned} &(q + cs - cv) \int_0^\infty \delta(t, q, s, v)\, dt \\ &\quad = \sum_{n=0}^{\infty} \delta(\tau_n + 0, q, s, v)[1 - e^{-(q + cs - cv)(\tau_{n+1} - \tau_n)}]. \end{aligned} \tag{230}$$

If $c \geqslant 0$ and $\tau_n < t < \tau_{n+1}$, then $\eta(t) = \eta(\tau_n + 0)$, $\xi(t) = \xi(\tau_n + 0) - c(t - \tau_n)$ and

$$\delta(t, q, s, v) = \delta(\tau_n + 0, q, s, v)e^{-(q + cs - cv)(t - \tau_n)}. \tag{231}$$

If we integrate (231) form τ_n to τ_{n+1} and add for $n = 0, 1, 2, \ldots$, then we get (230).

Let

$$U_n(s, q, v) = \mathbf{E}\{e^{-s\eta(\tau_n + 0) - q\tau_n - (v - s)\xi(\tau_n + 0)}\} \tag{232}$$

for $n \geqslant 0$, $\mathrm{Re}(s) = \mathrm{Re}(v) = 0$ and $\mathrm{Re}(q) > 0$, and

$$U(s, q, v, \rho) = \sum_{n=0}^{\infty} U_n(s, q, v)\rho^n \tag{233}$$

for $\mathrm{Re}(s) = \mathrm{Re}(v) = 0$, $\mathrm{Re}(q) > 0$, and $|\rho| \leqslant 1$.

If we form the expectation of (230), then we obtain that

$$(q + cs - cv) \int_0^\infty e^{-qt}\Phi(t, s, v)\,dt = U(s, q, v, 1)[1 - \varphi(q + cs - cv)] \tag{234}$$

for $\mathrm{Re}(s) = \mathrm{Re}(v) = 0$, $\mathrm{Re}(q) > 0$ and $c \geqslant 0$.

Since

$$\xi(\tau_{n+1} + 0) = \xi(\tau_n + 0) + \chi_{n+1} - c(\tau_{n+1} - \tau_n) \tag{235}$$

and

$$\eta(\tau_{n+1} + 0) - \xi(\tau_{n+1} - 0) = [\eta(\tau_n + 0) - \xi(\tau_n + 0) - \chi_{n+1} + c(\tau_{n+1} - \tau_n)]^+ \tag{236}$$

for $n \geqslant 0$ and $\eta(\tau_0 + 0) = \xi(\tau_0 + 0) = \eta_0$, it follows that

$$U_{n+1}(s, q, v) = \mathbf{T}\{U_n(s, q, v)\varphi(q + cs - cv)\psi(v - s)\} \tag{237}$$

for $n \geqslant 0$ and $U_0(s, q, v) = \mathbf{E}\{e^{-v\eta_0}\}$. In (237) $\mathbf{T}$ is defined by (63) and it operates on the variable s. Now the generating function (233) can be obtained by (71).

If $c < 0$, $\mathrm{Re}(q) > 0$ and $\mathrm{Re}(s) = \mathrm{Re}(v) = 0$, then for almost all realizations of the process $\{\xi(u), 0 \leqslant u < \infty\}$ we have

$$\begin{aligned}(q + cs - cv) &\int_0^\infty \delta(t, q, s, v)\,dt \\ &= \sum_{n=0}^{\infty} \delta(\tau_n + 0, q, s, v) + \frac{cs}{q - cv}\sum_{n=0}^{\infty} \delta\left(\tau_n + 0, q, v - \frac{q}{c}, v\right) \\ &\quad - \sum_{n=1}^{\infty} \delta(\tau_n - 0, q, s, v) - \frac{cs}{q - cv}\sum_{n=1}^{\infty} \delta\left(\tau_n - 0, q, v - \frac{q}{c}, v\right).\end{aligned} \tag{238}$$

The proof of (238) is based on the following identity: If $\alpha < \beta$, a and b are real numbers, s and $w \neq 0$ are complex (or real) numbers, then

$$(w+bs)\int_\alpha^\beta e^{-wt-s[a+bt]^+}\,dt = \{e^{-w\alpha-s[a+b\alpha]^+} - e^{-w\beta-s[\alpha+b\beta]^+}\} + \frac{bs}{w}\{e^{-w\alpha+w[a+b\alpha]^+/b} - e^{-w\beta+w[a+b\beta]^+/b}\}. \quad (239)$$

If $b = 0$, then the second expression on the right-hand side of (239) is 0.

If $c < 0$, and $\tau_n < t < \tau_{n+1}$, then $\xi(t) = \xi(\tau_n + 0) - c(t - \tau_n)$ and

$$\eta(t) - \xi(t) = [\eta(\tau_n + 0) - \xi(\tau_n + 0) + c(t - \tau_n)]^+ \quad (240)$$

for $n \geqslant 0$. If we integrate (229) from τ_n to τ_{n+1} and add for $n = 0, 1, 2, \ldots$, then we get (238).

Let

$$U_n(s, q, v) = \mathbf{E}\{e^{-s\eta(\tau_n+0)-q\tau_n-(v-s)\xi(\tau_n+0)}\} \quad (241)$$

for $n \geqslant 0$, $\mathrm{Re}(s) = \mathrm{Re}(v) = 0$ and $\mathrm{Re}(q) > 0$, and

$$V_n(s, q, v) = \mathbf{E}\{e^{-s\eta(\tau_n-0)-q\tau_n-(v-s)\xi(\tau_n-0)}\} \quad (242)$$

for $n \geqslant 1$, $\mathrm{Re}(s) = \mathrm{Re}(v) = 0$ and $\mathrm{Re}(q) > 0$. Furthermore, let

$$U(s, q, v, \rho) = \sum_{n=0}^{\infty} U_n(s, q, v)\rho^n, \quad (243)$$

$$V(s, q, v, \rho) = \sum_{n=1}^{\infty} V_n(s, q, v)\rho^n \quad (244)$$

and

$$Q(s, q, v) = U(s, q, v, 1) - V(s, q, v, 1) \quad (245)$$

for $\mathrm{Re}(s) = \mathrm{Re}(v) = 0$ and $\mathrm{Re}(q) > 0$.

If we form the expectation of (238), then we obtain that

$$(q + cs - cv)\int_0^\infty e^{-qt}\Phi(t, s, v)\,dt = Q(s, q, v) + \frac{cs}{q - cv}\,Q\left(v - \frac{q}{c}, q, v\right) \quad (246)$$

for $\mathrm{Re}(s) = \mathrm{Re}(v) = 0$, $\mathrm{Re}(q) > 0$ and $c < 0$.

Since

$$\xi(\tau_{n+1} - 0) = \xi(\tau_n - 0) + \chi_n - c(\tau_{n+1} - \tau_n) \quad (247)$$

and

$$\eta(\tau_{n+1} - 0) - \xi(\tau_{n+1} - 0) = [\eta(\tau_n - 0) - \xi(\tau_n - 0) - \chi_n + c(\tau_{n+1} - \tau_n)]^+ \quad (248)$$

for $n \geqslant 1$ and $\eta(\tau_1 - 0) = \xi(\tau_1 - 0) = \eta_0 - c\tau_1$, it follows that

$$V_{n+1}(s, q, v) = \mathbf{T}\{V_n(s, q, v)\varphi(q + cs - cv)\psi(v - s)\} \tag{249}$$

for $n = 1, 2, \ldots$ and $V_1(s, q, v) = \mathbf{E}\{e^{-v\eta_0}\}\varphi(q - cv)$. Thus the generating function (244) can be obtained by (71).

On the other hand

$$\xi(\tau_n + 0) = \xi(\tau_n - 0) + \chi_n \tag{250}$$

and

$$\eta(\tau_n + 0) - \xi(\tau_n + 0) = [\eta(\tau_n - 0) - \xi(\tau_n - 0) - \chi_n]^+ \tag{251}$$

for $n \geqslant 1$ imply that

$$U_n(s, q, v) = \mathbf{T}\{V_n(s, q, v)\psi(v - s)\} \tag{252}$$

for $n \geqslant 1$. Hence it follows that

$$U(s, q, v, \rho) = \mathbf{E}\{e^{-v\eta_0}\} + \mathbf{T}\{V(s, q, v, \rho)\psi(v - s)\} \tag{253}$$

for $\mathrm{Re}(s) = \mathrm{Re}(v) = 0$, $\mathrm{Re}(q) > 0$ and $|\rho| \leqslant 1$. By (245) we can write that

$$Q(s, q, v) = \mathbf{E}\{e^{-v\eta_0}\} + \mathbf{T}\{V(s, q, v, 1)\psi(v - s)\} - V(s, q, v, 1) \tag{254}$$

for $\mathrm{Re}(s) = \mathrm{Re}(v) = 0$ and $\mathrm{Re}(q) > 0$. This determines (246).

12. Semi-Markov Sequences

Now we suppose that $\xi_1, \xi_2, \ldots, \xi_n, \ldots$ is a semi-Markov sequence of real random variables. We say that $\{\xi_n\}$ is a semi-Markov sequence if there exists a homogeneous Markov chain $\sigma_0, \sigma_1, \ldots, \sigma_n, \ldots$ with a finite or a countably infinite state space I and transition probabilities $\mathbf{P}\{\sigma_{n+1} = j | \sigma_n = i\} = p_{ij}$ $(i \in I, j \in I)$ such that

$$\begin{aligned} \mathbf{P}\{\xi_{n+1} \leqslant x, \sigma_{n+1} = j | \sigma_0, \ldots, \sigma_n, \xi_1, \ldots, \xi_n\} \\ = \mathbf{P}\{\xi_{n+1} \leqslant x, \sigma_{n+1} = j | \sigma_n\} \end{aligned} \tag{255}$$

for $n \geqslant 0$, $x \in (-\infty, \infty)$, $j \in I$ with probability one, and

$$\mathbf{P}\{\xi_{n+1} \leqslant x, \sigma_{n+1} = j | \sigma_n = i\} = H_{ij}(x) \tag{256}$$

for $n \geqslant 0$, $x \in (-\infty, \infty)$, $i \in I$, $j \in I$, where $H_{ij}(\infty) = p_{ij}$.

Let

$$\psi_{ij}(s) = \int_{-\infty}^{\infty} e^{-sx}\, dH_{ij}(x) \tag{257}$$

and

$$\boldsymbol{\Psi}(s) = [\psi_{ij}(s)]_{i,\, j \in I} \tag{258}$$

for $\text{Re}(s) = 0$. Obviously, $\boldsymbol{\Psi}(s) \in \mathbf{R}_2$ and $\|\boldsymbol{\Psi}\| = 1$. We suppose that ζ_0 is a real random variable which may depend only on σ_0.

In this case the solutions of many fluctuation problems can be reduced to the solutions of the recurrence equations (95), (96), (110), and (111), where $\mathbf{L}$ is one of the transformations $\mathbf{T}$, $\mathbf{T}_0$, $\mathbf{T}_1$ defined in Section 7.

Now we shall consider the random variables introduced in Section 1 and describe the methods of finding their distributions.

The Maximal Partial Sum

The problem of finding the distribution of $\eta_n{}^*$ can be reduced to the problem of finding the distribution of $\eta_n{}^*$ in the case when $\mathbf{P}\{\zeta_0 = 0\} = 1$. Thus we assume that $\mathbf{P}\{\zeta_0 = 0\} = 1$ and write

$$U_{ij}(s, n) = \mathbf{E}\{e^{-s\eta_n{}^*}\delta(\sigma_n = j) | \sigma_0 = i\} \tag{259}$$

for $n \geqslant 0$ and $\text{Re}(s) \geqslant 0$, where $\eta_n{}^*$ is defined by (3). Write

$$\mathbf{U}_n(s) = [U_{ij}(s, n)]_{i,\, j \in I} \tag{260}$$

for $\text{Re}(s) \geqslant 0$.

By (3) we can write that

$$U_{ij}(s, n) = \sum_{r \in I} \mathbf{T}\{\psi_{ir}(s) U_{rj}(s, n-1)\} \tag{261}$$

for $n \geqslant 1$, $i \in I$, $j \in I$ and $\text{Re}(s) \geqslant 0$, where now $\mathbf{T}$ is defined by (63). Equivalently we can write that

$$\mathbf{U}_n(s) = \mathbf{T}\{\boldsymbol{\Psi}(s)\mathbf{U}_{n-1}(s)\} \tag{262}$$

for $n \geqslant 1$ and $\text{Re}(s) \geqslant 0$ and in this case $\mathbf{U}_0(s) = \mathbf{I}$. Now by (146) we obtain that

$$\mathbf{U}(s, \rho) = \sum_{n=0}^{\infty} \mathbf{U}_n(s)\rho^n \tag{263}$$

can be expressed as follows

$$\mathbf{U}(s, \rho) = [\boldsymbol{\Phi}^+(s, \rho)]^{-1}[\boldsymbol{\Phi}^-(0, \rho)]^{-1} \tag{264}$$

for $\text{Re}(s) \geqslant 0$ and $|\rho| < 1$.

If we define

$$U_{ij}(s, n) = \mathbf{E}\{e^{-s\eta_n}\delta(\sigma_n = j) | \sigma_0 = i\} \tag{265}$$

for $n \geqslant 0$ and $\text{Re}(s) = 0$, where η_n is given by (2) and if

$$\mathbf{U}_n(s) = [U_{ij}(s, n)]_{i,\, j \in I} \tag{266}$$

for $\text{Re}(s) = 0$, then by (2) we can write that

$$U_{ij}(s, n) = \sum_{r \in I} \mathbf{T}\{U_{ir}(s, n-1)\psi_{rj}(s)\} \tag{267}$$

for $n \geqslant 1$, $i \in I$, $j \in I$ and $\text{Re}(s) \geqslant 0$, where $\mathbf{T}$ is defined by (63). Equivalently we can write that

$$\mathbf{U}_n(s) = \mathbf{T}\{\mathbf{U}_{n-1}(s)\boldsymbol{\Psi}(s)\} \tag{268}$$

for $n \geqslant 1$ and $\text{Re}(s) \geqslant 0$ and in this case

$$\mathbf{U}_0(s) = [\delta_{ij}\mathbf{E}\{e^{-s\zeta_0}\delta(\sigma_0 = j)\}]_{i,\, j \in I}. \tag{269}$$

Now the generating function

$$\mathbf{U}(s, \rho) = \sum_{n=0}^{\infty} \mathbf{U}_n(s)\rho^n \tag{270}$$

is given by (139) and we have

$$\mathbf{U}(s, \rho) = \mathbf{T}\{\mathbf{U}_0(s)[\boldsymbol{\Psi}^-(s, \rho)]^{-1}\}[\boldsymbol{\Psi}^+(s, \rho)]^{-1} \tag{271}$$

for $\text{Re}(s) = 0$ and $|\rho| < 1$. If, in particular, $\mathbf{U}_0(s) = \mathbf{I}$, then (271) reduces to

$$\mathbf{U}(s, \rho) = [\boldsymbol{\Psi}^-(0, \rho)]^{-1}[\boldsymbol{\Psi}^+(s, \rho)]^{-1} \tag{272}$$

for $\text{Re}(s) \geqslant 0$ and $|\rho| < 1$.

First Passage Times

Now let

$$\Phi_{ij}(s, n) = \mathbf{E}\{e^{-s\zeta_n}\delta(\nu = n)\delta(\sigma_n = j) | \sigma_0 = i\} \tag{273}$$

for $n \geqslant 1$, $i \in I$, $j \in I$ and $\text{Re}(s) \leqslant 0$, where ν is defined by (4). Our aim is to find

$$\boldsymbol{\Phi}_n(s) = [\Phi_{ij}(s, n)]_{i,\, j \in I} \tag{274}$$

for $n \geqslant 1$ and $\text{Re}(s) \leqslant 0$. The matrix (274) is determined by the generating function

$$\boldsymbol{\Phi}(s, \rho) = \sum_{n=1}^{\infty} \boldsymbol{\Phi}_n(s)\rho^n \tag{275}$$

for $\text{Re}(s) \leqslant 0$ and $|\rho| < 1$.

The problem of finding (273), (274), and (275) can be reduced to the problem of finding the following functions

$$U_{ij}(s, n) = \mathbf{E}\{e^{-s\zeta_n}\delta(\nu > n)\delta(\sigma_n = j)|\sigma_0 = i\} \tag{276}$$

for $n \geqslant 0$, $i \in I$, $j \in I$ and $\mathrm{Re}(s) = 0$, or the matrices

$$\mathbf{U}_n(s) = [U_{ij}(s, n)]_{i,\, j \in I} \tag{277}$$

for $n \geqslant 0$ and $\mathrm{Re}(s) = 0$, or the generating function

$$\mathbf{U}(s, \rho) = \sum_{n=0}^{\infty} \mathbf{U}_n(s)\rho^n \tag{278}$$

for $\mathrm{Re}(s) = 0$ and $|\rho| < 1$.

Obviously, we have

$$\boldsymbol{\Phi}_n(s) = \mathbf{U}_{n-1}(s)\boldsymbol{\Psi}(s) - \mathbf{U}_n(s) \tag{279}$$

for $n \geqslant 1$ and $\mathrm{Re}(s) = 0$, and thus

$$\boldsymbol{\Phi}(s, \rho) = \mathbf{U}_0(s) - \mathbf{U}(s, \rho)[\mathbf{I} - \rho\boldsymbol{\Psi}(s)] \tag{280}$$

for $\mathrm{Re}(s) = 0$ and $|\rho| < 1$.

In a similar way as we proved (174) it follows that

$$\mathbf{U}_n(s) = \mathbf{T}_0\{\mathbf{U}_{n-1}(s)\boldsymbol{\Psi}(s)\} \tag{281}$$

for $n \geqslant 1$ and $\mathrm{Re}(s) \geqslant 0$. Accordingly, $\mathbf{U}(s, \rho)$ can be obtained by (138) and $\boldsymbol{\Phi}(s, \rho)$ by (280).

Define

$$\Phi_{ij}^*(s, n) = \mathbf{E}\{e^{-s\zeta_n}\delta(\nu^* = n)\delta(\sigma_n = j)|\sigma_0 = i\} \tag{282}$$

for $n \geqslant 1$ and $\mathrm{Re}(s) \leqslant 0$, where ν^* is defined by (5). Let

$$\boldsymbol{\Phi}_n{}^*(s) = [\Phi_{ij}^*(s, n)]_{i,\, j \in I} \tag{283}$$

for $n \geqslant 1$ and $\mathrm{Re}(s) \leqslant 0$ and

$$\boldsymbol{\Phi}^*(s, \rho) = \sum_{n=1}^{\infty} \boldsymbol{\Phi}_n{}^*(s)\rho^n \tag{284}$$

for $\mathrm{Re}(s) \leqslant 0$ and $|\rho| < 1$.

To find (282), (283), and (284) let us introduce the notations

$$U_{ij}^*(s, n) = \mathbf{E}\{e^{-s\zeta_n}\delta(\nu^* > n)\delta(\sigma_n = j)|\sigma_0 = i\} \tag{285}$$

for $n \geqslant 0$, $i \in I$, $j \in I$ and $\mathrm{Re}(s) = 0$,

$$\mathbf{U}_n{}^*(s) = [U_{ij}(s, n)]_{i,\, j \in I} \tag{286}$$

for $n \geqslant 0$ and $\mathrm{Re}(s) = 0$ and

$$\mathbf{U}^*(s, \rho) = \sum_{n=0}^{\infty} \mathbf{U}_n^*(s)\rho^n \tag{287}$$

for $\mathrm{Re}(s) = 0$ and $|\rho| < 1$.

Obviously, we have

$$\boldsymbol{\Phi}_n^*(s) = \mathbf{U}_{n-1}^*(s)\boldsymbol{\Psi}(s) - \mathbf{U}_n^*(s) \tag{288}$$

for $n \geqslant 1$ and $\mathrm{Re}(s) = 0$, and thus

$$\boldsymbol{\Phi}^*(s, \rho) = \mathbf{U}_0^*(s) - \mathbf{U}^*(s, \rho)[\mathbf{I} - \rho\boldsymbol{\Psi}(s)] \tag{289}$$

for $\mathrm{Re}(s) = 0$ and $|\rho| < 1$.

In a similar way as (183) we can prove that

$$\mathbf{U}_n^*(s) = \mathbf{T}_1\{\mathbf{U}_{n-1}^*(s)\boldsymbol{\Psi}(s)\} \tag{290}$$

for $n \geqslant 1$ and $\mathrm{Re}(s) \geqslant 0$. Accordingly, $\mathbf{U}^*(s, \rho)$ can be obtained by (130) and $\boldsymbol{\Phi}^+(s, \rho)$ by (289).

Positive Partial Sums

Let

$$U_{ij}(s, w, n) = \mathbf{E}\{e^{-s\zeta_n} w^{\Delta_n} \delta(\sigma_n = j) | \sigma_0 = i\} \tag{291}$$

for $n \geqslant 0$ and $\mathrm{Re}(s) = 0$, where Δ_n is defined in Section 1 and

$$\mathbf{U}_n(s, w) = [U_{ij}(s, w, n)]_{i, j \in I} \tag{292}$$

for $n \geqslant 0$ and $\mathrm{Re}(s) = 0$. The generating function

$$\mathbf{U}(s, w, \rho) = \sum_{n=0}^{\infty} \mathbf{U}_n(s, w)\rho^n \tag{293}$$

can be determined by (133) for $\mathrm{Re}(s) = 0$, $|\rho| < 1$ and $|\rho w| < 1$.

Now by (188) we have

$$\mathbf{U}_n(s, w) = w\mathbf{T}_0\{\mathbf{U}_{n-1}(s, w)\boldsymbol{\Psi}(s)\} + \mathbf{T}_0^*\{\mathbf{U}_{n-1}(s, w)\boldsymbol{\Psi}(s)\} \tag{294}$$

for $n \geqslant 1$. This equation is of type (110) and $\mathbf{U}(s, w, \rho)$ is given by (133).

Define

$$U_{ij}^*(s, w, n) = \mathbf{E}\{e^{-s\zeta_n} w^{\Delta_n^*} \delta(\sigma_n = j) | \sigma_0 = i\} \tag{295}$$

for $n \geqslant 0$ and $\mathrm{Re}(s) = 0$, where Δ_n^* is defined in Section 1 and

$$\mathbf{U}_n^*(s, w) = [U_{ij}(s, w, n)]_{i, j \in I} \tag{296}$$

for $n \geqslant 0$ and $\text{Re}(s) = 0$. The generating function

$$\mathbf{U}^*(s, w, \rho) = \sum_{n=0}^{\infty} \mathbf{U}_n^*(s, w)\rho^n \tag{297}$$

can be determined by (133) for $\text{Re}(s) = 0$, $|\rho| < 1$ and $|\rho w| < 1$.

Now we have

$$\mathbf{U}_n^*(s, w) = w\mathbf{T}_1\{\mathbf{U}_{n-1}^*(s, w)\boldsymbol{\Psi}(s)\} + \mathbf{T}_1^*\{\mathbf{U}_{n-1}^*(s, w)\boldsymbol{\Psi}(s)\} \tag{298}$$

for $n \geqslant 1$. This equation is of type (110) and $\mathbf{U}^*(s, w, \rho)$ is given by (133).

Ordered Partial Sums

In this section we assume that $\mathbf{P}\{\zeta_0 = 0\} = 1$. The general case can immediately be reduced to this particular case.

Define

$$\Phi_{ij}^{(n)}(s, v, w) = \sum_{k=0}^{n} \mathbf{E}\{e^{-s\eta_{n,k} - v\zeta_n}\delta(\sigma_n = j)\,|\,\sigma_0 = i\}w^k \tag{299}$$

for $n \geqslant 0$, $i \in I$, $j \in I$, $\text{Re}(s) = \text{Re}(v) = 0$, where $\eta_{n,k}$ is given by (6). Let

$$\boldsymbol{\Phi}_n(s, v, w) = [\Phi_{ij}^{(n)}(s, v, w)]_{i,\, j \in I} \tag{300}$$

and

$$\boldsymbol{\Phi}(s, v, w, \rho) = \sum_{n=0}^{\infty} \boldsymbol{\Phi}_n(s, v, w)\rho^n \tag{301}$$

for $\text{Re}(s) = \text{Re}(v) = 0$, $|\rho| < 1$ and $|\rho|[\min(|w|, 1) + |1 - w|] < 1$.

Now (193) holds unchangeably. If we multiply (193) by $e^{-v\zeta_{n+1}}\delta(\sigma_n = j)$ and form its conditional expectation given $\sigma_0 = i$, then we obtain that

$$\begin{aligned}\boldsymbol{\Phi}_{n+1}(s, v, w) = {} & w\mathbf{T}\{\boldsymbol{\Psi}(s + v)\boldsymbol{\Phi}_n(s, v, w)\} \\ & + \mathbf{T}^*\{\boldsymbol{\Psi}(s + v)\boldsymbol{\Phi}_n(s, v, w)\} + [\boldsymbol{\Psi}(v)]^{n+1}\end{aligned} \tag{302}$$

for $n \geqslant 0$ and $\boldsymbol{\Phi}_0(s, v, w) = \mathbf{I}$.

Let

$$\mathbf{U}_n(s, v, w) = \boldsymbol{\Phi}_n(s, v, w) - \boldsymbol{\Psi}(v)\boldsymbol{\Phi}_{n-1}(s, v, w) \tag{303}$$

for $n \geqslant 1$ and $\mathbf{U}_0(s, v, w) = \boldsymbol{\Phi}_0(s, v, w) = \mathbf{I}$. Then by (302) we get

$$\mathbf{U}_{n+1}(s, v, w) = w\mathbf{T}\{\boldsymbol{\Psi}(s + v)\mathbf{U}_n(s, v, w)\} + \mathbf{T}^*\{\boldsymbol{\Psi}(s + v)\mathbf{U}_n(s, v, w)\} \tag{304}$$

for $n \geqslant 1$. This equation is of type (111) and

$$\mathbf{U}(s, v, w, \rho) = \sum_{n=0}^{\infty} \mathbf{U}_n(s, v, w)\rho^n \tag{305}$$

is given by (135) for $\mathrm{Re}(s) = \mathrm{Re}(v) = 0$ and $|\rho|[\min(|w|, 1) + |1 - w|] < 1$. Finally, by (303) it follows that

$$\boldsymbol{\Phi}(s, v, w, \rho) = \mathbf{U}(s, v, w, \rho)[\mathbf{I} - \rho\boldsymbol{\Psi}(v)]^{-1} \tag{306}$$

for $\mathrm{Re}(s) = \mathrm{Re}(v) = 0$, $|\rho| < 1$ and $|\rho|[\min(|w|, 1) + |1 - w|] < 1$.

Changes of Sign

Define ω_n and ω_n^* as in the Introduction. Let

$$U_{ij}^{(n)}(s, z) = \mathbf{E}\{e^{-s\zeta_n} z^{\omega_n} \delta(\zeta_n > 0)\delta(\sigma_n = j) | \sigma_0 = i\} \tag{307}$$

and

$$V_{ij}^{(n)}(s, z) = \mathbf{E}\{e^{-s\zeta_n} z^{\omega_n} \delta(\zeta_n \leqslant 0)\delta(\sigma_n = j) | \sigma_0 = i\} \tag{308}$$

for $n \geqslant 0$ and $\mathrm{Re}(s) = 0$. Furthermore, let

$$\mathbf{U}_n(s, z) = [U_{ij}^{(n)}(s, z)]_{i, j \in I} \tag{309}$$

and

$$\mathbf{V}_n(s, z) = [V_{ij}^{(n)}(s, z)]_{i, j \in I} \tag{310}$$

for $\mathrm{Re}(s) = 0$.

In a similar way as (207) and (208) we can prove that

$$\mathbf{U}_n(s, z) = \mathbf{T}_0\{[\mathbf{U}_{n-1}(s, z) + z\mathbf{V}_{n-1}(s, z)]\boldsymbol{\Psi}(s)\} \tag{311}$$

and

$$\mathbf{V}_n(s, z) = \mathbf{T}_0^*\{[z\mathbf{U}_{n-1}(s, z) + \mathbf{V}_{n-1}(s, z)]\boldsymbol{\Psi}(s)\} \tag{312}$$

for $n \geqslant 1$ and $\mathrm{Re}(s) = 0$. Hence we can write that

$$\begin{bmatrix} \mathbf{U}_n(s, z) & \mathbf{V}_n(s, z) \\ \mathbf{0} & \mathbf{0} \end{bmatrix} = \mathbf{T}_0\left\{\begin{bmatrix} \mathbf{U}_{n-1}(s, z) & \mathbf{V}_{n-1}(s, z) \\ \mathbf{0} & \mathbf{0} \end{bmatrix}\begin{bmatrix} \boldsymbol{\Psi}(s) & \mathbf{0} \\ z\boldsymbol{\Psi}(s) & \mathbf{0} \end{bmatrix}\right\}$$
$$+ \mathbf{T}_0^*\left\{\begin{bmatrix} \mathbf{U}_{n-1}(s, z) & \mathbf{V}_{n-1}(s, z) \\ \mathbf{0} & \mathbf{0} \end{bmatrix}\begin{bmatrix} \mathbf{0} & z\boldsymbol{\Psi}(s) \\ \mathbf{0} & \boldsymbol{\Psi}(s) \end{bmatrix}\right\} \tag{313}$$

for $n \geqslant 1$ and $\mathrm{Re}(s) = 0$, where we form the transformations $\mathbf{T}_0$ and $\mathbf{T}_0^*$ element by element for the matrices occurring in (313).

In the space $\mathbf{S}_2$ the recurrence equation (313) is of the same type as (110) in the space $\mathbf{R}_2$ and it can be solved in the same way as (110) in $\mathbf{R}_2$.

Let

$$\bar{U}_{ij}^{(n)}(s, z) = \mathbf{E}\{e^{-s\zeta_n} z^{\omega_n^*} \delta(\zeta_n \geqslant 0)\delta(\sigma_n = j) | \sigma_0 = i\} \tag{314}$$

and

$$\bar{V}_{ij}^{(n)}(s, z) = \mathbf{E}\{e^{-s\zeta_n} z^{\omega_n^*} \delta(\zeta_n < 0)\delta(\sigma_n = j) | \sigma_0 = i\} \tag{315}$$

for $n \geqslant 0$ and $\text{Re}(s) = 0$. Furthermore, let

$$\mathbf{U}_n^*(s, z) = [\bar{U}_{ij}^{(n)}(s, z)]_{i, j \in I} \tag{316}$$

and

$$\mathbf{V}_n^*(s, z) = [\bar{V}_{ij}^{(n)}(s, z)]_{i, j \in I} \tag{317}$$

for $\text{Re}(s) = 0$.

In a similar way as (217) and (218) we can prove that

$$\mathbf{U}_n^*(s, z) = \mathbf{T}_1\{[\mathbf{U}_{n-1}^*(s, z) + z\mathbf{V}_{n-1}^*(s, z)]\boldsymbol{\Psi}(s)\} \tag{318}$$

and

$$\mathbf{V}_n^*(s, z) = \mathbf{T}_1^*\{[z\mathbf{U}_{n-1}^*(s, z) + \mathbf{V}_{n-1}^*(s, z)]\boldsymbol{\Psi}(s)\} \tag{319}$$

for $n \geqslant 1$ and $\text{Re}(s) = 0$.

Hence it follows that

$$\begin{bmatrix} \mathbf{U}_n^*(s, z) & \mathbf{V}_n^*(s, z) \\ \mathbf{0} & \mathbf{0} \end{bmatrix} = \mathbf{T}_1 \left\{ \begin{bmatrix} \mathbf{U}_{n-1}^*(s, z) & \mathbf{V}_{n-1}^*(s, z) \\ \mathbf{0} & \mathbf{0} \end{bmatrix} \begin{bmatrix} \boldsymbol{\Psi}(s) & \mathbf{0} \\ z\boldsymbol{\Psi}(s) & \mathbf{0} \end{bmatrix} \right\} + \mathbf{T}_1^* \left\{ \begin{bmatrix} \mathbf{U}_{n-1}^*(s, z) & \mathbf{V}_{n-1}^*(s, z) \\ \mathbf{0} & \mathbf{0} \end{bmatrix} \begin{bmatrix} \mathbf{0} & z\boldsymbol{\Psi}(s) \\ \mathbf{0} & \boldsymbol{\Psi}(s) \end{bmatrix} \right\} \tag{320}$$

for $n \geqslant 1$ and $\text{Re}(s) = 0$. In the space $\mathbf{S}_2$ the recurrence equation (320) is of the same type as (110) in the space $\mathbf{R}_2$ and it can be solved in the same way.

13. Compound Semi-Markov Processes

Let $\chi_1, \chi_2, \ldots, \chi_n, \ldots$ be a sequence of real random variables and $\tau_1 - \tau_0, \tau_2 - \tau_1, \ldots, \tau_n - \tau_{n-1}, \ldots$, where $\tau_0 = 0$, a sequence of positive random variables. We assume that there exists a homogeneous Markov chain $\sigma_0, \sigma_1, \ldots, \sigma_n, \ldots$ with a finite or a countably infinite state space I and transitions probabilities $\mathbf{P}\{\sigma_{n+1} = j \mid \sigma_n = i\} = p_{ij}$ $(i \in I, j \in I)$ such that

$$\mathbf{P}\{\chi_{n+1} \leqslant x, \tau_{n+1} - \tau_n \leqslant y, \sigma_{n+1} = j \mid \sigma_0, \ldots, \sigma_n, \chi_1, \ldots, \chi_n, \tau_1, \ldots, \tau_n\} = \mathbf{P}\{\chi_{n+1} \leqslant x, \tau_{n+1} - \tau_n \leqslant y, \sigma_{n+1} = j \mid \sigma_n\} \tag{321}$$

for $n \geqslant 0$ with probability one, and

$$\mathbf{P}\{\chi_{n+1} \leqslant x, \tau_{n+1} - \tau_n \leqslant y, \sigma_{n+1} = j \mid \sigma_n = i\} = H_{ij}(x, y) \tag{322}$$

for $n \geqslant 0$, $x \in (-\infty, \infty)$ $y \in (0, \infty)$, $i \in I$, $j \in I$, where $H_{ij}(\infty, \infty) = p_{ij}$.

Define

$$\chi(u) = \sum_{0<\tau_n \leqslant u} \chi_n \tag{323}$$

for $u \geqslant 0$. We say that the family of random variables $\{\chi(u), 0 \leqslant u < \infty\}$ forms a compound semi-Markov process.

Let

$$\psi_{ij}(s, q) = \int_{-\infty}^{\infty} \int_{0}^{\infty} e^{-sx-qy} \, d_x d_y \, H_{ij}(x, y) \tag{324}$$

for $\mathrm{Re}(s) = 0$ and $\mathrm{Re}(q) \geqslant 0$, and

$$\boldsymbol{\Psi}(s, q) = [\psi_{ij}(s, q)]_{i, j \in I}. \tag{325}$$

Obviously $\boldsymbol{\Psi}(s, q) \in \mathbf{R}_2$ and $\|\boldsymbol{\Psi}(s, q)\| < 1$ for $\mathrm{Re}(q) > 0$.

Let $\sigma(t) = \sigma_n$ for $\tau_n \leqslant t < \tau_{n+1}$ and

$$\Gamma_{ij}(t, s) = \mathbf{E}\{e^{-s\chi(t)}\delta(\sigma(t) = j) | \sigma_0 = i\} \tag{326}$$

for $\mathrm{Re}(s) = 0$, $i \in I$, $j \in I$ and

$$\boldsymbol{\Gamma}(t, s) = [\Gamma_{ij}(t, s)]_{i, j \in I} \tag{327}$$

for $\mathrm{Re}(s) = 0$.

By using the identity (225) we easily obtain that

$$q \int_0^\infty e^{-qt} \boldsymbol{\Gamma}(t, s) \, dt = [\mathbf{I} - \boldsymbol{\Psi}(s, q)]^{-1}[\mathbf{I} - \boldsymbol{\Psi}(0, q)] \tag{328}$$

for $\mathrm{Re}(s) = 0$ and $\mathrm{Re}(q) > 0$.

Now let

$$\xi(u) = \chi(u) - cu + \eta_0 \tag{329}$$

for $u \geqslant 0$, where c is a real constant and η_0 is a real random variable which may depend only on σ_0. We are interested in determining the distribution of the random variable

$$\eta(t) = \sup_{0 \leqslant u \leqslant t} \xi(u) \tag{330}$$

for $t \geqslant 0$.

Let us introduce the notations

$$\Phi_{ij}(t, s, v) = \mathbf{E}\{e^{-s\eta(t) - (v-s)\xi(t)}\} \tag{331}$$

and

$$\boldsymbol{\Phi}(t, s, v) = [\Phi_{ij}(t, s, v)]_{i, j \in I} \tag{332}$$

for $t \geqslant 0$ and $\mathrm{Re}(s) = \mathrm{Re}(v) = 0$,

$$U_{ij}^{(n)}(s, q, v) = \mathbf{E}\{e^{-s\eta(\tau_n + 0) - q\tau_n - (v-s)\xi(\tau_n + 0)}\}, \tag{333}$$

and

$$\mathbf{U}_n(s, q, v) = [U_{ij}^{(n)}(s, q, v)]_{i,\, j\in I} \tag{334}$$

for $n \geqslant 0$, $\mathrm{Re}(s) = \mathrm{Re}(v) = 0$ and $\mathrm{Re}(q) > 0$, and

$$\mathbf{U}(s, q, v, \rho) = \sum_{n=0}^{\infty} \mathbf{U}_n(s, q, v)\rho^n \tag{335}$$

for $\mathrm{Re}(s) = \mathrm{Re}(v) = 0$, $\mathrm{Re}(q) > 0$ and $|\rho| \leqslant 1$. Furthermore,

$$V_{ij}^{(n)}(s, q, v) = \mathbf{E}\{e^{-s\eta(\tau_n - 0) - q\tau_n - (v-s)\xi(\tau_n - 0)}\}, \tag{336}$$

and

$$\mathbf{V}_n(s, q, v) = [V_{ij}^{(n)}(s, q, v)]_{i,\, j\in I} \tag{337}$$

for $n \geqslant 1$, $\mathrm{Re}(s) = \mathrm{Re}(v) = 0$ and $\mathrm{Re}(q) > 0$, and

$$\mathbf{V}(s, q, v, \rho) = \sum_{n=1}^{\infty} \mathbf{V}_n(s, q, v)\rho^n \tag{338}$$

for $\mathrm{Re}(s) = \mathrm{Re}(v) = 0$, $\mathrm{Re}(q) > 0$ and $|\rho| \leqslant 1$.

In exactly the same way as we proved (234) we can prove that

$$(q + cs - cv)\int_0^{\infty} e^{-qt}\boldsymbol{\Phi}(t, s, v)\,dt = \mathbf{U}(s, q, v, 1)[\mathbf{I} - \boldsymbol{\Psi}(0, q + cs - cv)] \tag{339}$$

for $\mathrm{Re}(s) = \mathrm{Re}(v) = 0$, $\mathrm{Re}(q) > 0$ and $c \geqslant 0$, where $\mathbf{U}(s, q, v, \rho)$ is defined by (335). Since in this case

$$\mathbf{U}_{n+1}(s, q, v) = \mathbf{T}\{\mathbf{U}_n(s, q, v)\boldsymbol{\Psi}(v - s, q + cs - cv)\} \tag{340}$$

for $n \geqslant 0$, the generating function (335) is determined by (130).

In exactly the same way as we proved (246) we can prove that

$$(q + cs - cv)\int_0^{\infty} e^{-qt}\boldsymbol{\Phi}(t, s, v)\,dt = \mathbf{Q}(s, q, v) + \frac{cs}{q - cv}\mathbf{Q}\left(v - \frac{q}{c}, q, v\right) \tag{341}$$

for $\mathrm{Re}(s) = \mathrm{Re}(v) = 0$, $\mathrm{Re}(q) > 0$ and $c < 0$, where

$$\mathbf{Q}(s, q, v) = \mathbf{U}(s, q, v, 1) - \mathbf{V}(s, q, v, 1). \tag{342}$$

In this case we have

$$\mathbf{U}_n(s, q, v) = \mathbf{T}\{\mathbf{V}_n(s, q, v)\boldsymbol{\Psi}(v - s, 0)\} \tag{343}$$

for $n \geqslant 1$, which implies that

$$\mathbf{U}(s, q, v, \rho) = \mathbf{U}_0(s, q, v) + \mathbf{T}\{\mathbf{V}(s, q, v, \rho)\boldsymbol{\Psi}(v - s, 0)\} \tag{344}$$

for $\mathrm{Re}(s) = \mathrm{Re}(v) = 0$, $\mathrm{Re}(q) > 0$ and $|\rho| < 1$. Furthermore, we have

$$\mathbf{V}_{n+1}(s, q, v) = \mathbf{T}\{\mathbf{V}_n(s, q, v)\mathbf{\Psi}(v - s, q + cs - cv)\} \tag{345}$$

for $n = 1, 2, \ldots,$ where $\mathbf{V}_1(s, q, v) = \mathbf{U}_0(s, q, v)\mathbf{\Psi}(0, q - cv)$. Thus the generating function (338) can be obtained by (130). Accordingly, (342) is completely determined by the above formulas.

References

1. E. S. Andersen, On sums of symmetrically dependent random variables, *Skand. Aktuarietidskr.* **36** (1953), 123–138.
2. E. S. Andersen, On the fluctuations of sums of random variables, *Math. Scand.* **1** (1953), 263–285.
3. E. S. Andersen, On the fluctuations of sums of random variables, II. *Math. Scand.* **2** (1954), 195–223.
4. E. S. Andersen, An algebraic treatment of fluctuations of sums of random variables, *Proc. Fifth Berkeley Symp. Math. Statist. Probability*, Vol. 2, Part 1, *Probability Theory*, pp. 423–429, Univ. of California Press, Berkeley, 1967.
5. G. Baxter, An operator identity, *Pacific J. Math.* **8** (1958), 649–663.
6. G. Baxter, A two-dimensional operator identity with application to the change of sign in sums of random variables, *Trans. Amer. Math. Soc.* **96** (1960), 210–221.
7. G. Baxter, An analytic approach to finite fluctuation problems in probability, *J. Analyse Math.* **9** (1961), 31–70.
8. D. A. Darling, A unified treatment of finite fluctuation problems, *in* "Colloquium on Combinatorial Methods in Probability Theory," pp. 3–6, Matematisk Institut, Aarhus Universitet, Denmark, 1962.
9. J. H. A. De Smit, A simple analytic proof of the Pollaczek-Wendel identity for ordered partial sums, *Ann. Probability* **1** (1973), 348–351.
10. W. Feller, On combinatorial methods in fluctuation theory, "Probability and Statistics," The Harald Cramér Volume, pp. 79–91 (U. Grenander, ed.), Almqvist & Wiksell, Stockholm, 1959.
11. I. C. Gohberg and M. G. Krein, Systems of integral equations on a half line with kemels depending on the difference of arguments, *Uspekhi Mat. Nauk* **13**, No. 2 (1958), 3–72; English transl., *Amer. Math. Soc. Transl.* (2) **14** (1960), 217–287.
12. J. H. B. Kemperman, "The Passage Problem for a Stationary Markov Chain," Univ. of Chicago Press, Chicago, 1961.
13. J. H. B. Kemperman, Changes of sign in cumulative sums. I–II, *Proc. Nederl. Akad. Wetensch., Ser. A* **64** (1961), 291–303; 304–313.
14. J. F. C. Kingman, Spitzer's identity and its use in probability theory, *J. London Math. Soc.* **37** (1962), 309–316.
15. J. F. C. Kingman, The algebra of queues, *J. Appl. Probability* **3** (1966), 285–326.
16. P. Masani, The Laurent factorization of operator-valued functions, *Proc. London Math. Soc.* (3) **6** (1956), 59–69.
17. H. D. Miller, A matrix factorization problem in the theory of random variables defined on a finite Markov chain, *Proc. Cambridge Philos. Soc.* **58** (1962), 268–285.
18. H. D. Miller, Absorption probabilities for sums of random variables defined on a finite Markov chain, *Proc. Cambridge Philos. Soc.* **58** (1962), 286–298.

19. G. MORERA, Un teorema fondamentale nella teoria delle funzioni di una variable complessa, *Rend. Reale Ist. Lombardo Sci. Lettere Milano* (2) **19** (1886), 304–307.
20. W. F. OSGOOD," Functions of a Complex Variable," Hafner, New York, 1948.
21. F. POLLACZEK, Fonctions caractéristiques des certaines répartitions définies au moyen de la notion d' ordre. Application à la théorie des attentes, *C. R. Acad. Sci. Paris* **234** (1952), 2334–2336.
22. F. POLLACZEK, Sur la répartition des périodes d' occupation ininterrompue d' un guichet, *C. R. Acad. Sci. Paris* **234** (1952), 2042–2044.
23. F. POLLACZEK, Order statistics of partial sums of mutually independent random variables, *J. Appl. Probability* **12** (1975), 390–395.
24. S. C. PORT, An elementary probability approach to fluctuation theory, *J. Math. Anal. Appl.* **6** (1963), 109–151.
25. É. L. PRESMAN, Factorization methods and boundary problems for sums of random variables given on Markov chains, *Izv. Akad. Nauk SSSR Ser. Mat.* **33** (1969), 861–900; English transl., *Math. USSR Izv.* **3** (1969), 816–852.
26. G.-C. ROTA, Baxter algebras and combinatorial identities. I–II, *Bull. Amer. Math. Soc.* **75** (1969), 325–334.
27. F. SPITZER, A combinatorial lemma and its application to probability theory, *Trans. Amer. Math. Soc.* **82** (1956), 323–339.
28. L. TAKÁCS, On a linear transformation in the theory of probability, *Acta Sci. Math. (Szeged)* **33** (1972), 15–24.
29. L. TAKÁCS, On risk reserve processes, *Skand. Aktuarietidskr.* **53** (1970), 64–75.
30. L. TAKÁCS, On some recurrence equations in a Banach algebra, *Acta Sci. Math. (Szeged)* **38** (1976), 399–416.
31. J. G. WENDEL, Spitzer's formula: A short proof, *Proc. Amer. Math. Soc.* **9** (1958), 905–908.
32. J. G. WENDEL, Order statistics of partial sums. *Ann. Math. Statist.* **31** (1960), 1034–1044.

STUDIES IN PROBABILITY AND ERGODIC THEORY
ADVANCES IN MATHEMATICS SUPPLEMENTARY STUDIES, VOL. 2

Almost-Sure Invariance Principle for Branching Brownian Motion

L. G. GOROSTIZA[†]

Departamento de Matemáticas, Centro de Investigación del IPN, Mexico City, Mexico

AND

A. R. MONCAYO[‡]

Universidad Autónoma Metropolitana (Iztapalapa), Mexico City, Mexico

1. INTRODUCTION

Consider a supercritical age-dependent branching process, with one initial parent, and each element in the family tree producing at least one offspring. Associate to each branch of time-length T a random process with values in d-dimensional Euclidean space R^d. We call these processes *branch processes*; they may be thought of as obtained by piecing in order along the branches the *offspring processes*; the offspring processes may be interpreted as random motions that the offspring perform throughout their lifetimes. The offspring lifetimes are the durations of the corresponding offspring processes, but otherwise the branch processes are assumed to be independent of the family tree. Suppose that the branch processes all converge weakly to the same process V as $T \to \infty$, with some time and space normalizations. The *empirical distribution process* is defined, for each realization of the family tree and of the branch processes, by choosing at random a branch of time-length T and taking its corresponding branch process realization. The question that interests us is if there is an invariance principle stating that the empirical distribution process converges weakly as $T \to \infty$ to a process related to V, with probability one (i.e., for almost all realizations of the branching process and of the branch processes).

Examples of this model are branching diffusions, branching random walks, and branching transport processes. For the first two examples, limit theorems

[†] Research supported in part by CONACYT Grant PNCB 000060.
[‡] Research supported in part by CONACYT Grant PNCB 000050.

ISBN 0-12-599102-9

for the distribution of the offspring positions at time T have been proved by various authors [5–7, 9, 10, 13–16, 18–20, 22–24].

In [10] we proved an almost-sure invariance principle for the Galton–Watson case. The age-dependent case is significantly different, because in the Galton–Watson model there is precisely one generation living at each time, whereas in age-dependent models, at any given time there are members of many generations present, and no generation is notably dominant (see [21]); the new problems thus posed require new techniques.

As a first step towards a general age-dependent almost-sure invariance principle, we treat in this paper the case of a continuous-time Markov branching process where the offspring processes are Brownian motions. But since we are seeking a method of proof for general branch processes, we present here a general argument, and only at the end do we specialize to Brownian motion.

We use results from the following sources without special reference: Harris [11] and Athreya and Ney [1] for branching processes, Doob [8] for martingales, and Billingsley [3] for weak convergence.

2. Model, Notation, and Theorem

We assume the existence of a basic probability space $(\Omega, \mathfrak{F}, P)$ where everthing is defined. Ω contains all family histories, and $\mathfrak{F}$ is sufficiently large (see [11]).

We consider a continuous-time Markov branching process that starts out with one initial parent (the 0th generation), and such that each member produces at least one offspring (hence all branches are infinite), and the population is always finite. The offspring production law has mean $m > 1$, and we assume it has finite second moment. λ is the parameter of the exponential offspring lifetime distribution, and $\alpha = \lambda(m-1)$ is the Malthusian parameter.

For a positive time t, and a generation number k, let

$\Gamma_t =$ the population at time t,
$Z_t =$ the size of Γ_t,
$\Theta_k =$ the population of generation k,
$\xi_k =$ the size of Θ_k.

Recall that there are a.s. positive random variables W_1 and W_2 such that $\xi_k m^{-k} \to W_1$ a.s. as $k \to \infty$, and $Z_t e^{-\alpha t} \to W_2$ a.s. as $t \to \infty$.

The notation $\gamma \in \Gamma_t$ means that γ is the ancestry line of an offspring living at time t, and γ is also considered as a branch of time-length t.

For $t_1 < t_2$, and $\gamma \in \Gamma_{t_1}$, let

$\Gamma_{t_2}^{\gamma}$ = the population at time t_2 that descends from γ,
$Z_{t_2}^{\gamma}$ = the size of $\Gamma_{t_2}^{\gamma}$ (it is 1 if $\gamma \in \Gamma_{t_2}$ also).

For $k_1 < k_2$, and $\theta \in \Theta_{k_1}$, let

$\Theta_{k_2}^{\theta}$ = the population of generation k_2 that descends from θ,
$\xi_{k_2}^{\theta}$ = the size of $\Theta_{k_2}^{\theta}$.

$\Theta = \bigcup_{k=0}^{\infty} \Theta_k$ is the set of all nodes in the family tree. An element $\theta \in \Theta$ represents a point at which an offspring dies and reproduces (splits), and we denote $\tau(\theta)$ the time at which this occurs.

If $\theta \in \Theta_k$, $k > 0$, and γ is a branch on which θ is, then for $i = 0, 1, \ldots, k$, $\tau_i(\cdot)$ = the time of the ith split on the line of θ (for convenience of notation, $\cdot$ will be allowed to be θ or γ); in particular, $\tau_0(\cdot) = 0$, and $\tau_k(\cdot) = \tau(\theta)$; similarly, $\beta_i(\cdot) = \tau_{i+1}(\cdot) - \tau_i(\cdot)$ = the lifetime of the ith member on the given line. The lifetime of the offspring splitting at θ is written $\beta(\theta)$. Hence the $\beta(\theta)$ are i.i.d., exponentially distributed with parameter λ.

The population at time t that descends from a node θ is written Γ_t^{θ}, and defined as follows: if γ is the branch terminating at θ, then

$$\Gamma_t^{\theta} = \begin{cases} \Gamma_t^{\gamma} & \text{if } \tau(\theta) < t, \\ \{\gamma\} & \text{if } \tau(\theta) = t, \\ \phi & \text{if } \tau(\theta) > t, \end{cases}$$

and its size is

$$Z_t^{\theta} = \begin{cases} Z_t^{\gamma} & \text{if } \tau(\theta) < t, \\ 1 & \text{if } \tau(\theta) = t, \\ 0 & \text{if } \tau(\theta) > t. \end{cases}$$

$C[0, T]^d$ and $C[0, \infty)^d$ are the spaces of continuous functions from $[0, T]$ and $[0, \infty)$, respectively, to R^d, and C^d stands for $C[0, 1]^d$. Weak convergence of probability measures on these spaces is denoted $\Rightarrow$.

In general, for each infinite branch γ, let $X(\gamma) \equiv \{X(t, \gamma), t \geqslant 0\}$ be a random element of $C[0, \infty)^d$. These are the *branch processes*. If θ is the ith node on γ, then the *offspring process* corresponding to the offspring splitting at θ (i.e., the $(i-1)$th offspring on the line) is

$$Y_i(t, \cdot) = X(t, \cdot) - X(\tau_{i-1}(\cdot), \cdot), \qquad \tau_{i-1}(\cdot) \leqslant t \leqslant \tau_i(\cdot)$$

(again, $\cdot$ can be θ or γ). The offspring lifetimes are the durations of the respective offspring processes, but otherwise the branch processes are assumed to be independent of the family tree. The *maximal deviation* of $Y_i(\cdot)$ is

$$D_i(\cdot) = \sup_{\tau_{i-1}(\cdot) \leqslant t \leqslant \tau_i(\cdot)} \|Y_i(t, \cdot)\|,$$

where $\| \ \|$ is the Euclidean norm in R^d. For $\theta \in \Theta_i$, we write $D_i(\theta) = D(\theta)$.

For weak convergence, we introduce the following normalizations. For fixed $T > 0$, and $\gamma \in \Gamma_T$, we consider the branch process $X(\gamma)$ restricted to $[0, T]$ as a random element of $C[0, T]^d$. Then, dividing time by T, we define the process $X_T(\gamma)$ by

$$X_T(t, \gamma) = X(Tt, \gamma), \qquad 0 \leqslant t \leqslant 1.$$

Thus, for each $\gamma \in \Gamma_T$, the branch process $X_T(\gamma)$ is a random element of C^d. Space is normalized by dividing by $a_T > 0$, such that $a_T \to \infty$ as $T \to \infty$. We shall study the processes $a_T^{-1} X_T(\gamma)$, $\gamma \in \Gamma_T$, on C^d, as $T \to \infty$.

We will assume that all the branch processes have the same distribution, and we write X (resp. X_T) for a process having the common distribution of the $X(\gamma)$ (resp. $X_T(\gamma)$). We remark that a branch process includes the randomness of the (split times on the) branch to which it is associated; we are not talking about branch processes associated to realizations of branches. It is in this sense that all the branch processes are assumed to be identically distributed. In the case of branching transport processes, for example, the branch processes associated to different branch realizations are in general not identically distributed.

Our main assumption is that all the branch processes converge weakly to the same limit; hence X_T must satisfy the

CONVERGENCE HYPOTHESIS. *$a_T^{-1} X_T \Rightarrow V$ as $T \to \infty$, where V is a random element of C^d.*

In the case of Brownian motion offspring processes, the branch processes are clearly also Brownian motions. Let B denote (standard d-dimensional) Brownian motion, and $a_T = T^{1/2}$. Then $a_T^{-1} X_T(\gamma)$ is distributed as B on $[0, 1]$ (see [4]) for all $\gamma \in \Gamma_T$, and therefore the convergence hypothesis is trivially satisfied, the limit being B.

Notice that for Brownian motion branch processes, the assumption of identical distribution of all the branch processes is satisfied also for realizations of branches. This fact makes it possible to simplify several parts of the proof, by conditioning on realization of the family tree; but since we are interested in a more general point of view, we will omit mention of these points.

For technical reasons, we also introduce the following conditions on the branch processes:

INDEPENDENCE HYPOTHESIS. *$D(\theta)$, $\theta \in \Theta$, are independent.*

MOMENT HYPOTHESES. *$D(\theta)$, $\theta \in \Theta$, have the same mean ED, and uniformly bounded second moments. And*

$$E\left[\sup_{s \leqslant t \leqslant r} \|X(t) - X(s)\| \,\Big|\, \text{split times}\right] \leqslant K(1 + r - s)$$

for all $s < r$, where K is a constant.

These conditions are satisfied by Brownian motion and other processes.

For each $\omega \in \Omega$, we have a realization of the branching process, and realizations of the branch processes. We denote $\Gamma_T(\omega)$, $Z_T(\omega)$, $X_T(\gamma, \omega)$, etc., the "values" of the things defined above, corresponding to the sample point ω.

The *empirical distribution process* is the following. For each $\omega \in \Omega$, $(\Gamma_T(\omega), P_T(\omega))$ is the (random) probability space where the sample space is the set of branches $\Gamma_T(\omega)$, and $P_T(\omega)$ is the uniform probability measure over all subsets of $\Gamma_T(\omega)$ (i.e., the "random choice" of a branch of time-length T). The value of the empirical distribution process corresponding to the sample point $\gamma \in \Gamma_T(\omega)$ is $X_T(\gamma, \omega)$, i.e., the realization of the branch process of γ.

Under the above hypotheses, branching Brownian motion satisfies the following almost-sure invariance principle.

THEOREM.

$$a_T^{-1} X_T(\cdot, \omega) \Rightarrow B \quad as \quad T \to \infty,$$

for P-almost all ω, i.e.,

$$Z_T(\omega)^{-1} \sum_{\gamma \in \Gamma_T(\omega)} 1_{[a_T^{-1} X_T(\gamma, \omega) \in A]} \to B(A)$$

as $T \to \infty$, for P-almost all ω, where A is any Borel set in C^d whose boundary has B-measure 0 ($B(\cdot)$ is the Wiener measure, and $1_{[\cdot]}$ is the indicator of the set $[\cdot]$).

3. PROOF OF THE THEOREM

We will prove a proposition for general branch processes, and then apply it to Brownian motion.

The following *convention* will be in force: It should be clear at each step of the proof whether we are restricted to $[0, T]$ or not. When we are restricted to $[0, T]$, then, for example, if the last split on $\gamma \in \Gamma_T$ up to time T is the kth, we put $\tau_j(\gamma) = T$ for $j > k$; otherwise not.

Let k_T be a generation number depending on T, nondecreasing and growing to ∞ as $T \to \infty$. It will be adequately chosen throughout the proof. Firstly, it is necessary that k_T might be chosen so that with probability 1 for large T there be $\gamma \in \Gamma_T$ such that $\tau_{k_T}(\gamma) < T$. On the other hand, it will be necessary to have control over the elements of the k_Tth generation that live at times beyond T. Both of these requirements are satisfied by the following results. As $T \to \infty$,

$$k_T^{-1} \min_{\theta \in \Theta_{k_T}} \tau(\theta) \to x_1 \quad \text{and} \quad k_T^{-1} \max_{\theta \in \Theta_{k_T}} \tau(\theta) \to x_2$$

with probability one, where x_1 and x_2 are finite real numbers. This is obtained by a theorem that we learned from H. Kesten, which holds for general offspring lifetime distribution. In the case of exponential lifetime with parameter λ, x_1 and x_2 are the minimum and maximum roots of the equation $\lambda x \exp(1 - \lambda x) = m^{-1}$, $x \geqslant 0$; this can be obtained by the method of [2]. In this connection, see also [17]. In particular, we have $x_1 > 0$, and hence the first requirement mentioned above is satisfied if we keep k_T sufficiently smaller than T.

We have associated to each $\gamma \in \Gamma_T$ a random element $X_T(\gamma)$ of C^d. In terms of these, we define the following random functions for each $\gamma \in \Gamma_T$, and $i = 1, 2$,

$$X_T^{(i)}(t, \gamma) = \begin{cases} 0 & \text{if } 0 \leqslant t \leqslant \tau_T^{(i)}(\gamma)/T, \\ X_T(t, \gamma) - X_T(\tau_T^{(i)}(\gamma)/T, \gamma) & \text{if } \tau_T^{(i)}(\gamma)/T \leqslant t \leqslant 1, \end{cases}$$

where

$$\tau_T^{(1)}(\gamma) = \tau_{k_T}(\gamma) \quad \text{and} \quad \tau_T^{(2)}(\gamma) = \max_{\theta \in \Theta_{k_T}} \tau(\theta).$$

Notice that $\tau_T^{(2)}(\gamma) = \tau_T^{(2)}$ is independent of γ.

We write $X_T^{(i)}$ for a random function with the common distribution of the $X_T^{(i)}(\gamma)$, $\gamma \in \Gamma_T$, for each i.

Let f be a bounded, complex-valued function on C^d that satisfies the Lipschitz condition

$$|f(x) - f(y)| \leqslant M \sup_{0 \leqslant t \leqslant 1} \|x(t) - y(t)\|, \qquad x, y \in C^d,$$

where M is a constant. For such an f given, we define the following random variables:

$$\varphi_{1,T} = Z_T^{-1} \sum_{\gamma \in \Gamma_T} f(a_T^{-1} X_T(\gamma)),$$

$$\varphi_{2,T} = Z_T^{-1} \sum_{\gamma \in \Gamma_T} f(a_T^{-1} X_T^{(1)}(\gamma)),$$

$$\varphi_{3,T} = Z_T^{-1} \sum_{\theta \in \Theta_{k_T}} E\left[\sum_{\gamma \in \Gamma_T{}^\theta} f(a_T^{-1} X_T^{(1)}(\gamma)) \,\middle|\, \mathfrak{G}_T \right],$$

$$\varphi_{4,T} = Z_T^{-1} \sum_{\theta \in \Theta_{k_T}} E\left[\sum_{\gamma \in \Gamma_T{}^\theta} f(a_T^{-1} X_T^{(2)}(\gamma)) \,\middle|\, \mathfrak{G}_T \right],$$

where the $\mathfrak{G}_T$ in $\varphi_{3,T}$ and $\varphi_{4,T}$ denotes the σ-algebra generated by the family tree up to time $\tau_T^{(2)}$.

Since

$$\varphi_{1,T}(\omega) = \int_{\Gamma_T(\omega)} f(a_T^{-1} X_T(\omega))\, dP_T(\omega),$$

the theorem will be proved if we show that

$$\varphi_{1,T} \to \int_{C^d} f dB \quad \text{a.s.} \quad \text{as} \quad T \to \infty,$$

due to a general weak convergence result (see Lemma 5 of [10]).

We will prove the following statements in general:

1. $\varphi_{1,T} - \varphi_{2,T} \to 0$ a.s. as $T \to \infty$.
2. $\varphi_{2,T} - \varphi_{3,T} \to 0$ a.s. as $T \to \infty$.
3. $\varphi_{3,T} - \varphi_{4,T} \to 0$ a.s. as $T \to \infty$.

First we need a lemma.

LEMMA. *Let $U(\theta)$, $\theta \in \Theta$, be random variables such that $(U(\theta), \beta(\theta))$, $\theta \in \Theta$, are i.i.d., $U(\theta)$ depends on the tree only through $\beta(\theta)$, and $Ee^{-2\alpha\beta} U^2 < \infty$. Then there is a constant K such that*

$$e^{-2\alpha T} E \sum_{\theta \in \Theta_j} (Z_T{}^\theta)^2 U(\theta)^2 \leqslant K(m/(2m-1))^j,$$

for $j = 0, 1, \ldots.$

Proof. First, let

$$H_j = \sum_{\theta \in \Theta_j} e^{-2\alpha\tau(\theta)}, \qquad j = 0, 1, \ldots.$$

For $j \geqslant 1$, we have

$$H_j = \sum_{\theta \in \Theta_j} e^{-2\alpha \Sigma_{i=1}^{j} \beta_{i-1}(\theta)}$$
$$= \sum_{\theta \in \Theta_{j-1}} e^{-2\alpha \Sigma_{i=1}^{j-1} \beta_{i-1}(\theta)} \sum_{\theta' \in \Theta_j^{\theta}} e^{-2\alpha\beta_{j-1}(\theta')}.$$

Let $\mathfrak{T}_j$ denote the σ-algebra generated by the family tree up to generation j. Then

$$EH_j = EE[H_j | \mathfrak{T}_{j-1}]$$
$$= E\left\{ \sum_{\theta \in \Theta_{j-1}} \exp\left(-2\alpha \sum_{i=1}^{j-1} \beta_{i-1}(\theta) \right) E\left[\sum_{\theta' \in \Theta_j^{\theta}} \exp(-2\alpha\beta(\theta')) \middle| \mathfrak{T}_{j-1} \right] \right\};$$

but $e^{-2\alpha\beta(\theta')}$, $\theta' \in \Theta_j^{\theta}$, are i.i.d. and independent of $\mathfrak{T}_{j-1}$, so

$$E\left[\sum_{\theta' \in \Theta_j^{\theta}} e^{-2\alpha\beta(\theta')} \middle| \mathfrak{T}_{j-1} \right] = E \sum_{\theta' \in \Theta_j^{\theta}} e^{-2\alpha\beta(\theta')}$$
$$= m \int_0^{\infty} e^{-2\alpha x} \lambda e^{-\lambda x}\, dx = m\lambda/(2\alpha + \lambda) = m/(2m - 1);$$

hence $EH_j = EH_{j-1} m/(2m - 1)$, and consequently

$$EH_j = (m/(2m - 1))^j.$$

Now,

$$E \sum_{\theta \in \Theta_j} (Z_T^{\theta})^2 U(\theta)^2 = EE\left[\sum_{\theta \in \Theta_j} (Z_T^{\theta})^2 U(\theta)^2 \middle| \mathfrak{T}_j \right]$$
$$= E\left\{ \sum_{\theta \in \Theta_j} E[(Z_T^{\theta})^2 U(\theta)^2 | \mathfrak{T}_j] \right\}.$$

Since $U(\theta)$ depends on the tree only through the offspring lifetime $\beta(\theta)$, then $U(\theta)$ and Z_T^{θ} are independent conditioned upon $\mathfrak{T}_j$; so

$$E[(Z_T^{\theta})^2 U(\theta)^2 | \mathfrak{T}_j] = E[(Z_T^{\theta})^2 | \mathfrak{T}_j] E[U(\theta)^2 | \mathfrak{T}_j].$$

Since the offspring production law has finite second moment, there is a constant K_1 such that

$$E[(Z_T^{\theta})^2 | \mathfrak{T}_j] \leqslant K_1 e^{2\alpha(T - \tau(\theta))}.$$

Hence

$$E \sum_{\theta \in \Theta_j} (Z_T^{\theta})^2 U(\theta)^2 \leqslant K_1 E\left\{ \sum_{\theta \in \Theta_j} e^{2\alpha(T - \tau(\theta))} E[U(\theta)^2 | \mathfrak{T}_j] \right\}$$
$$= K_1 e^{2\alpha T} EE\left[\sum_{\theta \in \Theta_j} e^{-2\alpha\tau(\theta)} U(\theta)^2 \middle| \mathfrak{T}_j \right]$$
$$= K_1 e^{2\alpha T} E \sum_{\theta \in \Theta_j} e^{-2\alpha\tau(\theta)} U(\theta)^2.$$

Proceeding as in the beginning of the proof, we can write

$$
\begin{aligned}
E & \sum_{\theta\in\Theta_j} e^{-2\alpha\tau(\theta)}U(\theta)^2 \\
& = E\left\{\sum_{\theta\in\Theta_{j-1}} e^{-2\alpha\tau(\theta)}E\left[\sum_{\theta'\in\Theta_j{}^\theta} e^{-2\alpha\beta(\theta')}U(\theta')^2 \,\middle|\, \mathfrak{T}_{j-1}\right]\right\} \\
& = E\left\{\sum_{\theta\in\Theta_{j-1}} e^{-2\alpha\tau(\theta)}E\sum_{\theta'\in\Theta_j{}^\theta} e^{-2\alpha\beta(\theta')}U(\theta')^2\right\} \\
& = E\sum_{\theta\in\Theta_{j-1}} e^{-2\alpha\tau(\theta)}mEe^{-2\alpha\beta}U^2 \\
& = EH_{j-1}mEe^{-2\alpha\beta}U^2 = (m/(2m-1))^{j-1}mEe^{-2\alpha\beta}U^2 \\
& \leqslant K_2(m/(2m-1))^j,
\end{aligned}
$$

where K_2 is a constant.

The conclusion follows by putting things together.

COROLLARY.

$$\sup_T e^{-2\alpha T}\sum_{j=1}^{k_T} E\sum_{\theta\in\Theta_j}(Z_T{}^\theta)^2U(\theta)^2 < \infty.$$

Proof. $m > 1$ implies that $0 < m/(2m-1) < 1$.

Remark. Setting $U(\theta)\equiv 1$ for all θ, the above results express properties of continuous-time Markov branching processes.

Proof of Statement 1.

$$
\begin{aligned}
& |\varphi_{1,T}-\varphi_{2,T}| \\
& \quad \leqslant Z_T^{-1}\sum_{\gamma\in\Gamma_T}|f(a_T^{-1}X_T(\gamma)) - f(a_T^{-1}X_T^{(1)}(\gamma)| \\
& \quad \leqslant MZ_T^{-1}a_T^{-1}\sum_{\gamma\in\Gamma_T}\sup_{0\leqslant t\leqslant 1}\|X_T(t,\gamma) - X_T^{(1)}(t,\gamma)\| \\
& \quad = MZ_T^{-1}a_T^{-1}\sum_{\gamma\in\Gamma_T}\sup_{0\leqslant t\leqslant \tau_{k_T}(\gamma)/T}\|X_T(t,\gamma)\| \\
& \quad = MZ_T^{-1}a_T^{-1}\sum_{\gamma\in\Gamma_T}\sup_{0\leqslant t\leqslant \tau_{k_T}(\gamma)}\|X(t,\gamma)\| \\
& \quad \leqslant MZ_T^{-1}a_T^{-1}\sum_{\gamma\in\Gamma_T}\sum_{j=1}^{k_T}\sup_{\tau_{j-1}(\gamma)\leqslant t\leqslant\tau_j(\gamma)}\|X(t,\gamma) - X(\tau_{j-1}(\gamma),\gamma)\| \\
& \quad \leqslant MZ_T^{-1}a_T^{-1}\sum_{\gamma\in\Gamma_T}\sum_{j=1}^{k_T} D_j(\gamma) \\
& \quad = MZ_T^{-1}a_T^{-1}\sum_{j=1}^{k_T}\sum_{\gamma\in\Gamma_T} D_j(\gamma).
\end{aligned}
$$

In the last sum, $D_j(\gamma) = D(\theta)$, where θ is the jth node on the branch $\gamma \in \Gamma_T$. Then $D(\theta)$ is counted $Z_T{}^\theta$ times if the jth split on γ occurs (i.e., if $\tau(\theta) < T$), and it is counted once if the last split on γ before time T is the $(j-1)$th. Hence

$$\sum_{\gamma \in \Gamma_T} D_j(\gamma) = \sum_{\theta \in \Theta_j} D(\theta) V_T{}^\theta,$$

where, for $\theta \in \Theta_j$,

$$V_T{}^\theta = \begin{cases} Z_T{}^\theta & \text{if} \quad \tau(\theta) < T, \\ 1 & \text{if} \quad \tau_{j-1}(\theta) < T \leqslant \tau(\theta), \\ 0 & \text{if} \quad \tau_{j-1}(\theta) \geqslant T. \end{cases}$$

In any case, $V_T{}^\theta \leqslant Z_T{}^\theta + 1$, and therefore we have $|\varphi_{1,T} - \varphi_{2,T}| \leqslant A_{1,T} + A_{2,T}$, where

$$A_{1,T} = M Z_T^{-1} a_T^{-1} \sum_{j=1}^{k_T} \sum_{\theta \in \Theta_j} D(\theta),$$

and

$$A_{2,T} = M Z_T^{-1} a_T^{-1} \sum_{j=1}^{k_T} \sum_{\theta \in \Theta_j} D(\theta) Z_T{}^\theta.$$

To show that $A_{1,T} \to 0$ a.s., since $Z_T e^{-\alpha T} \to W_2$ a.s., it suffices to prove that

$$a_T^{-1} e^{-\alpha T} \sum_{j=1}^{k_T} \sum_{\theta \in \Theta_j} D(\theta) \to 0 \quad \text{a.s.}$$

For this, we write

$$a_T^{-1} e^{-\alpha T} \sum_{j=1}^{k_T} \sum_{\theta \in \Theta_j} D(\theta) \leqslant \frac{k_T}{a_T} \frac{\xi_{k_T}}{m^{k_T}} \frac{m^{k_T}}{e^{\alpha T}} Q_T,$$

where

$$Q_T = (k_T \xi_{k_T})^{-1} \sum_{j=1}^{k_T} \sum_{\theta \in \Theta_j} \xi_{k_T}^\theta D(\theta).$$

By the strong law for random variables on Galton–Watson trees (Theorem 2 of [10]), $Q_T \to ED$ a.s. Also, $\xi_{k_T} m^{-k_T} \to W_1$ a.s., and since $m^{k_T} e^{-\alpha T} = (m^{k_T/T} e^{-\alpha})^T$, then $m^{k_T} e^{-\alpha T} \to 0$ if $k_T = o(T)$; and finally it suffices to have $k_T = O(a_T)$ (we will need $k_T = o(a_T)$ later).

For $A_{2,T}$, again since $Z_T e^{-\alpha T} \to W_2$ a.s., and $a_T \to \infty$, it suffices to show that $a_T^{-1} Y_T \to 0$ a.s., where

$$Y_T = e^{-\alpha T} \sum_{j=1}^{k_T} \sum_{\theta \in \Theta_j} D(\theta) Z_T{}^\theta.$$

Let

$$U(\theta) = D(\theta) - E[D(\theta)|\beta(\theta)],$$

$$Y_{1,T} = e^{-\alpha T} \sum_{j=1}^{k_T} \sum_{\theta \in \Theta_j} U(\theta) Z_T^{\theta},$$

$$Y_{2,T} = e^{-\alpha T} \sum_{j=1}^{k_T} \sum_{\theta \in \Theta_j} E[D(\theta)|\beta(\theta)] Z_T^{\theta}.$$

Hence $Y_T = Y_{1,T} + Y_{2,T}$, and we will show that $a_T^{-1} Y_{1,T}$ and $a_T^{-1} Y_{2,T}$ both converge to 0 a.s.

For $Y_{1,T}$, it is sufficient to show that it is a.s. uniformly bounded, and we will do this by proving that $\{Y_{1,T}, \mathfrak{F}_T\}$ is an a.s. convergent martingale, where $\mathfrak{F}_T$ denotes the σ-algebra generated by the family tree up to time T, and by the branch processes of the branches $\gamma \in \Gamma_T$ up to the respective times $\max\{\tau_i(\gamma): \tau_i(\gamma) \leqslant t \text{ and } i \leqslant k_T\}$. Let $T_1 < T_2$; then

$$\begin{aligned} Y_{1,T_2} = e^{-\alpha T_2} &\sum_{j=1}^{k_{T_1}} \sum_{\theta \in \Theta_j,\, \tau(\theta) \leqslant T_1} U(\theta) Z_{T_2}^{\theta} \\ &+ e^{-\alpha T_2} \sum_{j=1}^{k_{T_1}} \sum_{\theta \in \Theta_j,\, \tau(\theta) > T_1} U(\theta) Z_{T_2}^{\theta} \\ &+ e^{-\alpha T_2} \sum_{j=k_{T_1}+1}^{k_{T_2}} \sum_{\theta \in \Theta_j} U(\theta) Z_{T_2}^{\theta}; \end{aligned}$$

hence

$$\begin{aligned} E[Y_{1,T_2}|\mathfrak{F}_{T_1}] = e^{-\alpha T_2} &\sum_{j=1}^{k_{T_1}} \sum_{\theta \in \Theta j,\, \tau(\theta) \leqslant T_1} U(\theta) E[Z_{T_2}^{\theta}|\mathfrak{F}_{T_1}] \\ &+ e^{-\alpha T_2} \sum_{j=1}^{k_{T_1}} E\left[\sum_{\theta \in \Theta_j,\, \tau(\theta) > T_1} E[U(\theta) Z_{T_2}^{\theta}|\mathfrak{F}_{T_1}] \middle| \mathfrak{F}_{T_1} \right] \\ &+ e^{-\alpha T_2} \sum_{j=k_{T_1}+1}^{k_{T_2}} E\left[\sum_{\theta \in \Theta_j} E[U(\theta) Z_{T_2}^{\theta}|\mathfrak{F}_{T_1}] \middle| \mathfrak{F}_{T_1} \right]. \end{aligned}$$

For θ such that $\tau(\theta) \leqslant T_1$,

$$Z_{T_2}^{\theta} = \sum_{\gamma \in \Gamma_{T_1}^{\theta}} Z_{T_2}^{\gamma},$$

so

$$E[Z_{T_2}^{\theta}|\mathfrak{F}_{T_1}] = \sum_{\gamma \in \Gamma_{T_1}^{\theta}} E[Z_{T_2}^{\gamma}|\mathfrak{F}_{T_1}];$$

but the residual life of $\gamma \in \Gamma_{T_1}$ after time T_1 has thc same distribution of the offspring lifetime; therefore

$$E[Z_{T_2}^{\gamma} | \mathfrak{F}_{T_1}] = e^{\alpha(T_2 - T_1)} \qquad \text{for all} \quad \gamma \in \Gamma_{T_1}^{\theta},$$

and hence

$$e^{-\alpha T_2} \sum_{j=1}^{k_{T_1}} \sum_{\theta \in \Theta_j,\, \tau(\theta) \leqslant T_1} U(\theta) E[Z_{T_2}^{\theta} | \mathfrak{F}_{T_1}]$$

$$= e^{-\alpha T_1} \sum_{j=1}^{k_{T_1}} \sum_{\theta \in \Theta_j} U(\theta) Z_{T_1}^{\theta} = Y_{1,T_1}$$

(the condition $\tau(\theta) \leqslant T_1$ is satisfied due to the definition of $Z_{T_1}^{\theta}$). We leave for the reader to verify that the remaining terms in the above expression for $E[Y_{1,T_2} | \mathfrak{F}_{T_1}]$ are zero (use conditional independence of $U(\theta)$ and $Z_{T_2}^{\theta}$, given $\beta(\theta)$ and $\tau(\theta)$). Hence $\{Y_{1,T}, \mathfrak{F}_T\}$ is a martingale. The a.s. convergence follows because the $Y_{1,T}$ have uniformly bounded second moments. Indeed, by first conditioning on realizations of the tree, and using the independence of the $U(\theta)$ and the fact that they have conditional mean zero, it follows that

$$E(Y_{1,T})^2 = e^{-2\alpha T} \sum_{j=1}^{k_T} E \sum_{\theta \in \Theta_j} U(\theta)^2 (Z_T{}^{\theta})^2,$$

and this, by the Corollary, is uniformly bounded.

Now, using one of the moment hypotheses, we have

$$a_T^{-1} Y_{2,T} \leqslant K e^{-\alpha T} Z_T k_T a_T^{-1} + K e^{-\alpha T} a_T^{-1} \sum_{j=1}^{k_T} \sum_{\theta \in \Theta_j} \beta(\theta) Z_T{}^{\theta}.$$

The first term on the right goes to 0 if $k_T = o(a_T)$. For the second term, since

$$\sum_{j=1}^{k_T} \sum_{\theta \in \Theta_j} \beta(\theta) Z_T{}^{\theta} \leqslant \sum_{\gamma \in \Gamma_T} \tau_{k_T}(\gamma) \leqslant Z_T \max_{\theta \in \Theta_{k_T}} \tau(\theta),$$

then

$$e^{-\alpha T} a_T^{-1} \sum_{j=1}^{k_T} \sum_{\theta \in \Theta_j} \beta(\theta) Z_T{}^{\theta} \leqslant e^{-\alpha T} Z_T k_T a_T^{-1} k_T^{-1} \max_{\theta \in \Theta_{k_T}} \tau(\theta),$$

and since $k_T^{-1} \max\{\tau(\theta), \theta \in \Theta_{k_T}\}$ has a positive limit a.s., it also suffices to have $k_T = o(a_T)$. So $a_T^{-1} Y_{2,T} \to 0$ a.s.

We have shown that $a_T^{-1} Y_T \to 0$ a.s., hence $A_{2,T} \to 0$ a.s.; and before we showed that $A_{1,T} \to 0$ a.s., so statement 1 is proved.

Proof of Statement 2. Splitting the sum in $\varphi_{2,T}$ into the terms with $\tau_{k_T}(\gamma) < T$, and the terms with $\tau_{k_T}(\gamma) = T$ (recall the convention), we have

$$\varphi_{2,T} - \varphi_{3,T} = e^{\alpha T} Z_T^{-1} (G_{1,T} + G_{2,T}),$$

where

$$G_{1,T} = e^{-\alpha T} \sum_{\gamma \in \Gamma_T,\ \tau_{k_T}(\gamma) = T} f(a_T^{-1} X_T^{(1)}(\gamma)),$$

and

$$G_{2,T} = e^{-\alpha T} \sum_{\theta \in \Theta_{k_T}} \left\{ \sum_{\gamma \in \Gamma_T{}^\theta} f(a_T^{-1} X_T^{(1)}(\gamma)) - E\left[\sum_{\gamma \in \Gamma_T{}^\theta} f(a_T^{-1} X_T^{(1)}(\gamma)) \middle| \mathfrak{G}_T \right] \right\}.$$

Since f is bounded, there is a constant K such that $|G_{1,T}| \leqslant K e^{-\alpha T} \xi_{k_T}$, and we have seen that $m^{k_T} e^{-\alpha T} \to 0$; hence $G_{1,T} \to 0$ a.s.

In $G_{2,T}$, conditioned upon $\mathfrak{G}_T$ the terms of the sum $\sum_\theta$ corresponding to different θ are independent, and all have (conditional) expectation zero; hence

$$E|G_{2,T}|^2 = e^{-2\alpha T} E \sum_{\theta \in \Theta_{k_T}} E\left[\left| \sum_{\gamma \in \Gamma_T{}^\theta} f(a_T^{-1} X_T^{(1)}(\gamma)) \right.\right.$$

$$\left.\left. - E\left[\sum_{\gamma \in \Gamma_T{}^\theta} f(a_T^{-1} X_T^{(1)}(\gamma)) \middle| \mathfrak{G}_T \right] \right|^2 \middle| \mathfrak{G}_T \right]$$

and since f is bounded, there is a constant K such that

$$E|G_{2,T}|^2 \leqslant K e^{-2\alpha T} E \sum_{\theta \in \Theta_{k_T}} E[(Z_T{}^\theta)^2 \,|\, \mathfrak{G}_T] = K e^{-2\alpha T} E \sum_{\theta \in \Theta_{k_T}} (Z_T{}^\theta)^2;$$

so, using the Lemma, there is a constant K_1 such that

$$E|G_{2,T}|^2 \leqslant K_1 \delta^{k_T},$$

where $\delta = m/(2m-1) < 1$. Now let k_T be constant for $T \in [n, n+1)$, n natural, and such that $\sum_n \delta^{k_n} < \infty$. Then, by Chebyshev's inequality and the Borel–Cantelli lemma, it follows that $G_{2,T} \to 0$ a.s.

Statement 2 is proved.

Proof of Statement 3.

$$|\varphi_{3,T} - \varphi_{4,T}|$$

$$\leqslant Z_T^{-1} \sum_{\theta \in \Theta_{k_T}} \left| E\left[\sum_{\gamma \in \Gamma_T{}^\theta} f(a_T^{-1} X_T^{(1)}(\gamma)) \middle| \mathfrak{G}_T \right] - E\left[\sum_{\gamma \in \Gamma_T{}^\theta} f(a_T^{-1} X_T^{(2)}(\gamma)) \middle| \mathfrak{G}_T \right] \right|$$

$$\leqslant Z_T^{-1} \sum_{\theta \in \Theta_{k_T}} E\left[\sum_{\gamma \in \Gamma_T{}^\theta} |f(a_T^{-1} X_T^{(1)}(\gamma)) - f(a_T^{-1} X_T^{(2)}(\gamma))| \middle| \mathfrak{G}_T \right]$$

$$\leqslant M Z_T^{-1} a_T^{-1} \sum_{\theta \in \Theta_{k_T}} E\left[\sum_{\gamma \in \Gamma_T{}^\theta} \sup_{0 \leqslant t \leqslant 1} \|X_T^{(1)}(t, \gamma) - X_T^{(2)}(t, \gamma)\| \middle| \mathfrak{G}_T \right]$$

$$= M Z_T^{-1} a_T^{-1} \sum_{\theta \in \Theta_{k_T}} E\left[\sum_{\gamma \in \Gamma_T{}^\theta} \sup_{\tau_T^{(1)}(\gamma) \leqslant t \leqslant \tau_T^{(2)}} \|X(t, \gamma) - X(\tau_T^{(1)}(\gamma), \gamma)\| \middle| \mathfrak{G}_T \right].$$

By conditioning on realizations of the trcc, and using one of the moment hypotheses, we obtain

$$E\left[\sum_{\gamma\in\Gamma_T{}^\theta}\sup_{\tau_T^{(1)}(\gamma)\leqslant t\leqslant\tau_T^{(2)}}\|X(t,\gamma)-X(\tau_T^{(1)}(\gamma),\gamma)\|\,\middle|\,\mathfrak{G}_T\right]$$
$$\leqslant KE\left[\sum_{\gamma\in\Gamma_T{}^\theta}(1+\tau_T^{(2)}-\tau_T^{(1)}(\gamma))\,\middle|\,\mathfrak{G}_T\right]$$
$$\leqslant KE\left[\sum_{\gamma\in\Gamma_T{}^\theta}(1+\tau_T^{(2)})\,\middle|\,\mathfrak{G}_T\right]$$
$$\leqslant K(1+\tau_T^{(2)})E[Z_T{}^\theta|\mathfrak{G}_T],$$

and since $E[Z_T{}^\theta|\mathfrak{G}_T]=e^{\alpha(T-\tau_T^{(2)})}Z^\theta_{\tau_T^{(2)}}$, we have

$$|\varphi_{3,T}-\varphi_{4,T}|\leqslant MKe^{\alpha T}Z_T^{-1}e^{-\alpha\tau_T^{(2)}}Z_{\tau_T^{(2)}}a_T^{-1}(1+\tau_T^{(2)}),$$

where the right-hand side goes to 0 a.s., because $k_T^{-1}\tau_T^{(2)}$ has a finite limit a.s., and $k_T=o(a_T)$.

So statement 3 is proved.

From statements 1, 2, and 3 it follows that $\varphi_{1,T}-\varphi_{4,T}\to 0$ a.s., so now it suffices to investigate convergence of $\varphi_{4,T}$. We will first write $\varphi_{4,T}$ in another way.

For $\theta\in\Theta_{k_T}$, $\tau(\theta)<T$,

$$E\left[\sum_{\gamma\in\Gamma_T{}^\theta}f(a_T^{-1}X_T^{(2)}(\gamma))\,\middle|\,\mathfrak{G}_T\right].$$
$$=E\left[\sum_{\gamma\in\Gamma^\theta_{\tau_T^{(2)}}}\sum_{\gamma'\in\Gamma_T{}^\gamma}f(a_T^{-1}X_T^{(2)}(\gamma'))\,\middle|\,\mathfrak{G}_T\right]$$
$$=Z^\theta_{\tau_T^{(2)}}E\left[\sum_{\gamma\in\Gamma_T^{\gamma_0}}f(a_T^{-1}X_T^{(2)}(\gamma))\,\middle|\,\mathfrak{G}_T\right],$$

where γ_0 is a fixed arbitrary branch in $\Gamma_{\tau_T^{(2)}}$. The last equality is due to the fact that the terms $\sum_{\gamma'\in\Gamma_T{}^\gamma}$, $\gamma\in\Gamma^\theta_{\tau_T^{(2)}}$, are identically distributed, because the residual times of the $\gamma\in\Gamma^\theta_{\tau_T^{(2)}}$ after time $\tau_T^{(2)}$ all have the same distribution. Hence

$$\varphi_{4,T}=Z_T^{-1}Z_{\tau_T^{(2)}}E\left[\sum_{\gamma\in\Gamma_T^{\gamma_0}}f(a_T^{-1}X_T^{(2)}(\gamma))\,\middle|\,\mathfrak{G}_T\right]$$

(by the convention, $\tau_T^{(2)}$ is interpreted here as $\max\{\tau(\theta):\theta\in\Theta_{k_T},\tau(\theta)<T\}$). Notice that in the last expression we may replace $\mathfrak{G}_T$ by $\tau_T^{(2)}$.

Let

$$\psi_T=e^{-\alpha(T-\tau_T^{(2)})}E\left[\sum_{\gamma\in\Gamma_T^{\gamma_0}}f(a_T^{-1}X_T^{(2)}(\gamma))\,\middle|\,\tau_T^{(2)}\right].$$

Since $T - \tau_T^{(2)} \to \infty$ a.s., because $\tau_T^{(2)} = o(T)$ a.s., then

$$(Z_T^{-1} e^{\alpha T})(Z_{\tau_T^{(2)}} e^{-\alpha \tau_T^{(2)}}) \to 1 \text{ a.s.},$$

and therefore $\varphi_{4,T} - \psi_T \to 0$ a.s.

In conclusion, we have proved the following proposition for general branch processes.

PROPOSITION. *$\varphi_{1,T} - \psi_T \to 0$ a.s. as $T \to \infty$.*

The significance of this is essentially that the empirical distribution of the branch processes $\varphi_{1,T}$ behaves asymptotically as the quotient of means ψ_T.

This proposition is the basic result.

It remains to show that ψ_T converges a.s. in general, as a consequence of the convergence hypothesis for the individual branch processes, and identify the limit of ψ_T, or to do it for special branch processes of interest.

In the case of Brownian motion, $f(a_T^{-1} X_T^{(2)}(\gamma))$, $\gamma \in \Gamma_T^{\gamma_0}$, are identically distributed conditioned upon $Z_T^{\gamma_0}$, and therefore

$$E\left[\sum_{\gamma \in \Gamma_T^{\gamma_0}} f(a_T^{-1} X_T^{(2)}(\gamma)) \middle| \tau_T^{(2)}\right] = E[Z_T^{\gamma_0} f(a_T^{-1} X_T^{(2)}(\gamma_1)) | \tau_T^{(2)}],$$

where γ_1 is a fixed arbitrary element of $\Gamma_T^{\gamma_0}$. Moreover, conditioned upon $\tau_T^{(2)}$, the random variables $Z_T^{\gamma_0}$ and $f(a_T^{-1} X_T^{(2)}(\gamma_1))$ are independent, hence

$$E[Z_T^{\gamma_0} f(a_T^{-1} X_T^{(2)}(\gamma_1)) | \tau_T^{(2)}] = e^{\alpha(T - \tau_T^{(2)})} E[f(a_T^{-1} X_T^{(2)}(\gamma_1) | \tau_T^{(2)}],$$

and

$$\psi_T = E[f(a_T^{-1} X_T^{(2)}(\gamma_1)) | \tau_T^{(2)}].$$

Now, let h_T be constants such that $x_0 k_T < h_T < T$, with $x_0 > x_2$ (recall that $k_T^{-1} \tau_T^{(2)} \to x_2$ a.s.), and $h_T = o(a_T)$. Then, for each ω in a subset of Ω of probability one, $\tau_T^{(2)}(\omega) < h_T$ for all sufficiently large T. Let

$$X_T^{(3)}(t) = \begin{cases} 0 & \text{if } 0 \leqslant t \leqslant h_T/T, \\ X_T(t) - X_T(h_T/T) & \text{if } h_T/T \leqslant t \leqslant 1. \end{cases}$$

Then

$$\begin{aligned} &|E[f(a_T^{-1} X_T^{(2)}(\gamma_1)) | \tau_T^{(2)}] - E[f(a_T^{-1} X_T^{(3)}(\gamma_1)) | \tau_T^{(2)}]| \\ &\leqslant M a_T^{-1} E\left[\sup_{\tau_T^{(2)} \leqslant t \leqslant h_T} \|X(t) - X(\tau_T^{(2)})\| \middle| \tau_T^{(2)}\right] \\ &\leqslant MK a_T^{-1}(1 + h_T) \qquad \text{(a.s. for all sufficiently large } T), \end{aligned}$$

but for each ω in a set of probability one,

$$E[f(a_T^{-1} X_T^{(3)}(\gamma_1)) | \tau_T^{(2)}](\omega) = E f(a_T^{-1} X_T^{(3)})$$

for all sufficiently large T; and, similarly as above,

$$|Ef(a_T^{-1}X_T^{(3)}) - Ef(a_T^{-1}X_T)| \leqslant MKa_T^{-1}(1 + h_T).$$

In conclusion, $\psi_T - Ef(a_T^{-1}X_T) \to 0$ a.s., and therefore $\varphi_{1,T} - Ef(a_T^{-1}X_T) \to 0$ a.s. In this case $a_T^{-1}X_T$ is Brownian motion B, so

$$\varphi_{1,T} \to Ef(B) \quad \text{a.s.} \quad \text{as} \quad T \to \infty,$$

and the theorem is proved.

4. Example

Let A be a Borel set of C^d, and for $T > 1$, let A_T be a Borel set of $C[0, T]^d$ such that for all $t \in [0, 1]$ the projections of A at t and of A_T at Tt coincide (i.e., A_T is obtained from A by stretching time); then it follows from the theorem that

$$Z_T^{-1} \sum_{\gamma \in \Gamma_T} 1_{[\{X(t,\gamma),\, 0 \leqslant t \leqslant T\} \in T^{1/2}A_T]} \to B(A) \qquad \text{a.s.}$$

as $T \to \infty$. In particular, if A is a the cylinder determined by D at $t = 1$, where D is a Borel set of R^d, then

$$Z_T^{-1} \sum_{\gamma \in \Gamma_T} 1_{[X(T,\gamma) \in T^{1/2}D]} \to (2\pi)^{-d/2} \int_D e^{-\|x\|^2/2}\, dx \qquad \text{a.s.}$$

as $T \to \infty$, which coincides with the result obtained by Watanabe (see [1], p. 245).

Other limit results for branching Brownian motion can be obtained by using special functionals f (see [12], p. 237).

Acknowledgments

We thank Norman Kaplan for valuable conversations on some points of the proof, and the hospitality of the Centre de Recherches Mathématiques of the University of Montreal.

References

1. K. B. Athreya and P. Ney, "Branching Processes," Springer-Verlag, Berlin and New York, 1973.
2. R. R. Bahadur and R. Ranga-Rao, On deviations of the sample mean, *Ann. Math. Statist.* **31** (1960), 1015–1027.
3. P. Billingsley, "Convergence of Probability Measures," Wiley, New York, 1968.

4. L. Breiman, "Probability," Addison-Wesley, Reading, Mass., 1968.
5. H. E. Conner, Asymptotic behavior of averaging processes for a branching process of restricted Brownian particles, *J. Math. Anal. Appl.* **20** (1967), 464–479.
6. A. W. Davis, Branching diffusion processes with no absorbing boundaries, I, *J. Math. Anal. Appl.* **18** (1967), 276–296.
7. A. W. Davis, Branching diffusion processes with no absorbing boundaries, II, *J. Math. Anal. Appl.* **19** (1967), 1–25.
8. J. L. Doob, "Stochastic Processes,.. Wiley, New York, 1953.
9. L. G. Gorostiza and A. R. Moncayo, Invariance principle for sums of independent random variables on a binary tree, *Bol. Soc. Mat. Mexicana* **19** (1974), 44–47.
10. L. G. Gorostiza and A. R. Moncayo, Invariance principle for random processes on Galton-Watson trees, *J. Math. Anal. Appl.* **60**, No. 2 (1977), 461–476.
11. T. E. Harris, "The Theory of Branching Processes," Springer-Verlag, Berlin and New York, 1963.
12. D. L. Iglehart, Weak convergence in applied probability, *Stochastic Processes Appl.* **2** (1974), 211–241.
13. A. Joffe and A. R. Moncayo, On sums of independent random variables defined on a binary tree, *Bol. Soc. Mat. Mexicana* **18**, No. 1 (1973), 50–54.
14. A. Joffe and A. R. Moncayo, Random variables, trees, and branching random walks, *Advances in Math.* **10**, No. 3 (1973), 401–416.
15. N. Kaplan and S. Asmussen, Branching random walks II, *Stochastic Processes Appl.* **4** (1976), 15–31.
16. B. P. Kharlamov, The number of generations in a branching process with an arbitrary set of particle types, *Theor. Probability Appl.* **14** (1969), 432–449.
17. J. F. C. Kingman, The first birth problem for an age-dependent branching process, *Ann. Probability* **3**, No. 5 (1975), 790–801.
18. A. N. Kolmogorov, Über das logarithmisch normale Verteilungsgesetz der Dimensionen der Teilchen bei Zerstüchelung, *Dokl. Akad. Nauk SSSR* **31** (1941), 99–101.
19. P. E. Ney, The limit distribution of a binary cascade process, *J. Math. Anal. Appl.* **10** (1965), 30–36.
20. P. E. Ney, The convergence of a random distribution function associated with branching processes, *J. Math. Anal. Appl.* **12** (1965), 316–327.
21. M. L. Samuels, Distribution of the branching process population among generations, *J. Appl. Probability* **8** (1971), 655–667.
22. A. J. Stam, On a conjecture by Harris, *Z. Wahrscheinlichkeitstheorie und Verw. Gebiete* **5** (1966), 202–206.
23. S. Watanabe, On the branching process for Brownian particles with an absorbing boundary, *J. Math. Kyoto Univ.*, **4**, No. 2 (1965), 385–398.
24. S. Watanabe, Limit theorems for a class of branching processes (Markov processes and potential theory), *Proc. Symp. Math. Res. Center, Madison, Wis.*, pp. 205–232, Wiley, New York, 1967.

AMS (MOS) 1970 subject classifications: 60J80, 60F05.

STUDIES IN PROBABILITY AND ERGODIC THEORY
ADVANCES IN MATHEMATICS SUPPLEMENTARY STUDIES, VOL. 2

On Operator Inequalities and Projections

M. HASEEB RIZVI[†]

Department of Statistics, Stanford University, Stanford, California,
and
Department of Mathematics, Sir George Williams University, Montreal, Canada

AND

R. W. SHORROCK[‡]

Centre de Recherches Mathématiques, Université de Montréal, Montréal, Canada

A Cauchy–Schwarz inequality for random commuting operators is given, and some recently obtained operator inequalities are shown to be simple consequences of positiveness of $A \bar{\otimes} A$, the norm-reducing property of orthogonal projections, and a basic formula for a generalized inverse.

1. INTRODUCTION

This paper exploits the fact that Kronecker products $A \bar{\otimes} A$ map positive semi-definite operators into positive semi-definite operators and the fact that orthogonal projections are norm-reducing to obtain some operator inequalities. A Cauchy–Schwarz inequality for operator-valued random variables that commute almost surely is proved in Section 2 making use of the positiveness of $A \bar{\otimes} A$. In Section 3 we give a brief exposition leading to a known formula for the orthogonal projection on the image of a bounded linear operator on a Hilbert space. This formula, at least for finite dimensional spaces, is given, for instance, in Albert [1]. In view of this formula, we are able, in Section 4, to show that certain recent results of Lieb and Ruskai [2] are immediate consequences of the norm-reducing property of projections and the positiveness of $A \bar{\otimes} A$.

For random vectors X taking values in a Hilbert space $\mathscr{X}$ with inner product $\langle \cdot, \cdot \rangle$ we define the expectation EX by

$$\langle \xi, EX \rangle = E\langle \xi, X \rangle, \qquad \forall \xi \in \mathscr{X}$$

[†] Present address: Committee on National Statistics, National Academy of Sciences, 2101 Constitution Avenue, Washington, D.C. 20418.

[‡] Present address: Management Sciences Division, Bell Canada, 620 Belmont, Montreal, Canada.

ISBN 0-12-599102-9

whenever $E|\langle \xi, X \rangle| \leqslant c\|X\|$. For random operators A we define EA by

$$(EA)\xi = E(A\xi)$$

whenever the right-hand side is defined for all ξ in the domain of A. For example,

$$E(A \bar{\otimes} B)(C) = E(ACB'),$$

for A and B random and C fixed. For a map $X:(\Omega, \mathscr{A}) \to \mathscr{X}$ to be called a random variable we require that it be at least "weakly measurable", that is,

$$\omega \to \langle \xi, X(\omega) \rangle$$

is to be measurable for all $\xi \in \mathscr{X}$. Similarly we require of a random operator $A:(\Omega, \mathscr{A}) \to \mathscr{B}(\mathscr{X}, \mathscr{Y})$, where $\mathscr{B}(\mathscr{X}, \mathscr{Y})$ denotes the set of all bounded linear operators from $\mathscr{X}$ to $\mathscr{Y}$, that

$$\omega \to \langle A(\omega)x, y \rangle$$

be measurable for each $x \in \mathscr{X}$ and $y \in \mathscr{Y}$. When we write $E(A \bar{\otimes} B)$ for A, B random, we tacitly assume that all random operators $\omega \to A(\omega)CB'(\omega)$ are measurable for C fixed. This follows if, for example, the Hilbert spaces are all separable, so that the inner products $\omega \to \langle X(\omega), Y(\omega) \rangle$ of random vectors X and Y are always measurable.

2. A Cauchy–Schwarz Inequality for Commuting Operators

Let X and Y be random bounded linear operators on a Hilbert space $\mathscr{X}$ such that the expectations $E(X \bar{\otimes} X)$, $E(Y \bar{\otimes} Y)$, and $E(X \bar{\otimes} Y)$ all exist, where we define $X \bar{\otimes} Y$ as a linear operator mapping $\mathscr{B}(\mathscr{X})$ into $\mathscr{B}(\mathscr{X})$ by $X \bar{\otimes} Y(C) = XCY'$; here $\mathscr{B}(\mathscr{X})$ is the set of all bounded linear operators on $\mathscr{X}$. Suppose $[X_1, Y_1]$ and $[X_2, Y_2]$ are independent copies of $[X, Y]$. We assume that X and Y commute in the sense that

$$P\{X_1 Y_2 = Y_2 X_1\} = 1, \tag{1}$$

where P denotes probability; that is, $X_1 Y_2$ equals $Y_2 X_1$ a.s. Observe that for any bounded linear operators A and C we have $A \bar{\otimes} A(C) \geqslant 0$ whenever $C \geqslant 0$ $[C \geqslant 0 \Leftrightarrow \langle x, Cx \rangle \geqslant 0$ for all $x \in \mathscr{X}]$. We shall write $A \bar{\otimes} A \geqslant 0$ or $[0 \leqslant A \bar{\otimes} A]$ to denote that the operator $A \bar{\otimes} A$ is order-preserving in this sense. Then we have the following theorem.

Theorem. *Assuming* (1),

$$E(X \bar{\otimes} X) \circ E(Y \bar{\otimes} Y) - E(X \bar{\otimes} Y) \circ E(Y \bar{\otimes} X) \geqslant 0. \tag{2}$$

Proof. From the preceding remarks it is clear that

$$\begin{aligned} 0 &\leqslant (X_1 Y_2 - X_2 Y_1) \bar{\otimes} (X_1 Y_2 - X_2 Y_1) \\ &\overset{\text{a.s.}}{=} X_1 Y_2 \bar{\otimes} X_1 Y_2 + X_2 Y_1 \bar{\otimes} X_2 Y_1 \\ &\quad - X_2 Y_1 \bar{\otimes} Y_2 X_1 - X_1 Y_2 \bar{\otimes} Y_1 X_2 \\ &= (X_1 \bar{\otimes} X_1) \circ (Y_2 \bar{\otimes} Y_2) + (X_2 \bar{\otimes} X_2) \circ (Y_1 \bar{\otimes} Y_1) \\ &\quad - (X_2 \bar{\otimes} Y_2) \circ (Y_1 \bar{\otimes} X_1) - (X_1 \bar{\otimes} Y_1) \circ (Y_2 \bar{\otimes} X_2). \end{aligned}$$

Taking expectations,

$$0 \leqslant E(X \bar{\otimes} X) \circ E(Y \bar{\otimes} Y) - E(X \bar{\otimes} Y) \circ E(Y \bar{\otimes} X).$$

COROLLARY 1. *Let C be any positive operator. Then for any random operator X we have*

$$E(XCX') - (EX)C(EX)' \geqslant 0.$$

Proof. Take $Y = I$ and apply (2) to C. [This corollary just restates the obvious fact that $E(X - EX)C(X - EX)' \geqslant 0$. Taking $C = I$ and assuming $X = X'$ a.s., it states the well-known fact that the function $X \to X^2$ is operator convex].

COROLLARY 2. *If X and Y are self-adjoint a.s. and commute in the sense of* (1), *then*

$$E(X^2)\, E(Y^2) - E\{XE(XY)Y\} \geqslant 0.$$

Proof. Apply (2) to I.

3. GENERALIZED INVERSES

Suppose H and $\mathscr{X}$ are Hilbert spaces with inner products both denoted by $\langle \cdot, \cdot \rangle$. Let $A: H \to \mathscr{X}$ be a bounded linear operator with closed range. Consider the problem of finding the orthogonal projection P_A on the image of A. It is well known that if A is a partial isometry, then $P_A = AA'$, where A' is the adjoint map of A defined by

$$\langle x, A\xi \rangle = \langle A'x, \xi \rangle.$$

When A is not a partial isometry, we can nevertheless define a new inner product

$$(\xi, \eta) = \langle \xi, C\eta \rangle$$

which makes A a partial isometry as explained below. We would then have $P_A = AA^*$, where A^* is the adjoint of A with respect to the inner product

$(\cdot, \cdot)$ and is defined by

$$\langle \xi, CA^*x \rangle = (\xi, A^*x) = \langle A\xi, x \rangle = \langle \xi, A'x \rangle.$$

Thus $CA^* = A'$ and $A^* = C^{-1}A'$ when C is invertible.

We now obtain, in terms of A and A', a class of such operators C. Let

$$H = H_0 \oplus H_1, \qquad \mathscr{X} = \mathscr{X}_0 \oplus \mathscr{X}_1$$

be orthogonal direct sum decompositions where H_0 is the kernel (null-space) of A and $\mathscr{X}_1$ is the image of A and is closed by assumption. With respect to these decompositions, A and C have the matrix representations

$$A = \begin{bmatrix} 0 & 0 \\ 0 & A_1 \end{bmatrix}, \qquad C = \begin{bmatrix} C_0 & 0 \\ 0 & C_1 \end{bmatrix}.$$

Then A is a partial isometry with respect to $(\cdot, \cdot)$ iff $A_1 : H_1 \to \mathscr{X}_1$ is an isometry with respect to $(\cdot, \cdot)$ iff $C_1 = A_1{}'A_1$. Since A_1 is bijective and bounded and H_1 and $\mathscr{X}_1$ are Hilbert spaces, $(A_1{}'A_1)^{-1}$ is also bounded. For any invertible (and positive and self-adjoint) C_0, we have

$$A^* = C^{-1}A' = \begin{bmatrix} 0 & 0 \\ 0 & (A_1{}'A_1)^{-1}A_1{}' \end{bmatrix}.$$

We observe that

$$\begin{aligned} (A'A + \epsilon^2 I)^{-1}A' &= \begin{bmatrix} \epsilon^2 I & 0 \\ 0 & A_1{}'A_1 + \epsilon^2 I \end{bmatrix}^{-1} \begin{bmatrix} 0 & 0 \\ 0 & A_1{}' \end{bmatrix} \\ &= \begin{bmatrix} 0 & 0 \\ 0 & (A_1{}'A_1 + \epsilon^2 I)^{-1}A_1{}' \end{bmatrix} \\ &\to \begin{bmatrix} 0 & 0 \\ 0 & (A_1{}'A_1)^{-1}A_1{}' \end{bmatrix} = A^*, \end{aligned}$$

as $\epsilon \to 0$ in the uniform operator topology. (Uniform operator convergence follows since for D positive, self-adjoint, and invertible,

$$\begin{aligned} \|(D + \epsilon^2 I)^{-1} - D^{-1}\| &= \epsilon^2 \|D^{-1}(D + \epsilon^2 I)^{-1}\| \\ &\leqslant \epsilon^2 \|D^{-1}\| \, \|(D + \epsilon^2 I)^{-1}\| \\ &\leqslant \epsilon^2 \|D^{-1}\|^2.) \end{aligned}$$

We finally have

$$P_A = AA^* = \lim_{\epsilon \to 0} A(A'A + \epsilon^2 I)^{-1}A',$$

and A^* is the Moore–Penrose generalized inverse of A.

4. Operator Convexity of $A'S^{-1}A$

Theorem (Lieb and Ruskai). *Suppose H and $\mathscr{X}$ are Hilbert spaces and $A:H \to \mathscr{X}$ and $S:\mathscr{X} \to \mathscr{X}$ are random bounded linear operators, and assume that $S = S' \geqslant 0$ and that S^{-1} exists a.s. Then*

$$E(A'S^{-1}A) - (EA)'(ES)^{-1}(EA) \geqslant 0$$

provided these expectations exist.

Proof. Write $\mu = EA$ and $\Sigma = ES$. Then

$$\begin{aligned} 0 &\leqslant E(A - S\Sigma^{-1}\mu)'S^{-1}(A - S\Sigma^{-1}\mu) \\ &= E(A'S^{-1}A) - \mu'\Sigma^{-1}\mu. \end{aligned}$$

Remark. Lieb and Ruskai [2] actually show in their Theorem 1 that the theorem holds even when S is not invertible, provided $A'S^{-1}A$ can be defined by

$$A'S^{-1}A = \operatorname*{slim}_{\epsilon \to 0} A'(S + \epsilon^2 I)^{-1}A$$

where "slim" denotes limit in the strong operator topology. This follows by applying the theorem stated above to $S + \epsilon^2 I$ and letting $\epsilon \to 0$. With this definition of $A'S^{-1}A$ we have, by Section 3,

$$P_A = A(A'A)^{-1}A'$$

where P_A denotes the orthogonal projection on the image of A. Suppose $C:H \to \mathscr{X}$ is another bounded linear operator. Since it is clear that $P_A \leqslant I$ and $C \mathbin{\bar{\otimes}} C$ preserves order, we have

$$CA(A'A)^{-1}A'C' \leqslant CC'.$$

Making the notational changes $A \to BS$, $C \to S'A'$, and setting $\varphi(X) = S'XS$, the preceding inequality becomes

$$\varphi(A'B)[\varphi(B'B)]^{-1}\varphi(B'A) \leqslant \varphi(A'A)$$

where now A, $B:H \to H$ and $S:\mathscr{X} \to H$.
This is precisely Theorem 2 of Lieb and Ruskai [2].

Acknowledgments

We would like to thank Marek Kanter for bringing reference [2] to our attention, and Ingram Olkin for making some useful comments.

References

1. A. Albert, "Regression and the Moore–Penrose Pseudoinverse," Academic Press, New York, 1972.
2. E. H. Lieb and M. B. Ruskai, Some operator inequalities of the Schwarz type, *Advances in Math.* **12** (1974), 269–273.

AMS (MOS) 1970 subject classifications: primary, 47D99; secondary, 15A09.

STUDIES IN PROBABILITY AND ERGODIC THEORY
ADVANCES IN MATHEMATICS SUPPLEMENTARY STUDIES, VOL. 2

Boundary Behavior of Laplace–Stieltjes Transforms with Applications to Uniformly Distributed Sequences

JEFFREY D. VAALER

Department of Mathematics, University of Texas at Austin, Austin, Texas

Let $\mu(y)$ be nonnegative and increasing on $[0, \infty)$ and let $f(s) = \int_{0-}^{\infty} e^{-sy}\, d\mu(y)$ be its Laplace–Stieltjes transform. If $f(\sigma)$ has regular growth as $\sigma \to 0+$ then sufficient conditions are given for (*) $\lim_{\sigma\to 0+} f(\sigma)^{-1} f(\sigma + it) = 0$ at almost all real t. An example is also produced to show that regular growth alone does not imply (*). By using a Tauberian theorem which relates (*) to the condition $\lim_{x\to\infty} \mu(x)^{-1} \int_{0-}^{x+} e^{-ity}\, d\mu(y) = 0$, a metric theorem for uniformly distributed sequences is obtained.

1. INTRODUCTION

Let $s = \sigma + it$ be a complex variable and let

$$f(s) = \int_{0-}^{\infty} e^{-sy}\, d\mu(y)$$

define an analytic function in the half plane $\sigma > 0$. We shall assume that $d\mu(y)$ is a Borel–Stieltjes measure on $[0, \infty)$ determined by a nondecreasing, right continuous function $\mu(y)$. Also, we assume that $\mu(y) = 0$ for $y < 0$ so that $\mu(x) = \int_{0-}^{x+} d\mu(y)$.

We say that a positive real valued function $L(x)$ is *slowly oscillating* if for each $\beta > 0$

$$\lim_{x\to\infty} L(x)^{-1} L(\beta x) = 1.$$

A well known Tauberian theorem of Karamata [5, 6] and its Abelian converse (Feller [3]) states that

$$f(\sigma) \sim \sigma^{-\kappa} L(\sigma^{-1}) \qquad \text{as} \quad \sigma \to 0+ \tag{1.1}$$

if and only if

$$\mu(x) \sim \{\Gamma(\kappa + 1)\}^{-1} x^{\kappa} L(x) \qquad \text{as} \quad x \to \infty. \tag{1.2}$$

ISBN 0-12-599102-9

Here κ is a positive constant and $L(x)$ is a slowly oscillating function. We say that $f(s)$ or $\mu(x)$ has *regular growth* whenever one of the two equivalent conditions (1.1) or (1.2) holds.

In [11] we show that if $f(s)$ has regular growth then for any real t,

$$\lim_{\sigma \to 0+} f(\sigma)^{-1} f(\sigma + it) = 0 \tag{1.3}$$

if and only if

$$\lim_{x \to \infty} \mu(x)^{-1} \int_{0-}^{x+} e^{-ity}\, d\mu(y) = 0. \tag{1.4}$$

Our aim in the present paper is to give sufficient conditions for (1.3) and (1.4) to hold at almost all t, in the sense of Lebesgue measure. As a special case of our results we obtain a metric theorem for sequences which are uniformly distributed mod 1 with weights.

In Theorem 5.1 we prove that if $f(s)$ has regular growth then

$$\liminf_{\sigma \to 0+} f(\sigma)^{-1} |f(\sigma + it)| = 0 \tag{1.5}$$

for almost all t. Thus it is natural to ask if regular growth alone is a sufficient condition for (1.3) to hold at almost all t. We answer this in the negative by constructing an example to show that, in general, $\liminf_{\sigma \to 0+}$ in (1.5) cannot be replace by $\lim_{\sigma \to 0+}$.

2. Preliminary Lemmas

Lemma 2.1. *Let $f(s)$ have regular growth. Then as $\sigma \to 0+$ we have* $-f'(\sigma) \sim \kappa\sigma^{-1} f(\sigma)$.

Proof. Let $\epsilon > 0$. Since

$$-f'(\sigma) = \int_{0-}^{\infty} e^{-\sigma y} y\, d\mu(y)$$

is decreasing we have

$$\begin{aligned} -\epsilon\sigma f'(\sigma) &\geqslant -\int_{\sigma}^{\sigma(1+\epsilon)} f'(\omega)\, d\omega \\ &= f(\sigma)\{1 - f(\sigma)^{-1} f(\sigma + \epsilon\sigma)\}. \end{aligned}$$

It follows from (1.1) that

$$\liminf_{\sigma \to 0+} -\sigma f'(\sigma) f(\sigma)^{-1} \geqslant \epsilon^{-1}\{1 - (1+\epsilon)^{-\kappa}\}.$$

Letting $\epsilon \to 0+$, we obtain

$$\liminf_{\sigma \to 0+} -\sigma f'(\sigma) f(\sigma)^{-1} \geqslant \kappa.$$

Similarly we can show that

$$\limsup_{\sigma \to 0+} -\sigma f'(\sigma) f(\sigma)^{-1} \leqslant \kappa. \qquad \text{Q.E.D.}$$

Since the following result is Lemma 4A of [1] we state it without proof. We remark that it can be proved under weaker hypothese (cf. [7]).

LEMMA 2.2. *Let $\kappa > 0$ and let $L(x)$ be slowly oscillating. Let $I(x)$ be a positive nondecreasing function on $[1, \infty)$ such that $I(x) \sim x^{\kappa} L(x)$ as $x \to +\infty$. Then for each $\epsilon > 0$ there exist positive numbers C_ϵ and D_ϵ such that*

$$\frac{1}{C_\epsilon} \left(\frac{x}{y}\right)^{\kappa - \epsilon} \leqslant \frac{I(x)}{I(y)} \leqslant D_\epsilon \left(\frac{x}{y}\right)^{\kappa + \epsilon}$$

whenever $1 \leqslant y \leqslant x$.

3. MAIN RESULTS

Let $\mathscr{G}(s)$ be analytic in the infinite strip $0 < \sigma < \delta$, for some $\delta > 0$. Also, let the boundary values

$$\lim_{\sigma \to 0+} \mathscr{G}(\sigma + it) = \mathscr{G}(it) \tag{3.1}$$

exist for almost all t and be nonzero for almost all t. We then have the following sufficient condition.

THEOREM 3.1. *Let $f(s)$ have regular growth. If*

$$\int_0^\sigma \sigma^{-1} f(\sigma)^{-2} \int_{-T}^{T} |f(\sigma + it)\mathscr{G}(\sigma + it)|^2 \, dt \, d\sigma < \infty$$

for each $T > 0$, then (1.3) *and* (1.4) *hold for almost all t.*

Remark: The hypotheses in Theorem 3.1 can be weakened in various ways. For example, the exponent 2 can be replaced by an arbitrary positive exponent p. We would then assume only that $|\mathscr{G}(\sigma + it)|^p$ is integrable on $[-T, T]$ for each σ and continuous in σ for each t. None of these generalizations will concern us here however.

Proof. For $0 < \sigma < \delta$, let

$$J(\sigma) = f(\sigma)^{-2} \int_{-T}^{T} |f(\sigma + it)\mathscr{G}(\sigma + it)|^2 \, dt$$

and

$$\lambda(\sigma) = \left\{ \int_0^\sigma \omega^{-1} J(\omega) \, d\omega \right\}^{-1/2}.$$

We observe that λ is positive, decreasing, unbounded, and that

$$\int_0^\delta \sigma^{-1} \lambda(\sigma) J(\sigma) \, d\sigma < \infty. \tag{3.2}$$

Next, let u_0 be a point in $(0, \delta)$ such that $\lambda(u_0) > 1$. We then define a sequence $\{u_0, u_1, \ldots\}$ of positive real numbers by

$$\begin{aligned} u_n &= u_{n-1}\left(1 - \frac{1}{\lambda(u_{n-1})}\right) \\ &= u_0 \prod_{K=0}^{n-1} \left(1 - \frac{1}{\lambda(u_K)}\right). \end{aligned}$$

It follows immediately that

$$u_0 > u_1 > \cdots > u_n > \cdots, \tag{3.3}$$

$$\lim_{n \to \infty} u_n = 0, \tag{3.4}$$

and that

$$\lim_{n \to \infty} u_n (u_{n-1})^{-1} = 1. \tag{3.5}$$

On the interval $[u_n, u_{n-1}]$ let $J(\sigma)$ take its minimum at σ_n, $n = 1, 2, 3, \ldots$. We note that σ_n exists since J is continuous. Then

$$\begin{aligned} J(\sigma_n) &\leqslant (u_{n-1} - u_n)^{-1} \int_{u_n}^{u_{n-1}} J(\sigma) \, d\sigma \\ &\leqslant u_{n-1}(u_{n-1} - u_n)^{-1} \int_{u_n}^{u_{n-1}} \sigma^{-1} J(\sigma) \, d\sigma \\ &= \lambda(u_{n-1}) \int_{u_n}^{u_{n-1}} \sigma^{-1} J(\sigma) \, d\sigma \\ &\leqslant \int_{u_n}^{u_{n-1}} \sigma^{-1} \lambda(\sigma) J(\sigma) \, d\sigma. \end{aligned}$$

Summing on n and using (3.2), we obtain

$$\sum_{n=1}^{\infty} f(\sigma_n)^{-2} \int_{-T}^{T} |f(\sigma_n + it)\mathscr{G}(\sigma_n + it)|^2 \, dt < \infty. \tag{3.6}$$

Since each of the terms $J(\sigma_n)$ is nonnegative we may interchange the sum and integral in (3.6). It follows that

$$\sum_{n=1}^{\infty} f(\sigma_n)^{-2}|f(\sigma_n + it)\mathscr{G}(\sigma_n + it)|^2 \tag{3.7}$$

converges for almost all t in $[-T, T]$. But since T is arbitrary, the sum in (3.7) must converge for almost all real t. Because of our hypothesis concerning the boundary values of $\mathscr{G}(s)$, the sequence $|\mathscr{G}(\sigma_n + it)|^2$ approaches a positive limit as $n \to \infty$ for almost all t. Thus

$$\sum_{n=1}^{\infty} f(\sigma_n)^{-2}|f(\sigma_n + it)|^2$$

converges for almost all t and in particular

$$\lim_{n\to\infty} f(\sigma_n)^{-2}|f(\sigma_n + it)|^2 = 0 \tag{3.8}$$

for almost all t.

Now suppose that $\sigma_{n+1} \leqslant \sigma \leqslant \sigma_n$. Then

$$\begin{aligned}\frac{|f(\sigma + it)|}{f(\sigma)} &\leqslant \frac{|f(\sigma + it) - f(\sigma_n + it)| + |f(\sigma_n + it)|}{f(\sigma)} \\ &\leqslant \frac{f(\sigma) - f(\sigma_n)}{f(\sigma)} + \frac{|f(\sigma_n + it)|}{f(\sigma_n)}.\end{aligned}$$

By the mean value theorem,

$$\frac{|f(\sigma + it)|}{f(\sigma)} \leqslant \frac{-f'(\sigma^*)(\sigma_n - \sigma)}{f(\sigma)} + \frac{|f(\sigma_n + it)|}{f(\sigma_n)} \tag{3.9}$$

for some σ^*, $\sigma \leqslant \sigma^* \leqslant \sigma_n$. From Lemma 2.1 there exists a positive number M, depending only on f, such that for all n

$$-f'(\sigma^*)/f(\sigma) \leqslant -f'(\sigma^*)/f(\sigma^*) \leqslant M/\sigma^* \leqslant M/\sigma.$$

It follows from (3.9) that

$$\begin{aligned}\frac{|f(\sigma + it)|}{f(\sigma)} &\leqslant M\left(\frac{\sigma_n}{\sigma} - 1\right) + \frac{|f(\sigma_n + it)|}{f(\sigma_n)} \\ &\leqslant M\left(\frac{u_{n-1}}{u_{n+1}} - 1\right) + \frac{|f(\sigma_n + it)|}{f(\sigma_n)}.\end{aligned}$$

Combining (3.3), (3.4), (3.5), and (3.8), we obtain (1.3) for almost all t. Since (1.3) and (1.4) are equivalent, the theorem is proved. Q.E.D.

Let $\Delta = \{s \in \mathbf{C} \mid \sigma > 0\}$. We recall that $\mathscr{G}(s)$ is in the Hardy class $H^2(\Delta)$ if it is analytic on Δ and if

$$\sup_{0<\sigma<\infty} \int_{-\infty}^{\infty} |\mathscr{G}(\sigma + it)|^2 \, dt$$

is finite. Moreover, if $\mathscr{G}(s) \in H^2(\Delta)$ and is not identically zero then for almost all t the boundary values (3.1) exist and are not zero ([4], pp. 128–133). Thus as a special case of Theorem 3.1 we have the following.

COROLLARY 3.2. *Let $f(s)$ have regular growth and let $\mathscr{G}(s)$ be a function in $H^2(\Delta)$ which is not identically zero. If*

$$\int_0^1 \sigma^{-1} f(\sigma)^{-2} \int_{-\infty}^{\infty} |f(\sigma + it)\mathscr{G}(\sigma + it)|^2 \, dt \, d\sigma < \infty, \tag{3.10}$$

then (1.3) *and* (1.4) *hold for almost all t.*

We now show that (3.10) can be replaced by a condition which depends more directly on $\mu(y)$.

Let $g(y)$ be a measurable complex valued function defined for $0 \leqslant y < \infty$ and integrable on $[0, x]$ for each $x \geqslant 0$. By Tonelli's theorem we have

$$\begin{aligned} \int_0^x \int_{0-}^{y+} |g(y - \omega)| \, d\mu(\omega) \, dy &= \int_{0-}^{x+} \int_{\omega}^{x} |g(y - \omega)| \, dy \, d\mu(\omega) \\ &\leqslant \left(\int_0^x |g(y)| \, dy \right) \mu(x). \end{aligned} \tag{3.11}$$

Now let the convolution $g^*\mu$ be defined by

$$g^*\mu(y) = \int_{0-}^{y+} g(y - \omega) \, d\mu(\omega).$$

From (3.11) we see that $g^*\mu(y)$ is integrable on $[0, x]$ and in particular finite for almost all y.

Our reason for introducing $g(y)$ is to apply the following form of the Paley–Wiener theorem: $\mathscr{G}(s) \in H^2(\Delta)$ if and only if

$$\mathscr{G}(s) = \int_0^{\infty} e^{-sy} g(y) \, dy \tag{3.12}$$

for some $g(y) \in L^2([0, \infty))$.

THEOREM 3.3. *Let $f(s)$ have regular growth and let $\mathscr{G}(s) \in H^2(\Delta)$ be given by* (3.12). *Then* (3.10) *is equivalent to*

$$\int_c^{\infty} \left| \frac{g^*\mu(y)}{\mu(y)} \right|^2 dy < \infty. \tag{3.13}$$

Here $c \geqslant 0$ is any constant such that $\mu(c) > 0$.

Proof. For $x \geqslant 0$ we have

$$\mathscr{G}(s)e^{-sx} = \int_x^\infty e^{-sy} g(y-x)\,dy.$$

Thus by Fubini's theorem

$$f(s)\mathscr{G}(s) = \int_0^\infty e^{-sy} g^*\mu(y)\,dy.$$

By Parseval's identity, with $\sigma > 0$,

$$\int_{-\infty}^{\infty} |f(\sigma+it)\mathscr{G}(\sigma+it)|^2\,dt = 2\pi \int_0^\infty e^{-2\sigma y}|g^*\mu(y)|^2\,dy.$$

It follows, using Tonelli's theorem, that

$$\int_0^1 \sigma^{-1} f(\sigma)^{-2} \int_{-\infty}^{\infty} |f(\sigma+it)\mathscr{G}(\sigma+it)|^2\,dt\,d\sigma$$
$$= 2\pi \int_0^\infty \left(\int_0^1 \sigma^{-1} f(\sigma)^{-2} e^{-2\sigma y}\,d\sigma \right) |g^*\mu(y)|^2\,dy. \qquad (3.14)$$

In particular *each side of* (3.14) *is finite if and only if the other side is.* Thus to prove the equivalence of (3.10) and (3.13) it suffices to show that

$$\mu(y)^2 \int_0^1 \sigma^{-1} f(\sigma)^{-2} e^{-2\sigma y}\,d\sigma \qquad (3.15)$$

approaches a positive limit as $y \to +\infty$.

Since (1.1) and (1.2) are both assumed to hold, we have

$$\mu(y)^2 \int_0^1 \sigma^{-1} f(\sigma)^{-2} e^{-2\sigma y}\,d\sigma$$
$$\sim \Gamma(\kappa+1)^{-2} \int_0^y \omega^{-1}\{f(y^{-1})^2 f(\omega y^{-1})^{-2}\} e^{-2\omega}\,d\omega,$$

as $y \to +\infty$. Now let $\epsilon > 0$ and apply Lemma 2.2 with $I(x) = f(x^{-1})$. We obtain

$$f(y^{-1})^2 f(\omega y^{-1})^{-2} \leqslant C_\epsilon^{\,2} \omega^{2\kappa - 2\epsilon} \qquad (0 < \omega \leqslant 1),$$
$$f(y^{-1})^2 f(\omega y^{-1})^{-2} \leqslant D_\epsilon^{\,2} \omega^{2\kappa + 2\epsilon} \qquad (1 \leqslant \omega \leqslant y).$$

Thus if we choose $\epsilon < \kappa$ we may apply the dominated convergence theorem to show that

$$\lim_{y\to\infty} \Gamma(\kappa+1)^{-2} \int_0^y \omega^{-1}\{f(y^{-1})^2 f(\omega y^{-1})^{-2}\} e^{-2\omega}\,d\omega$$
$$= \Gamma(\kappa+1)^{-2} \int_0^\infty \omega^{2\kappa-1} e^{-2\omega}\,d\omega = \Gamma(\kappa+1)^{-2} 4^{-\kappa} \Gamma(2\kappa).$$

Hence (3.15) approaches a positive limit as $y \to +\infty$ and this proves the theorem. Q.E.D.

For completeness we give the result for $\mu(x)$ which corresponds to Corollary 3.2.

COROLLARY 3.4. *Let $\mu(x)$ have regular growth and let $g(y) \in L^2([0, \infty))$ with $\int_0^\infty |g(y)|^2 \, dy > 0$. If*

$$\int_c^\infty \left| \frac{g^*\mu(y)}{\mu(y)} \right|^2 dy < \infty,$$

where $c \geqslant 0$ is such that $\mu(c) > 0$, then (1.3) *and* (1.4) *hold for almost all t.*

Proof. If $\mathcal{G}(s) \in H^2(\Delta)$ is given by (3.12) and $\int_0^\infty |g(y)|^2 \, dy > 0$ then $\mathcal{G}(s)$ is not identically zero. Now apply Theorem 3.3 and Corollary 3.2. Q.E.D.

4. APPLICATIONS

Let $\{x_n\}$, $n = 1, 2, 3, \ldots$, be a sequence of real numbers and let $a(n)$ be a sequence of positive real numbers. Also, let $S(N) = \sum_{n=1}^N a(n)$. Tsuji [10] has defined the sequence $\{x_n\}$ to be *$a(n)$-uniformly distributed mod* 1 if for each u and v, $0 \leqslant u < v \leqslant 1$, we have

$$\lim_{N\to\infty} S(N)^{-1} \sum_{\substack{n=1 \\ u \leqslant \langle x_n \rangle < v}}^{N} a(n) = v - u.$$

Here $\langle y \rangle = y - [y]$ is the fractional part of y. By a simple generalization of the Weyl criterion it follows that $\{x_n\}$ is $a(n)$-uniformly distributed mod 1 if and only if

$$\lim_{N\to\infty} S(N)^{-1} \sum_{n=1}^{N} a(n) e^{2\pi i J x_n} = 0 \tag{4.1}$$

for each integer $J \neq 0$.

Now let α be a real parameter. From the criterion (4.1) we see that $\{\alpha x_n\}$ is $a(n)$-uniformly distributed mod 1 for almost all real α if and only if

$$\lim_{N\to\infty} S(N)^{-1} \sum_{n=1}^{N} a(n) e^{-itx_n} = 0$$

for almost all t. In order to apply Corollary 3.4 we must also assume that the sequence $\{x_n\}$ is nonnegative and strictly increasing.

THEOREM 4.1. *Let* $0 \leqslant x_1 < x_2 < \cdots < x_n < \cdots$ *and let the function* $A(y) = \sum_{x_n \leqslant y} a(n)$ *satisfy* $A(y) \sim y^\kappa L(y)$ *as* $y \to +\infty$, *for some* $\kappa > 0$ *and some slowly oscillating function* $L(y)$. *If there exists a function* $g(y) \in L^2([0, \infty))$ *such that* $\int_0^\infty |g(y)|^2 \, dy > 0$ *and*

$$\int_c^\infty \left| \frac{g^* A(y)}{A(y)} \right|^2 dy < \infty,$$

then $\{\alpha x_n\}$ *is* $a(n)$*-uniformly distributed mod 1 for almost all real* α.

Proof. Apply Corollary 3.4 with $\mu(x) = A(x)$. Q.E.D.

We say that the sequence $\{x_n\}$ is *separated* if there exists $\delta > 0$ such that $\delta \leqslant x_{n+1} - x_n$ for each $n = 1, 2, 3, \ldots$.

THEOREM 4.2. *Let* $\{x_n\}$ *and* $A(y)$ *be as in Theorem* 4.1. *Also let the sequence* $\{x_n\}$ *be separated. If*

$$\sum_{n=1}^\infty \left(\frac{a(n)}{S(n)} \right)^2 < \infty, \tag{4.2}$$

then $\{\alpha x_n\}$ *is* $a(n)$*-uniformly distributed mod 1 for almost all real* α.

Proof. Let $0 < \delta \leqslant x_{n+1} - x_n$ for each $n = 1, 2, 3, \ldots$. We apply Theorem 4.1 with $c = x_1$ and

$$g(y) = \begin{cases} 1 & \text{if } 0 \leqslant y < \frac{1}{2}\delta \\ 0 & \text{if } \frac{1}{2}\delta \leqslant y. \end{cases}$$

We then have

$$g^* A(y) = A(y) - A(y - \tfrac{1}{2}\delta) = \begin{cases} a(n) & \text{if } x_n \leqslant y < x_n + \frac{1}{2}\delta \\ 0 & \text{otherwise.} \end{cases}$$

It follows that

$$\int_{x_1}^\infty \left| \frac{g^* A(y)}{A(y)} \right|^2 dy = \tfrac{1}{2}\delta \sum_{n=1}^\infty \left(\frac{a(n)}{S(n)} \right)^2. \qquad \text{Q.E.D.}$$

We remark that the example of the next section shows that $A(y) \sim y^\kappa L(y)$ does not imply the boundedness condition (4.2). Other conditions for metric theorems involving $a(n)$-uniform distribution mod 1 can be found in [9] and in Chapter 3, Section 4 of [8].

5. Consequences of Regular Growth

If $f(s)$ is only assumed to have regular growth then, as we shall see, we cannot in general conclude that (1.3) holds for almost all t. However we can prove the following weaker result.

THEOREM 5.1. *Let $f(s)$ have regular growth. Then*

$$\liminf_{\sigma\to 0+} f(\sigma)^{-1}|f(\sigma+it)| = 0$$

for almost all t.

Proof. Let $\epsilon > 0$ be given. From (1.2) we have

$$\lim_{y\to\infty} \mu(y)^{-1}\{\mu(y)-\mu(y-1)\} = 0. \tag{5.1}$$

It follows from (1.1) and (5.1) that there exists $T > 0$ such that

$$\{\mu(y)-\mu(y-1)\} \leqslant \epsilon f(y^{-1}) \tag{5.2}$$

whenever $y \geqslant T$. By Lemma 2.1 we may also assume that

$$\left|\kappa + \frac{1}{y} f\left(\frac{1}{y}\right)^{-1} f'\left(\frac{1}{y}\right)\right| \leqslant \tfrac{1}{2}\kappa \tag{5.3}$$

for $y \geqslant T$. Now let $0 < \sigma \leqslant \frac{1}{2}\kappa T^{-1}$ and define

$$b(y) = e^{-\sigma y} f(y^{-1})$$

for $T \leqslant y < \infty$. It follows easily from (5.3) that the derivative $b'(y)$ can be zero only if

$$(\tfrac{1}{2})\kappa\sigma^{-1} \leqslant y \leqslant (\tfrac{3}{2})\kappa\sigma^{-1}.$$

Thus we have the upper bound

$$\begin{aligned}\sup_{T\leqslant y<\infty} \{e^{-\sigma y} f(y^{-1})\} &\leqslant \sup_{T\leqslant y\leqslant (3/2)\kappa\sigma^{-1}} \{e^{-\sigma y} f(y^{-1})\} \\ &\leqslant \sup_{T\leqslant y\leqslant (3/2)\kappa\sigma^{-1}} \{f(y^{-1})\} = f(\tfrac{2}{3}\kappa^{-1}\sigma).\end{aligned} \tag{5.4}$$

For $\sigma > 0$ a simple computation shows that

$$f(s)s^{-1}(1-e^{-s}) = \int_0^\infty e^{-sy}\{\mu(y)-\mu(y-1)\}\,dy.$$

Thus by Fatou's lemma and Parseval's identity we have

$$\int_{-\infty}^{\infty}\left(\liminf_{\sigma\to 0+} f(\sigma)^{-2}|f(\sigma+it)|^2\right)\left(\frac{2\sin\{\frac{1}{2}t\}}{t}\right)^2 dt$$

$$\leqslant \liminf_{\sigma\to 0+} f(\sigma)^{-2}\int_{-\infty}^{\infty}\left|f(\sigma+it)\left\{\frac{1-e^{-\sigma-it}}{\sigma+it}\right\}\right|^2 dt$$

$$= \liminf_{\sigma\to 0+}(2\pi)f(\sigma)^{-2}\int_T^{\infty} e^{-2\sigma y}\{\mu(y)-\mu(y-1)\}^2\,dy. \tag{5.5}$$

If $0<\sigma\leqslant(\frac{1}{2})\kappa T^{-1}$ then using (5.2) and (5.4) we obtain

$$\int_T^{\infty} e^{-2\sigma y}\{\mu(y)-\mu(y-1)\}^2\,dy \leqslant \epsilon\int_T^{\infty} e^{-\sigma y}\{\mu(y)-\mu(y-1)\}\{e^{-\sigma y}f(y^{-1})\}\,dy$$

$$\leqslant \epsilon f(\tfrac{2}{3}\kappa^{-1}\sigma)f(\sigma)\sigma^{-1}(1-e^{-\sigma}). \tag{5.6}$$

Combining (5.5) and (5.6) we have

$$\int_{-\infty}^{\infty}\left(\liminf_{\sigma\to 0+} f(\sigma)^{-2}|f(\sigma+it)|^2\right)\left(\frac{2\sin\{\frac{1}{2}t\}}{t}\right)^2 dt$$

$$\leqslant \lim_{\sigma\to 0+}(2\pi\epsilon)f(\sigma)^{-1}f(\tfrac{2}{3}\kappa^{-1}\sigma) = 2\pi\epsilon(\tfrac{3}{2}\kappa)^{\kappa}.$$

Since $\epsilon>0$ was arbitrary it follows that

$$\liminf_{\sigma\to 0+} f(\sigma)^{-2}|f(\sigma+it)|^2 = 0$$

for almost all t. Q.E.D.

In the remainder of this paper we shall construct an example of a measure $d\mu(y)$ such that $\mu(y)$ has regular growth but

$$\limsup_{x\to\infty}\mu(x)^{-1}\int_{0-}^{x+} e^{-ity}\,d\mu(y) \geqslant (2\pi)^{-1} \tag{5.7}$$

for almost all t. In fact in our example $d\mu(y)$ is a pure point mass measure and thus has the form

$$\mu(y) = A(y) = \sum_{x_n\leqslant y} a(n)$$

considered in Section 4.

For $0<y<\infty$ let $\beta(y)=[(\log y)/(\log 2)]$. Then let $\{x_n\}$, $n=1,2,3,\ldots,$ be the strictly increasing sequence of positive integers $\{2^m(1+k2^{-\beta(m)})\}$, where $m=1,2,3,\ldots$ and $k=0,1,2,\ldots,2^{\beta(m)}-1$. If $x_n=2^m(1+k2^{-\beta(m)})$ we define $a(n)=2^{m-\beta(m)}$.

LEMMA 5.2. *As $y\to\infty$ we have $A(y)=\sum_{x_n\leqslant y} a(n) = y + O(y\log^{-1} y)$.*

Proof. Let $2^N \leqslant y < 2^{N+1}$. The set of x_n's with $2^N \leqslant x_n \leqslant y$ consists of the integers $2^N(1 + k2^{-\beta(N)})$, where $k = 0, 1, 2, \ldots, [2^{\beta(N)-N}(y - 2^N)]$. Thus

$$A(y) = \sum_{m=1}^{N-1} \sum_{k=0}^{2^{\beta(m)}-1} 2^{m-\beta(m)} + 2^{N-\beta(N)}\{1 + [2^{\beta(N)-N}(y - 2^N)]\}$$
$$= 2^N - 1 + y - 2^N + O(2^{N-\beta(N)})$$
$$= y + O(yN^{-1}) = y + O(y \log^{-1} y). \qquad \text{Q.E.D.}$$

Now let $0 \leqslant \xi < 1$ and let

$$\xi = \sum_{J=1}^{\infty} \epsilon_J(\xi)2^{-J}, \qquad \epsilon_J(\xi) = 0 \quad \text{or} \quad 1, \tag{5.8}$$

be the binary expansion of ξ. To achieve uniqueness we assume that expansions ending in infinitely many consecutive ones are not allowed. For each positive integer N let $E(N) \subseteq [0, 1)$ be the set of all ξ such that

$$\epsilon_{N-\beta(N)+1}(\xi) = \epsilon_{N-\beta(N)+2}(\xi) = \cdots = \epsilon_{N+1}(\xi) = 0.$$

If λ is Lebesgue measure on $[0, 1)$ it is clear that $\lambda\{E(N)\} = 2^{-1-\beta(N)}$. Also, let

$$\mathscr{E} = \limsup_{N\to\infty} E(N) = \bigcap_{M=1}^{\infty} \bigcup_{N=M}^{\infty} E(N).$$

We note that $\xi \in \mathscr{E}$ if and only if $\xi \in E(N)$ for infinitely many N. In order to establish (5.7) we require the following probabilistic result.

LEMMA 5.3. *The Lebesgue measure of $\mathscr{E}$ is* 1.

Proof. Let $K \geqslant 2$ be a fixed integer. Then the sets

$$\{2^K + (m-2)\beta(2^K) + m, 2^K + (m-2)\beta(2^K) + m + 1, \ldots, 2^K + (m-1)\beta(2^K) + m\}, \tag{5.9}$$

where $m = 1, 2, 3, \ldots, [(2^K - 1)/(K + 1)]$, are disjoint. We note that for each $m = 1, 2, 3, \ldots, [(2^K - 1)/(K + 1)]$ we have

$$2^K + (m - 1)\beta(2^K) + m < 2^{K+1} - \beta(2^{K+1}) + 1.$$

Thus the sets (5.9) are disjoint for $K = 1, 2, 3, \ldots,$ and $m = 1, 2, 3, \ldots, [(2^K - 1)/(K + 1)]$.

Let W be the set of positive integers of the form $\{2^K + (m - 1)K + (m - 1)\}$, where $K = 1, 2, 3, \ldots,$ and $m = 1, 2, 3, \ldots, [(2^K - 1)/(K + 1)]$. Since $\beta(2^K) = K$ it follows that each of the sets (5.9) has the form

$$\{N - \beta(N) + 1, N - \beta(N) + 2, \ldots, N + 1\} \tag{5.10}$$

for some $N \in W$. Now since the sets (5.10) are disjoint for $N \in W$, the sets $E(N)$, $N \in W$, are independent with respect to the "probability" measure λ. That is

$$\lambda\left\{\bigcap_{N \in W_0} E(N)\right\} = \prod_{N \in W_0} \lambda\{E(N)\},$$

for every finite subset $W_0 \subseteq W$.

We define

$$\mathscr{E}^* = \limsup_{\substack{N\to\infty \\ N \in W}} E(N) = \bigcap_{M=1}^{\infty} \bigcup_{\substack{N=M \\ N \in W}}^{\infty} E(N),$$

and observe that

$$\sum_{N \in W} \lambda\{E(N)\} = \sum_{K=2}^{\infty} \left[\frac{2^K - 1}{K+1}\right] 2^{-1-\beta(2^K)} = \sum_{K=2}^{\infty} \left[\frac{2^K - 1}{K+1}\right] 2^{-1-K} = +\infty.$$

Thus by the Borel–Cantelli lemma ([2], Theorem 4.3.1) we have $\lambda\{\mathscr{E}^*\} = 1$. Since $\mathscr{E}^* \subseteq \mathscr{E}$ the lemma is proved. Q.E.D.

THEOREM 5.4. *For almost all t,*

$$\limsup_{y\to\infty} A(y)^{-1} \left|\sum_{x_n \leqslant y} a(n) e^{-itx_n}\right| \geqslant (2\pi)^{-1}.$$

Proof. Since $\{x_n\}$, $n = 1, 2, 3, \ldots$ is a sequence of integers and $A(y) \sim y$, it suffices to show that

$$\limsup_{y\to\infty} y^{-1} \left|\sum_{x_n \leqslant y} a(n) e(\xi x_n)\right| \geqslant (2\pi)^{-1}$$

for almost all ξ in $[0, 1)$, where $e(x) = e^{2\pi i x}$. Next, since

$$\begin{aligned}(2^{N+1} - 1)^{-1} &\left|\sum_{2^N \leqslant x_n \leqslant 2^{N+1}-1} a(n) e(\xi x_n)\right| \\ &\leqslant (2^{N+1} - 1)^{-1} \left|\sum_{x_n \leqslant 2^{N+1}-1} a(n) e(\xi x_n)\right| + (2^N - 1)^{-1} \left|\sum_{x_n \leqslant 2^N - 1} a(n) e(\xi x_n)\right|,\end{aligned}$$

the theorem will follow if we can show that

$$\limsup_{N\to\infty} (2^{N+1} - 1)^{-1} \left|\sum_{2^N \leqslant x_n \leqslant 2^{N+1}-1} a(n) e(\xi x_n)\right| \geqslant \pi^{-1} \tag{5.11}$$

for almost all ξ in $[0, 1)$.

Let $\|x\|$ be the distance from x to the nearest integer. Then for each positive integer N,

$$\begin{aligned}(2^{N+1}-1)^{-1}&\left|\sum_{2^N \leqslant x_n \leqslant 2^{N+1}-1} a(n)e(\xi x_N)\right|\\ &= (2^{N+1}-1)^{-1}2^{N-\beta(N)}\left|\sum_{k=0}^{2^{\beta(N)}-1} e(\xi 2^{N-\beta(N)}k)\right|\\ &\geqslant 2^{-1-\beta(N)}\left|\frac{\sin(\pi\{\xi 2^N\})}{\sin(\pi\{\xi 2^{N-\beta(N)}\})}\right|\\ &\geqslant 2^{-\beta(N)}\left(\frac{\|\xi 2^N\|}{\pi\|\xi 2^{N-\beta(N)}\|}\right).\end{aligned} \tag{5.12}$$

If $\xi \in E(N)$ and (5.8) is the binary expansion of ξ, then

$$\|\xi 2^N\| = \left\|\sum_{J=N+1}^{\infty} \epsilon_J(\xi)2^{N-J}\right\| = \sum_{J=N+2}^{\infty} \epsilon_J(\xi)2^{N-J},$$

while

$$\begin{aligned}2^{\beta(N)}\|\xi 2^{N-\beta(N)}\| &= 2^{\beta(N)}\left\|\sum_{J=N-\beta(N)+1}^{\infty} \epsilon_J(\xi)2^{N-\beta(N)-J}\right\|\\ &= \sum_{J=N+2}^{\infty} \epsilon_J(\xi)2^{N-J}.\end{aligned}$$

Thus if $\xi \in \mathscr{E}$ then it follows from (5.12) that (5.11) holds. By Lemma 5.3, $\lambda(\mathscr{E}) = 1$ and so (5.11) holds for almost all ξ in $[0, 1)$. Q.E.D.

Finally, we remark that since (1.3) and (1.4) are equivalent it follows from Theorem 5.4 that

$$\limsup_{\sigma \to 0+} \sigma\left|\sum_{n=1}^{\infty} a(n)e^{-\sigma x_n - itx_n}\right| > 0$$

for almost all t.

References

1. P. T. Bateman and H. G. Diamond, Asymptotic distribution of Beurling's generalized prime numbers, *in* "Studies in Number Theory" (W. J. Le Veque, ed.), M.A.A. Studies Vol. 6, pp. 152–210, Prentice–Hall, Englewood Cliffs, New Jersey, 1969.
2. K. L. Chung, "A Course in Probability Theory," Harcourt, New York, 1968.
3. W. Feller, On the classical Tauberian theorems, *Arch. Math.* (*Basel*) **14** (1963), 317–322.
4. K. Hoffman, "Banach Spaces of Analytic Functions," Prentice-Hall, Englewood Cliffs, New Jersey 1962.

5. J. KARAMATA, Neuer Beweis und Verallgemeinerung der Tauberschen Sätz, welche die Laplacesche und Stieltjessche Transformation betreffen, *J. Reine Angew. Math.* **164** (1931), 27–39.
6. J. KARAMATA, Neuer Beweis und Verallgemeinerung einiger Tauberian-Sätze, *Math. Z.* **33** (1931), 294–299.
7. J. KOREVAAR, T. VAN AARDENNE-EHRENFEST, AND N. G. DE BRUIJN, A note on slowly oscillating functions, *Nieuw Arch. Wisk.* (2) **23** (1949), 77–86.
8. L. KUIPERS AND H. NIEDERREITER, "Uniform Distribution of Sequences," Wiley, New York, 1974.
9. W. PHILIPP, Über einen Satz von Davenport-Erdös-LeVeque, *Monatsh. Math.* **68** (1964), 52–58.
10. M. TSUJI, On the uniform distribution of numbers mod 1, *J. Math. Soc. Japan* **4** (1952), 313–322.
11. J. VAALER, A Tauberian theorem related to Weyl's criterion, *J. Number Theory* **9** (1977), 71–78.

STUDIES IN PROBABILITY AND ERGODIC THEORY
ADVANCES IN MATHEMATICS SUPPLEMENTARY STUDIES, VOL. 2

Regularities of Distribution†

LEONARD SHAPIRO

Mathematical Sciences Department, North Dakota State University, Fargo, North Dakota

Given an arbitrary sequence $\{\xi_j\}$ in $[0, 1)$ and a subinterval A of $[0, 1)$, define $N(n, A)$ to be the number of elements ξ_j with $0 \leqslant j < n$ which belong to A. We study the set of $\alpha \in [0, 1)$ such that for some interval A, $n\alpha$-$N(n, A)$ is bounded in n. Using methods of topological dynamics we prove that this set is countable, and we describe it for certain $\{\xi_j\}$. This extends previous work of Kesten and Schmidt. We obtain similar results in spaces more general than $[0, 1)$.

0. SUMMARY OF RESULTS

Suppose $\{\xi_j : j = 0, 1, 2, \ldots\}$ is a sequence in the unit interval $[0, 1)$, and A is a subinterval of $[0, 1)$. If the limit

$$\alpha = \lim_{n\to\infty} \frac{1}{n} \sum_{j=0}^{n-1} \chi_A(\xi_j)$$

exists, where χ_A denotes the characteristic function of A, then we might call α the "frequency" with which the sequence $\{\xi_j\}$ visits the set A. Concepts such as ergodicity and uniform distribution deal with this frequency. The ergodic theorem, for example, implies that if T is an ergodic measure preserving transformation on $[0, 1)$ then for almost all points $\xi \in [0, 1)$ the sequence $\xi_j = T^j\xi$ visits A with frequency equal to the measure of A.

Under any reasonable definition of frequency, one would expect that if $\{\xi_j\}$ visits A with frequency α then the quantity

$$\sum_{j=0}^{n-1} \chi_A(\xi_j),$$

which tells how many times $\{\xi_j\}$ has visited A, should be close to $n\alpha$ in some

† Supported by NSF Contract GP32306X-1.

ISBN 0-12-599102-9

sense. The above definition says that their difference, which we denote by D,

$$D(n, \alpha, A) = \left(\sum_{j=0}^{n-1} \chi_A(\xi_j) \right) - n\alpha,$$

is $o(n)$. In this paper we will be concerned with a stronger notion than frequency, namely we ask when $D(n, \alpha, A)$ is bounded in n.

For the purposes of this paper we will call a pair (α, A) *admissible* if $\alpha \in [0, 1)$, A is an interval, and $D(n, \alpha, A)$ is bounded in n.

We will show that, for any sequence $\{\xi_j\}$, there are at most countably many α for which (α, A) is an admissible pair for some A (Theorem 4.1). This result is still valid for sequences in topological spaces more general then $[0, 1)$, when the sequence is minimal, and with A allowed to range over a family including all open and closed sets (Theorems 2.4 and 5.1). We will also construct a sequence $\{\xi_j\}$ which is dense in $[0, 1)$ but for which the only admissible pairs are $(0, \phi)$ and $(1, [0, 1))$ (Example 6.3). When $\{\xi_j\}$ is the Van der Corput sequence, which has appeared in previous studies of admissible pairs, it is possible to identify all the α's which can appear in admissible pairs (Example 6.4).

In most studies of frequency or admissible pairs it is assumed that α is equal to the length (or measure) of A, but we do not need to make this assumption to obtain our results.

1. History of the Subject and Outline of the Paper

Previous investigations of such admissible pairs have fallen into two categories: Finding for specific sequences $\{\xi_j\}$ exactly which pairs are admissible, and determining for arbitrary sequences the size of the set of α's which can appear in an admissible pair.

The only nontrivial sequences for which the set of all admissible pairs have been found are those of the form $\{\xi_j = \langle b + j\xi \rangle : j = 0, 1, \ldots\}$ where ξ is irrational, ξ and $b \in [0, 1)$, and $\langle b + j\xi \rangle$ denotes $b + j\xi$ reduced mod 1. It is well known [8] that such a sequence $\{\xi_j\}$ is uniformly distributed, that is, $\{\xi_j\}$ visits every interval A with frequency $\mu(A)$ (= Lebesgue measure of A). Thus (α, A) is admissible only if $\alpha = \mu(A)$. Hecke ([6], reproduced in [9]) showed by a rather simple argument that (α, A) is admissible if $\alpha = \mu(A) = \langle j\xi \rangle$ for some integer j. Subsequently Kesten, in an ingenious proof using partial fractions [7], showed the converse: (α, A) is admissible only if $\alpha = \mu(A) = \langle j\xi \rangle$ for some integer j. Simpler proofs of Kesten's result are given in [4] and [9]. The proof in [4] is also contained in this paper. These later proofs used an idea originally due to Furstenberg: Consider the transformation T defined on $[0, 1)$ by $Tx = x + \xi \pmod 1$, and notice that $\xi_j = T^j b$ for $j \geqslant 0$. If (α, A) is an admissible pair, $\exp(2\pi i\alpha)$ turns out to be an eigenvalue of the (minimal and ergodic) transformation group determined by T (see definitions below), and the only such eigenvalues are $\exp(2\pi ij\xi)$ for integers j.

Thus this line of investigation showed that for certain special sequences $\{\xi_j\}$, the only α's which could appear in admissible pairs were related to eigenvalues of a certain minimal transformation group.

In [3] Erdös asked whether for every sequence $\{\xi_j\}$ one could find an α, $0 < \alpha < 1$, for which the pair $(\alpha, [0, \alpha))$ was not admissible.

In an impressive series of papers, Schmidt has studied this and related questions. He showed [11] that the α's for which $(\alpha, [0, \alpha))$ is not admissible formed a winning strategy in a modified Banach–Mazur game. (His proof uses generalizations of some integral inequalities due to Roth.) This implies that the set of α's for which $(\alpha, [0, \alpha))$ is admissible form a set of first (Baire) category.

In a subsequent paper [12] Schmidt showed that this set was actually countable. Here he used a rather topological proof, and he used a similar technique in [13] to prove that there are at most countably many α's for which (α, A) is admissible for some interval A of length α (not necessarily $A = [0, \alpha)$). The latter paper also contains a proof of the analogous statement with $[0, 1)$ replaced by the k-dimensional torus (the product of k copies of $[0, 1)$) and intervals A replaced by finite unions of products of k intervals ("boxes"), but Schmidt comments that "the generalization to arbitrary intervals causes considerable difficulties . . . generalization to arbitrary dimension and . . . unions of boxes are easy". It should be noted that Schmidt's results include, and in fact are based on, describing the size of the set E_κ of α's such that for some A with $\alpha =$ measure of A,

$$\left| \sum_{j=0}^{n-1} \chi_A(\xi_j) - n\alpha \right| < \kappa \qquad \text{for all} \quad n,$$

where κ is a fixed positive integer.

Hence this line of investigation shows that for *any* sequence $\{\xi_j\}$, the set of α's which appear in admissible sequences (with the additional requirement that $\alpha =$ measure of A) form a countable set.

In this paper we will unify and and extend these results. We will unify them by showing that for any sequence $\{\xi_j\}$, most α's which occur in admissible pairs are related to eigenvalues of a minimal transformation group constructed from $\{\xi_j\}$, and this will imply that the set of such α's is countable. We will extend the results by removing the restriction that $\alpha =$ measure of A.

Here is a very rough outline, using notation defined in Section 2, of the proof of our main result (Theorem 4.1) that the set of α appearing in admissible pairs is countable: Begin by constructing a compact space X, a homeomorphism T of X, and a function $f: X \to [0, 1)$ such that for some $\xi \in X$, $\{f(T^j\xi)\}$ is the given sequence $\{\xi_j\}$. Assume (α, A) is admissible. Defining the function g from X to $\mathbb{R}$ by $g(x) = \chi_A(f(x)) - \alpha$, it follows from the definition of admissibility that

$$\sum_{j=0}^{n-1} g(T^j\xi)$$

is bounded in n. By a crucial result from topological dynamics (Theorem 2.3) we can find a continuous $h: X \to \mathbb{R}$ such that

$$g(x) = h(Tx) - h(x), \qquad x \in X.$$

Applying $\exp(2\pi i\cdot)$ to each side of this equation and using the definition of g and the fact that $\exp(2\pi i\chi_A(\cdot))$ is always 1, we obtain

$$\exp(-2\pi i\alpha) = \exp(2\pi ih(Tx))/\exp(2\pi ih(x)).$$

This says that $\exp(2\pi i\alpha)$ is an eigenvalue of the transformation group (X, T), and (X, T) has only countably many eigenvalues (Theorem 2.4). Thus there can be only countably many α's.

In following this outline one encounters several technical difficulties. Theorem 2.3 can only be applied if X is minimal, so one must restrict attention to a minimal subset M of X. The function g must also be continuous to apply 2.3, and this necessitates constructing an auxiliary space Y with a map $\pi: Y \to X$, then defining g on Y, where it is continuous. Corresponding to M there is a minimal set N in Y with $\pi(N) = M$. Then, as in the outline, one obtains an eigenfunction $\exp(2\pi ih(x))$ defined on N (Theorem 3.4). But since N and Y change with A, we need the eigenfunction to be defined on M. So we make an assumption (3.1#) on A which assures that $\pi: N \to M$ is one-to-one at one point, and this implies that the eigenfunction can be defined on M. There is one more complication: since $D(N, \alpha, A)$ is bounded for positive n, we must pay special attention to the positive semi-orbits of points in X and Y, in constructing M and N.

The steps described in the previous paragraph are carried out in Section 3, where Theorem 3.1 is proved. The rough outline is then almost complete except that some intervals A may not satisfy 3.1#. In fact Example 6.1 shows that one can find uncountably many intervals which do not satisfy (3.1#). In Section 4 we show that intervals which do not satisfy (3.1#) give rise to only countably many α's, thus proving our main result (4.1). The method of Section 4 is limited in that it is essentially a counting argument and it does not say anything about the α's; it merely shows they form a countable set. We partially remedy this in Section 5 by showing that if $\{\xi_j\}$ is minimal (defined in Section 2) then one can dispense with condition (3.1 #): $\pi: Y \to X$ is always one-to-one at some point. Thus (Theorem 5.1) if $\{\xi_j\}$ is minimal and (α, A) is admissible then $\exp(2\pi i\alpha)$ is an eigenvalue of (X, T). We then use this to identify the set of α which appear in admissible pairs for certain minimal sequences, Examples 6.3 and 6.4.

The proofs in Section 2 and 3 may be easier to follow on first reading if one assumes that $\xi_j = j\xi (\text{mod } 1)$ for some irrational $\xi \in [0, 1)$. Then X is homeomorphic to $[0, 1)$, $T: X \to X$ is the map $x \to x + \xi (\text{mod } 1)$, and $M = X$. The map $\pi: Y \to X$ is one-to-one except on the orbits of the two points in ∂A, where the map is two-to-one. The space Y is minimal, so $N = Y$, and Condition (3.1 #) holds for all intervals A. With these assumptions in force, Sections 2 and 3 become a proof of Kesten's result mentioned above.

Our methods, similar to those in [4] and [14], are borrowed from topological dynamics, but we have tried to make the exposition self-contained. We have chosen to state our theorems in terms of sequences $\{\xi_j\}$ in $[0, 1)$ and intervals A. However, many of our results are valid in a more general setting (typically, for a compact Hausdorff space replacing $[0, 1)$ and A satisfying various topological conditions) and we include remarks to this effect where applicable. We do not obtain the results of Schmidt mentioned above, concerning the sets E_κ and the k-torus. (Added in proof: It is now possible to obtain Schmidt's results concerning the k-torus, by methods similar to those in this paper.)

2. Notions from Topological Dynamics

In this section we collect some standard definitions and results from topological dynamics. Most of these can be found, for example, in [1] and [5], but we include complete proofs here.

We call a pair (X, T) a *transformation group* if X is a Hausdorff space and T is a homeomorphism of X. A subset Y of X is called *invariant* if $TY = Y$; in this case T restricted to Y is a homeomorphism and we have the transformation group (Y, T). If X is compact we say (X, T) is *minimal* if the only closed invariant subsets of X are X and the empty set. If $x \in X$, the *orbit* of x is $\{T^i x : i \in \mathbb{Z}\}$. The set $\{T^i x : i = 0, 1, 2, \ldots\}$, denoted $\mathcal{O}^+(x)$, is called the *positive semi-orbit* of x. At times we will regard $\mathcal{O}^+(x)$ as a sequence: $\{x, Tx, T^2x, \ldots\}$.

2.1 Theorem. *Suppose (X, T) is a transformation group, and $x \in X$. If $\mathcal{O}^+(x)$ has compact closure then the set W of limit points of the sequence $\mathcal{O}^+(x)$*

is nonempty, compact, and invariant (NCI). Any NCI set contains an NCI subset M such that (M, T) is minimal.

Proof. We note that W is usually called the omega-limit set of x. W is closed since it is the set of limit points of a sequence and it is nonempty since $\mathcal{O}^+(x)$ has compact closure. Also W is invariant since TW is the set of limit points of the sequence $\{Tx, T^2x, T^3x, \ldots\}$, which is equal to the set of limit points of $\{x, Tx, T^2x, \ldots\}$. Since W is a closed subset of the (compact) closure of $\mathcal{O}^+$, W is also compact.

Let W be any NCI set and define $\mathcal{I}$ to be the set of *NCI* subsets of W, ordered by inclusion. Since W is compact Zorn's lemma applies to yield a minimal element M of $\mathcal{I}$. Then (M, T) is minimal, for if not then there would be a nonempty closed invariant proper subset of M, contradicting the minimality of M in $\mathcal{I}$.

2.2 Corollary. *A transformation group (X, T) with X compact is minimal iff $\mathcal{O}^+(x)$ is dense in X for every $x \in X$.*

Proof. Suppose $\mathcal{O}^+(x)$ is dense in X for every $x \in X$. If Y is a nonempty closed invariant subset of X then pick some $y \in Y$ and note that $\mathcal{O}^+(y) \subseteq Y$ and $\mathcal{O}^+(y)$ is dense in X, so $X = Y$. Thus (X, T) is minimal. The converse follows from the theorem.

2.3 Theorem (cf. [5, Theorem 14.11] and [2, Theorem I.1]. *Suppose (X, T) is a minimal transformation group, and g is a continuous real-valued function defined on X. Then there exists a continuous real-valued function h on X satisfying*

$$g(x) = h(Tx) - h(x) \qquad \text{for all} \quad x \in X$$

if and only if there is some point $\omega \in X$ such that

$$\sum_{i=0}^{n-1} g(T^i\omega) \tag{*}$$

is bounded for $n \geqslant 0$.

Proof. If $g(x) = h(Tx) - h(x)$ then (*) is a telescoping sum and (*) = $h(T^n\omega) - h(\omega)$. Since X is compact, h is bounded on X. Hence $h(T^n\omega)$ is bounded in n, so (*) is also bounded in n.

Conversely, assume (*) is bounded in n. Define a homeomorphism S of $X \times \mathbb{R}$ ($\mathbb{R}$ = real numbers) onto itself by

$$S(x, \lambda) = (Tx, \lambda + g(x)), \qquad x \in X, \quad \lambda \in \mathbb{R}.$$

One can show by induction that for $n \geqslant 0$,

$$S^n(x, \lambda) = \left(T^n x, \lambda + \sum_{i=0}^{n-1} g(T^i x) \right).$$

By our hypothesis that (*) is bounded for $n \geqslant 0$, the positive semi-orbit

$$\mathcal{O}^+(\omega, 0) = \left\{ \left(T^n \omega, \sum_{i=0}^{n-1} g(T^i \omega) \right) : n = 0, 1, 2, \ldots \right\}$$

has compact closure. Denote by M the set of limit points of $\mathcal{O}^+(\omega, 0)$. By 2.1, M is nonempty, closed, and invariant. (Actually M is minimal, but we do not need that fact.)

We claim that given $x \in X$ there is a unique $\lambda \in \mathbb{R}$ such that $(x, \lambda) \in M$. There is at least one such λ for each x because the set

$$\{x \in X : \text{there is some } \lambda \in \mathbb{R} \text{ with } (x, \lambda) \in M\}$$

is a closed, nonempty, invariant subset of X and since X is minimal it must be all of X.

We will show that uniqueness of λ in two steps. First we show that there is only one λ with $(\omega, \lambda) \in M$, namely $\lambda = 0$. For if not, let $\delta \neq 0$ with $(\omega, \delta) \in M$. Since M consists of limit points of $\mathcal{O}^+(\omega, 0)$, there must be a sequence $n(j) \to \infty$ such that

$$\lim_{j \to \infty} S^{n(j)}(\omega, 0) = (\omega, \delta).$$

By the formula for $S^n(\omega, \lambda)$ given above, this implies that

$$\lim_{j \to \infty} T^{n(j)} \omega = \omega \quad \text{and} \quad \lim_{j \to \infty} \sum_{i=0}^{n(j)-1} g(T^i \omega) = \delta.$$

Now we compute, again using the formula for $S^n(\omega, \lambda)$:

$$\lim_{j \to \infty} S^{n(j)}(\omega, \delta) = \left(\lim_{j \to \infty} T^{n(j)} \omega, \delta + \lim_{j \to \infty} \sum_{i=0}^{n(j)-1} g(T^i \omega) \right) = (\omega, \delta + \delta).$$

Since M is invariant and $(\omega, \delta) \in M$ we know $S^{n(j)}(\omega, \delta) \in M$ for each $n(j)$. Since M is closed, $\lim_{j \to \infty} S^{n(j)}(\omega, \delta) = (\omega, 2\delta)$ is in M. Continuing in this way we can show that for $k \geqslant 0$, $(\omega, k\delta) \in M$, contradicting the compactness of M. This completes the first step.

Second, we pick any $x \in X$ and suppose (x, δ) and (x, δ') are in M for some δ, δ' in $\mathbb{R}$. By the minimality of X there is a sequence $n(j) \to \infty$ such that

$$\lim_{j \to \infty} T^{n(j)} x = \omega.$$

Since M is compact we can refine the sequence $n(j)$ if necessary so the sequences

$$S^{n(j)}(x, \delta) \quad \text{and} \quad S^{n(j)}(x, \delta')$$

both converge to points in M. The first coordinate of $\lim_{j\to\infty} S^{n(j)}(x, \delta)$ must be $\lim_{j\to\infty} T^{n(j)}x = \omega$, and since $(\omega, 0)$ is the only point in M with first coordinate ω, we must have

$$\lim_{j\to\infty} (x, \delta) = (\omega, 0).$$

Considering second coordinates, this implies

$$\lim_{j\to\infty} \left[\delta + \sum_{i=0}^{n(j)-1} g(T^i x) \right] = 0.$$

The same reasoning applied to (x, δ') yields

$$\lim_{j\to\infty} \left[\delta' + \sum_{i=0}^{n(j)-1} g(T^i x) \right] = 0,$$

hence $\delta = \delta'$, as was to be proved.

Now we can define $h(x)$, for any $x \in X$, to be the unique point in $\mathbb{R}$ satisying $(x, h(x)) \in M$, so h is a function from X to $\mathbb{R}$. The graph of h is the compact set M, so h is continuous. Since M is invariant, $S(x, h(x)) = (Tx, h(x) + g(x)) \in M$, so

$$h(Tx) = h(x) + g(x) \qquad \text{for all} \quad x \in X. \qquad \text{Q.E.D.}$$

If (X, T) is a transformation group we call a continuous function H from X into the complex numbers of absolute value 1 an *eigenfunction* of (X, T) if

$$H(Tx) = \gamma H(x) \qquad \text{for all} \quad x \in X,$$

for some complex number γ. Note that $|\gamma| = 1$. We call γ the *eigenvalue* of (X, T) corresponding to H.

2.4 THEOREM. *If X is a compact metric space, there are at most countably many eigenvalues of the transformation group (X, T).*

Proof. We call a set of eigenvalues $\{\exp(2\pi i a_\lambda) | \lambda \in \Lambda\}$ independent if $\{1\} \cup \{a_\lambda | \lambda \in \Lambda\}$ is linearly independent over the rationals. To prove the theorem it suffices to prove that any independent set of eigenvalues of (X, T) is countable. Since X is compact metric, $C(X)$ (=continuous functions on X, with supremum norm) is separable. Thus it suffices to show that if α and β are independent eigenvalues of (X, T), corresponding to eigenfunctions

H_α and H_β, respectively, then

$$\sup_{x \in X} |H_\alpha(x) - H_\beta(x)| \geqslant 1.$$

For any integer k and any x,

$$|H_\alpha(T^k x) - H_\beta(T^k x)| = |\alpha^k H_\alpha(x) - \beta^k H_\beta(x)|.$$

Since α and β are independent the two-dimensional Kronecker theorem [8] applies to show that the set $\{(\alpha^k, \beta^k): k \geqslant 1\}$ is dense in $\{z: |z| = 1\} \times \{z: |z| = 1\}$. Since $|H_\alpha(x)| = |H_\beta(x)| = 1$ we can make the quantity $|\alpha^k H_\alpha(x) - \beta^k H_\beta(x)|$ greater than or equal to one for suitable k.

3. Admissible Pairs and Eigenvalues

Recall the sequence $\{\xi_j\}$ in $[0, 1)$ which is under study, and the definition of admissible pair given in Section 1. In this section we define a transformation group (X, T) related to the sequence $\{\xi_j\}$, and we show that with one extra assumption, every α appearing in an admissible pair (α, A) is related to an eigenvalue of a certain minimal subset of (X, T).

We give $[0, 1)$ the topology of the reals mod 1, so it becomes a compact metric space. The space $X' = \prod_{-\infty}^{\infty} [0, 1)$, with the product topology, is then a compact metric space. If $x \in X'$ we write x_j to denote the jth coordinate of x, so we can consider X' as the set of bisequences $(\ldots, x_{-1}, x_0, x_1, x_2, \ldots)$ with elements in $[0, 1)$. Define the "shift" transformation T on X' by $(Tx)_j = x_{j+1}$, so T shifts the bisequence x one coordinate to the left, and (X, T) is a transformation group.

We extend the sequence $\{\xi_j\}$ to a bisequence by defining ξ_j arbitrarily for $j < 0$, and denote the bisequence $\{\xi_j\}$ in X' by ξ. We denote by X the closure of the orbit of ξ. Since X' is compact we can apply Theorem 2.1 to obtain a compact subset M of X such that (M, T) is minimal and every point of M is a limit point of the sequence $\mathcal{O}^+(\xi)$.

The remainder of this section is devoted to proving:

3.1 Theorem. *If (α, A) is an admissible pair and there is a point $m \in M$ satisfying*

$$m_j \notin \partial A \quad (=\text{boundary of } A) \qquad \text{for all} \quad j, \tag{3.1\#}$$

then there is a continuous nonzero function H on M such that

$$H(Tx) = H(x)\exp(2\pi i\alpha) \qquad \text{for all} \quad x \in M.$$

Remark. This theorem is true with $[0, 1)$ replaced by any compact Hausdorff space I and A replaced by any subset of I. The proof given below goes over verbatim in this more general case.

We now fix $\alpha \in [0, 1]$ and $A \subseteq [0, 1)$, although we need not yet assume that the pair (α, A) is admissible.

Define a point ω in the space $Y' = \prod_{j=-\infty}^{\infty} [0, 1) \times \{0, 1\}$ of bisequences in $[0, 1) \times \{0, 1\}$ by

$$\omega_j = (\xi_j, \chi_A(\xi_j)),$$

where χ_A denotes the characteristic function of A. Let S be the shift in Y', $(Sy)_j = y_{j+1}$, and Y the closure of the orbit of ω. Define $\pi: Y \to X$ to be the projection on the first coordinate, so if $y_j = (x_j, \lambda_j)$ then $(\pi y)_j = x_j$. Define a subset C of Y by

$$C = \{y \in Y : y_0 = (x_0, 1) \text{ for some } x_0\}.$$

Finally, define the continuous function $f: X \to [0, 1)$ by

$$f(x) = x_0.$$

3.2 Proposition. *The objects defined above satisfy*

(a) *the map $\pi: Y \to X$ is continuous and equivariant, i.e., $\pi \circ S = T \circ \pi$,*

(b) *C is an open and closed subset of Y,*

(c) *$S^n\omega \in C$ iff $\xi_n \in A$,*

(d) *If $\chi_{f^{-1}(A)}: X \to \{0, 1\}$ is continuous on the orbit of x then $\pi^{-1}(x)$ is a singleton.*

Proof. Parts (a), (b), and (c) are clear. For part (d), suppose that $\chi_{f^{-1}(A)}$ is continuous at $T^j x$ for all j. Let y be an element of $\pi^{-1}(x)$, so $y_j = (x_j, \lambda_j)$, where each $\lambda_j \in \{0, 1\}$. We claim $\lambda_j = \chi_{f^{-1}(A)}(T^j x)$ for all j, which will show that y is the only element of $\pi^{-1}(x)$. Since $y \in Y$ there is a sequence $k(i)$ with $y = \lim_i S^{k(i)}\omega$. Considering each coordinate separately, this means that for each j,

$$x_j = \lim_i \xi_{j+k(i)} \quad \text{and} \quad \lambda_j = \lim_i \chi_A(\xi_{j+k(i)}).$$

This implies that for each j,

$$T^j x = \lim_i T^{j+k(i)}\xi \quad \text{and} \quad \lambda_j = \lim_i \chi_{f^{-1}(A)}(T^{j+k(i)}\xi).$$

Since $\chi_{f^{-1}(A)}$ is continuous at all $T^j x$, it follows that

$$\lambda_j = \chi_{f^{-1}(A)}(T^j x). \qquad \text{Q.E.D.}$$

Recall the minimal set M, contained in the set of limit points of the sequence $\mathcal{O}^+(\xi)$, which was chosen earlier. Since M is closed and T-invariant, $\pi^{-1}(M)$ is closed (hence compact), and $\pi^{-1}(M)$ is S-invariant, by 3.2a. Let V denote the set of limit points of $\mathcal{O}^+(\omega)$, which is compact and invariant by 2.1. By 2.1 there is a minimal set N in $V \cap \pi^{-1}(M)$.

We now assume that (α, A) is an admissible pair, i.e.,

$$D(n, \alpha, A) = \sum_{j=0}^{n-1} \chi_A(\xi_j) - n\alpha$$

is bounded in n. As remarked above, in this section we need not assume that A is a subinterval. By 3.2c,

$$D(n, \alpha, A) = \sum_{j=0}^{n-1} \chi_C(S^j\omega) - n\alpha.$$

Define the continuous (because of 3.2b) function g on Y by

$$g(y) = \chi_C(y) - \alpha.$$

Then it follows that

$$\sum_{j=0}^{n-1} g(S^j\omega) \text{ is bounded in } n.$$

We cannot yet apply Theorem 2.3 since ω may not be in the minimal set N. We need the following Proposition.

3.3 PROPOSITION. *For every $y \in N$, $\sum_{j=0}^{n-1} g(S^jy)$ is bounded in n.*

Proof. We know that for some B,

$$\left| \sum_{j=0}^{n-1} g(S^j\omega) \right| < B$$

for all n. Then for any $k > 0$,

$$\left| \sum_{j=k}^{k+n-1} g(S^j\omega) \right| < 2B.$$

Let $y \in N$, then since $N \subseteq V$, the limit points of $\mathcal{O}^+(\omega)$, there is a sequence $k(i) \to \infty$ with $\lim_i S^{k(i)}\omega = y$. Thus for each $n > 0$,

$$\left| \sum_{j=0}^{n-1} g(S^jy) \right| = \lim_i \left| \sum_{j=0}^{n-1} g(S^{j+k(i)}\omega) \right|$$

$$= \lim_i \left| \sum_{j=k(i)}^{k(i)+n-1} g(S^j\omega) \right| \leqslant 2B. \qquad \text{Q.E.D.}$$

We apply 2.3 to 3.3: there is a continuous function h on N such that

$$g(y) = h(Sy) - h(y) \qquad \text{for} \quad y \in N.$$

Now recalling that $g(y) = \chi_C(y) - \alpha$, we obtain

$$e^{-2\pi i\alpha} = e^{2\pi i(\chi_C(y)-\alpha)} = e^{2\pi ig(y)} = e^{2\pi ih(Sy)}/e^{2\pi ih(y)},$$

or, setting $K(y) = e^{-2\pi ih(y)}$, we have proved:

3.4 THEOREM. *There is a continuous function $K(y)$ defined on N such that*

$$|K(y)| = 1 \quad \text{and} \quad K(Sy) = \exp(2\pi i\alpha)K(y) \qquad \text{for} \quad y \in N.$$

Now we have almost proved Theorem 3.1, except that the function K is defined on N instead of M. To obtain an eigenfunction defined on M (this is not always possible—see Examples 1 and 2 in Section 6) we must use the assumption (3.1 #) about a point $m \in M$.

3.5 LEMMA. *Suppose there is some $m \in M$ such that $m_j \notin \partial A$ for all j. Then $T^j m \notin \partial f^{-1}(A)$ for all j.*

Proof. Recall that $m_j = f(T^j m)$. Thus if $m_j \notin \partial A$ then $T^j m \notin f^{-1}(\partial A)$. The inclusion $\partial f^{-1}(A) \subseteq f^{-1}(\partial A)$ holds in general for continuous functions f, so $T^j m \notin \partial f^{-1}(A)$.

3.6 PROPOSITION. *Suppose there is some $m \in M$ such that $T^j m \notin \partial f^{-1}(A)$ for all integers j. Then the function K defined in 3.4 is constant on the sets $\pi^{-1}(x)$, $x \in M$.*

Proof. First we note that $\chi_{f^{-1}(A)}$ is continuous at $T^j m$ for all j, for in general the characteristic function of a set is continuous off the boundary of that set. Now apply 3.2d to conclude that there is a unique point, which we denote by $\bar{m}$, in $\pi^{-1}(m)$.

Let $y, y' \in N$ with $\pi(y) = \pi(y')$. We will prove $K(y) = K(y')$. Since N is minimal there is a sequence $k(l)$ such that $\lim_l S^{k(l)}y = \bar{m}$ and $S^{k(l)}y'$ converges. We claim that also $\lim_l S^{k(l)}y' = \bar{m}$ (cf. the 5th paragraph of 2.3 for a similar argument). Using 3.2a we see that

$$m = \pi(\bar{m}) = \pi\left(\lim_l S^{k(l)}y\right) = \lim_l T^{k(l)}\pi(y)$$

and since $\pi(y) = \pi(y')$ we have

$$m = \lim_l T^{k(l)}\pi(y) = \lim_l T^{k(l)}\pi(y') = \pi\left(\lim_l S^{k(l)}y'\right).$$

Now since $\bar{m}$ is the only point satisfying $\pi(\bar{m}) = m$, we have proved our claim that $\lim_l S^{k(l)}y' = \bar{m}$.

By changing to a subsequence if necessary we can assume that $\exp(2\pi ik(l)\alpha)$ converges, say to γ. Then

$$K(m) = \lim_l K(S^{k(l)}y) = \lim_l \exp(2\pi ik(l)\alpha)H(y) = \gamma K(y).$$

Similarly, $K(m) = \gamma K(y')$. Since $\gamma \neq 0$ we conclude that $K(y) = K(y')$. Q.E.D.

To obtain 3.1 from 3.5 and 3.6 we need only define H on M by $H(x) = K(y)$ for any $y \in \pi^{-1}(x)$, so Theorem 3.1 is proved.

4. Countability of α's Appearing in Admissible Pairs

This section is devoted to proving:

4.1 Theorem. *The set of α's such that (α, A) is an admissible pair for some subinterval A of $[0, 1)$, is countable.*

Recall that we have given $[0, 1)$ the (compact) topology of the reals mod 1. By a subinterval of $[0, 1)$ we mean a connected subset A of $[0, 1)$ such that ∂A is two points. Actually, simple extensions of our arguments could prove 4.1 for finite unions of subintervals.

We fix some $m \in M$ and set $E = \{m_j | j \text{ is an integer}\}$. Combining 3.1 and 2.4 yields the following theorem.

4.2 Theorem. *The set of α's such that (α, A) is an admissible pair for some subinterval A with $\partial A \cap E = \varnothing$, is countable.*

4.3 Lemma. *The set of α's such that (α, A) is an admissible pair for some subinterval A with $\partial A \subseteq E$, is countable.*

Proof. The set E is countable, so the set of finite subsets of E is countable. For each finite subset F of E there are finitely many subsets A of $[0, 1)$ such that $\partial A = F$. Thus the set of intervals A with $\partial A \subseteq E$, is countable.

Thus we now need only prove the following to complete the proof of 4.1.

4.4 Theorem. *The set of α's such that (α, A) is an admissible pair for some A such that $E \cap \partial A$ is a single point, is countable.*

Proof. Since E is countable we can fix $a \in E$ and prove only that the set of α such that (α, A) is admissible for some A with $E \cap \partial A = \{a\}$, is countable. For each $b \in [0, 1)$ there are only eight intervals with $\partial A = \{a, b\}$; arguments for all eight are identical except in notation so we will focus on intervals of the form $[a, b]$. We define

$$F_a = \{\alpha : (\alpha, [a, b]) \text{ is admissible for some } b \in [0, 1) - E\}.$$

The proof will be complete if we show that F_a is countable. (Note that if $a > b$ then $[a, b]$ denotes $\{\alpha \mid 0 \leqslant x \leqslant b \text{ or } a \leqslant x < 1\}$, while if $a < b$ then $[a, b]$ has the usual meaning.) We pause for a lemma.

4.5 LEMMA. *Suppose* $A_1 = A_2 \cup A_3$ *and* $A_2 \cap A_3 = \varnothing$ *for subsets* A_i *of* $[0, 1)$, *and* (α_1, A_1), (α_2, A_2) *are admissible. Then* $(\alpha_1 - \alpha_2, A_3)$ *is admissible.*

Proof. Since $\chi_{A_1} = \chi_{A_2} + \chi_{A_3}$ we have

$$\left[\sum_{j=0}^{n-1} \chi_{A_1}(\xi_j) - n\alpha_1\right] = \left[\sum_{j=0}^{n-1} \chi_{A_2}(\xi_j) - n\alpha_2\right] + \left[\sum_{j=0}^{n-1} \chi_{A_3}(\xi_j) - n(\alpha_1 - \alpha_2)\right].$$

Since the first two bracketed expressions are bounded in n, so is the third. By the definition of admissible, $(\alpha_1 - \alpha_2, A_3)$ is admissible.

Now to return to the proof of 4.4. Suppose there is some $c \in [0, 1) - E$ and a β such that $(\beta, [a, c])$ is admissible (if there is no such c then F_a is empty and we are done). We claim that

$$F_a \subseteq (\beta + \Gamma) \cup (\beta - \Gamma),$$

where Γ is the set of α's such that (α, A) is admissible for some A with $\partial A \cap E = \varnothing$. For suppose $b \notin E$ and $(\alpha, [a, b])$ is admissible. First consider the case $b \in [a, c]$, and set $A_1 = [a, c]$, $A_2 = [a, b]$ and $A_3 = (b, c]$. Then (β, A_1) and (α, A_2) are admissible and the Lemma applies to show that $(\beta - \alpha, A_3)$ is admissible. But $\partial A_3 = \{b, c\}$ so $E \cap \partial A_3 = \varnothing$. Therefore $\beta - \alpha \in \Gamma$, so $\alpha \in \beta - \Gamma$. For the case $c \in [a, b]$, uses the same argument with $A_1 = [a, b]$, $A_2 = [a, c]$, $A_3 = (c, b]$ to get $\alpha - \beta \in \Gamma$, so $\alpha \in \beta + \Gamma$. We have proved $F_a \subseteq (\beta + \Gamma) \cup (\beta - \Gamma)$. By 4.2 Γ is countable, so $(\beta + \Gamma) \cup (\beta - \Gamma)$ is countable. The proof is completed.

5. REGULARITIES OF DISTRIBUTION FOR MINIMAL SEQUENCES

Suppose $\xi = \{\xi_j : j \in \mathbb{Z}\}$ is a bisequence in $[0, 1)$. As in Section 3, ξ is a point in the compact space of bisequences $X' = \prod_{-\infty}^{\infty} [0, 1)$, and T is the shift in X', $(Tx)_j = x_{j+1}$. We let X denote the closure of the orbit of ξ. If (X, T) is minimal (defined in Section 2) then we call ξ a *minimal bisequence.* If $\{\xi_j : j = 0, 1, \ldots\}$ is a sequence we call it a *minimal sequence* if we can define

ξ_j for $j < 0$ in such a way that the bisequence $\{\xi_j : j \in \mathbb{Z}\}$ is a minimal bisequence.

We will be concerned with sets A satisfying the condition

$$\begin{array}{l} \textit{If } B \subseteq \partial A \textit{ and } B \textit{ satisfies } B \subseteq cls(B \cap A) \cap cls(B \cap A^c) \\ (\textit{where } A^c \textit{ denotes the complement of } A), \textit{ then } B \textit{ is the empty set.} \end{array} \tag{5.1*}$$

Note that any closed or open set A satisfies (5.1*), since then either $B \cap A$ or $B \cap A^c$ is empty if $B \subseteq \partial A$. Any subinterval A, (or more generally any set A such that ∂A is finite) will satisfy (5.1*), since $cls(B \cap A) = B \cap A$ and $cls(B \cap A^c) = B \cap A^c$ whenever B is finite.

The proof of the following theorem parallels that of Theorem 3.1. As in that theorem, the results are valid for $[0, 1)$ replaced by any compact Hausdorff space I and A replaced by any subset of I.

5.1 THEOREM. *Suppose* $\{\xi_j : j = 0, 1, \ldots\}$ *is a minimal sequence. If* A *satisfies condition* (5.1*) (*in particular, if* A *is an interval*) *and* (α, A) *is an admissible pair, then* $\exp(2\pi i\alpha)$ *is an eigenvalue of* (X, T).

Proof. The proof proceeds exactly as the proof of 3.1 in Section 3 except for two changes: First, define ξ_j for $j < 0$ in such a way that $\xi = \{\xi_j : j \in \mathbb{Z}\}$ is a minimal bisequence. Note that this implies that X is minimal, so $M = X$. Second, replace Lemma 3.5 by:

5.2 LEMMA. *Suppose* ξ *is a minimal bisequence and* A *satisfies the condition* (5.1*). *Then there is some* $m \in M$ *such that* $T^j m \notin \partial f^{-1}(A)$ *for all* j.

Proof. First we claim that $\partial f^{-1}(A)$ has no interior. For suppose U were a nonempty open set with $U \subseteq \partial f^{-1}(A)$, and set $B = f(U)$. Let $x \in B$, and suppose V is an open set with $x \in V$. Then $f^{-1}(V) \cap U$ is a nonempty and open subset of $\partial f^{-1}(A)$. By the definition of boundary there are points x', x'' in $f^{-1}(V) \cap U$ with

$$x' \in f^{-1}(A), \qquad x'' \in M - f^{-1}(A) = f^{-1}([0, 1) - A).$$

Thus both $f(x')$ and $f(x'')$ are in $f(f^{-1}(V) \cap U) \subseteq V \cap B$, and $f(x') \in A$, $f(x'') \in [0, 1) - A$. We have shown that V has nonempty intersection with both $A \cap B$ and $B \cap ([0, 1) - A)$, so

$$x \in \mathrm{cls}(B \cap A) \cap \mathrm{cls}(B \cap ([0, 1) - A)).$$

Since A satisfies (5.1*) we conclude that $B = \varnothing$. Thus $\partial f^{-1}(A)$ has no interior and it follows from Baire's theorem that

$$\bigcup_{j=-\infty}^{\infty} T^j \partial f^{-1}(A)$$

is not the whole of M. Any point m outside this set will satisfy $T^j m \notin \partial f^{-1}(A)$ for all j. This completes the proof of the Lemma, and of Theorem 5.1.

As we have mentioned, the proof of 5.1 is valid for spaces I more general than $[0, 1)$, in particular for $I = \prod_{j=1}^{k} [0, 1)$, the k-torus. In that setting, any subset A of I which is a product of k intervals (i.e., a "box") satisfies (5.1*). For suppose $B \subseteq \partial A$ and

$$B \subseteq \text{cls}(B \cap A) \cap \text{cls}(B \cap A^c).$$

Then

$$B \subseteq \text{cls}(\partial A \cap A) \cap \text{cls}(\partial A \cap A^c).$$

Also, ∂A is a k-1 dimensional simplicial complex and $cls(\partial A \cap A) \cap cls(\partial A \cap A^c)$ is a complex of dimension at most k-2, so B is contained in a k-2 dimensional complex. Proceeding by induction we see that B must be finite (a 0 dimensional complex), and so B is empty as noted at the beginning of this section. This generalization of 5.1, together with 2.4, shows that for minimal sequences $\{\xi_j\}$ we can recover the results of Schmidt [13], again without requiring that α = volume of A, namely we can prove:

5.3 THEOREM. *Suppose $\{\xi_j\}$ is a minimal sequence in the k-torus. Then there are only countably many $\alpha \in [0, 1]$ for which there exists a finite union A of products of k subintervals of $[0, 1)$, such that*

$$\sum_{j=0}^{n-1} \chi_A(\xi_j) - n\alpha$$

is bounded in n. Moreover, for each such α, $\exp(2\pi i\alpha)$ is an eigenvalue of the minimal transformation group generated by $\{\xi_j\}$.

6. EXAMPLES

In this section we present examples of sequences $\{\xi_j\}$ which illustrate some interesting features and provide counterexamples to natural conjectures.

EXAMPLE 6.1. In Section 3 we constructed a minimal transformation group M and showed that some α's appearing in admissible pairs are related to eigenvalues of M. To see that not all such α's arise in this way, define

$$\xi_j = \tfrac{1}{2} + (-\tfrac{1}{2})^{j+1}$$

for $j = 0, 1, 2, \ldots$. Then (using Section 3's notation) there is only one minimal set M which consists of limit points in X of $\mathcal{O}^+(\xi)$, and that is the fixed point m with $m_j = \frac{1}{2}$ for all j. Such a minimal transformation group has no nontrivial eigenvalues, yet for any λ with $0 \leqslant \lambda < \frac{1}{2}$, the pair $(\frac{1}{2}, [\lambda, \frac{1}{2}])$ is admissible for the sequence $\{\xi_j\}$ we have defined. We commented in Section 1 that there may be uncountably many intervals A appearing in admissible pairs, which do not satisfy (3.1#). In fact the intervals $[\lambda, \frac{1}{2}]$ with $0 \leqslant \lambda < \frac{1}{2}$ do not satisfy (3.1#). For if they did, then by Theorem 3.1, $\exp(\pi i) = -1$ would be an eigenvalue of (M, T), which is impossible since M is one point.

EXAMPLE 6.2. We have restricted our attention to *subintervals* A in admissible pairs (α, A) for good reason: the result about countability of the set of α appearing in admissible pairs is false for general subsets of $[0, 1)$, even for sets which are closures of their interior (although, as we have mentioned, it is true for sets A with finite boundaries). In this example, for any α we will construct a set A_α which is the closure of its interior and for which (α, A_α) is admissible. Define $\xi_j = 1/(j+2)$ for $j = 0, 1, 2, \ldots$. If $\alpha \in (0, 1)$ then ([6], reproduced in [9])

$$\sum_{j=0}^{n-1} \chi_{[0,\alpha)}(\langle j\alpha \rangle) - n\alpha$$

is bounded in n, where $\langle j\alpha \rangle$ denotes $j\alpha$ reduced mod 1, so $\langle j\alpha \rangle \in [0, 1)$. One can easily find a set $A_\alpha \subseteq [0, 1)$ which is the closure of its interior and such that $\langle j\alpha \rangle \in [0, \alpha)$ iff $\xi_j \in A_\alpha$, e.g.,

$$A_\alpha = \{0\} \cup \{x : |x - \xi_j| \leqslant \tfrac{1}{2}|\xi_j - \xi_{j+1}| \quad \text{for some } j \text{ such that } \langle j\alpha \rangle \in [0, \alpha)\}.$$

Then

$$\chi_{[0,\alpha)}(\langle j\alpha \rangle) = \chi_{A_\alpha}(\xi_j),$$

so

$$\sum_{j=0}^{n-1} \chi_{A_\alpha}(\xi_j) - n\alpha$$

is bounded in n. Therefore (α, A_α) is admissible, for arbitrary $\alpha \in [0, 1]$. Of course the results of Section 3 do not apply in this example since again there is only one candidate for M, namely the fixed point $m_j = 0$ for all j, and the hypothesis of 3.1 is not satisfied for any of the sets A_α since $0 \in \partial A_\alpha$.

One might expect that if the set $\{\xi_j\}$ were dense in $[0, 1)$ then there would be at least one $\alpha \in (0, 1)$ and some A for which (α, A) is an admissible pair. This is not true, as our next example shows.

EXAMPLE 6.3. Let $\beta, \gamma \in (0, 1)$ with γ irrational and $\beta \neq n\gamma \pmod 1$ for all integers n. Let $\langle \cdot \rangle$ denote $\cdot$ mod 1. Define the sequence $\{\xi_j\}$ by taking the sequence $\{\langle k\gamma\rangle : k = 0, 1, \ldots\}$ and doubling every term $\langle k\gamma\rangle$ with $\langle k\gamma\rangle \in [0, \beta)$. To be specific, define $\{\xi_j\}$ inductively by $\xi_0 = \xi_1 = 0$, and for $j \geqslant 1$, $\xi_{j+1} = \langle \xi_j + \gamma\rangle$ unless $\xi_j \neq \xi_{j-1}$ and $\xi_j \in [0, \beta)$ in which case $\xi_{j+1} = \xi_j$. Since the set of numbers $\{\xi_j\}$ is equal to the set $\{\langle k\gamma\rangle : k = 0, 1, \ldots\}$, and the latter is dense in $[0, 1)$ since γ is irrational, $\{\xi_j\}$ is dense in $[0, 1)$. It is shown in [10] that the minimal set M generated by $\{\xi_j\}$ has no nontrivial (i.e., $\neq 1$) eigenvalues.[1] Thus 5.1 shows that there are no admissible pairs (α, A) with $\alpha \neq 0, 1$. In fact there are no admissible pairs other than $(0, \varnothing)$ and $(1, [0, 1))$. For if (α, A) is another admissible pair then since M is ergodic (even uniquely ergodic [10]) we must have α = length of A. Since $A \neq \varnothing$ or $[0, 1)$ we must have $0 < \alpha < 1$. By 5.1 $\exp(2\pi i\alpha)$ is a nontrivial eigenvalue of M, contradicting the quoted result.

EXAMPLE 6.4. This is the Van der Corput sequence discussed in [12]. We denote by D the group of 2-adic integers, so the elements in D can be viewed as formal power series

$$\sum_{k=0}^{\infty} \lambda_k 2^k, \qquad \lambda_k \in \{0, 1\},$$

and addition is the usual power-series addition. We can also view D as the sequences $\{\lambda_k\}$ with elements in $\{0, 1\}$. We define a metric on D by $d(\{\lambda_k\}, \{\lambda_k'\}) = 1/(n+1)$, where n is the least integer with $\lambda_n \neq \lambda_n'$. Then D is a compact group. Note that the nonnegative integers $\mathbb{Z}_+$ are contained in D: they correspond to the sequences which are eventually 0. Also, $\mathbb{Z}_+$ is dense in D.

Define the continuous function $f: D \to [0, 1)$ by

$$f(\{\lambda_k\}) = \sum_{k=0}^{\infty} \lambda_k 2^{-k-1} \pmod 1.$$

Then we can define the Van der Corput sequence by

$$\xi_j = f(j) \qquad \text{for} \quad j \in \mathbb{Z}_+.$$

We claim that $\{\xi_j\}$ is a minimal sequence. Extend $\{\xi_j\}$ to a bisequence ξ by defining $\xi_{-j} = f(j^{-1})$ for $j > 0$, where j^{-1} represents the inverse in D of $j \in \mathbb{Z}_+$. Let X denote the orbit closure of ξ under the shift T, and define the maps $S: D \to D$ and $\varphi: D \to X$ by

$$Sy = y + 1, \qquad (\varphi x)_j = f(S^j x).$$

[1] M is a homomorphic image of the symbolic flow χ'' defined in the last paragraph of page 383 of [10]. Since χ'' has no nontrivial eigenvalues, neither does M.

Then (D, S) is a transformation group. Also φ is a homeomorphism. (It is surjective since $\varphi(0) = \xi$; continuity and injectivity are easy to prove, and since D is compact this is sufficient). Finally, φ clearly satisfies $\varphi \circ S = T \circ \varphi$. Thus the transformation group (X, T) is isomorphic to (D, S). The transformation group (D, S) is minimal by 2.2 since $\mathcal{O}^+(0) = \{S^n 0 \mid n = 0, 1, \ldots\} = \mathbb{Z}_+$ is dense in D and $\mathcal{O}^+(y) = y + \mathcal{O}^+(0)$ for any $y \in D$. Thus (X, T) is minimal and $\{\xi_j\}$ is a minimal sequence.

In order to make use of 5.1, let us determine the eigenvalues of (D, S). Since D is a compact group and S is addition by an element of D, an eigenfunction of (D, S) is a group homomorphism into the circle group (of complex numbers with norm one). Since D is the inverse limit of cyclic groups of order 2^k for $k = 1, 2, \ldots$, the only homomorphic images of D are D itself (which is not a closed subgroup of the circle group) and the cyclic groups. The latter correspond to eigenvalues $\exp(2\pi i l/2^k)$ for $l, k \geqslant 0$.

Since the only eigenvalues of (D, S) are $\exp(2\pi i l/2^k)$ for $l, k \geqslant 0$ Theorem 5.1 implies that (α, A) is an admissible pair for $\{\xi_j\}$ only if $\alpha = l/2^k$ for some $k \geqslant 0$, $0 \leqslant l \leqslant 2^k$. In [12] Schmidt shows the converse: each of these α's does appear in an admissible pair.

EXAMPLE 6.5. One might conjecture from the above examples that the converse of 5.1 is true. Namely, if $\exp(2\pi i\alpha)$ is an eigenvalue for (M, T) and $0 \leqslant \alpha \leqslant 1$ then there is some interval (or perhaps just a set) A such that (α, A) is admissible. This is not true even for minimal sequences $\{\xi_j\}$. Define $\{\xi_j\}$ to be the sequence of period four $\{0, 0, 0, \frac{1}{2}, \ldots\}$. Then the minimal set M which it generates consists of four points, and the shift $T: M \to M$ permutes them cyclically. $\mathrm{Exp}(2\pi i\frac{1}{2})$ is an eigenvalue for (M, T) but the only admissible pairs for $\{\xi_j\}$ are those of the form $(\frac{1}{4}, A_1)$, $(\frac{3}{4}, A_2)$, $(0, A_3)$ and $(1, A_4)$ where $A_1 \cap \{0, \frac{1}{2}\} = \{\frac{1}{2}\}$, $A_2 \cap \{0, \frac{1}{2}\} = \{0\}$, $A_3 \cap \{0, \frac{1}{2}\} = \varnothing$, and $A_4 \cap \{0, \frac{1}{2}\} = \{0, \frac{1}{2}\}$. These statements remain valid even if we allow the A's to range over arbitrary subsets of $[0, 1)$.

REFERENCES

1. R. ELLIS, "Lectures in Topological Dynamics," Benjamin, New York, 1969; MR **42**, No. 2463.
2. R. ELLIS, Cocycles in topological dynamics, unpublished notes.
3. P. ERDÖS, Problems and results on diophantine approximations, *Compositio Math.* **16** (1964), 52–56.
4. H. FURSTENBERG, H. KEYNES, AND L. SHAPIRO, Prime flows in topological dynamics, *Israel J. Math.* **14** (1973), 26–38.
5. W. GOTTSCHALK AND G. HEDLUND, "Topological Dynamics," Amer. Math. Soc. Colloq. Publ. Vol. 36, American Mathematical Society Providence, R. I., 1955.

6. E. HECKE, Analytische Funktionen und die Verteilung von Zahlen mod Eins, *Abh. Math. Sem. Hamburg* **1** (1922), 54–76.
7. H. KESTEN, On a conjecture of Erdös and Szüsz related to uniform distribution mod 1, *Acta Arith.* **12** (1966, 1967), 193–212.
8. L. KUIPERS AND H. NIEDERREITER, "Uniform Distribution of Sequences," Wiley, New York, 1974.
9. K. PETERSEN, On a series of cosecants related to a problem in ergodic theory, *Compositio Math.* **26** (1973), 313–317.
10. K. PETERSEN AND L. SHAPIRO, Induced flows, *Trans. Amer. Math. Soc.* **177** (1973), 375–390.
11. W. SCHMIDT, Irregularities of distribution, *Quart. J. Math. Oxford Ser.* **19** (1958), 181–191.
12. W. SCHMIDT, Irregularities of distribution VI, *Compositio Math.* **24** (1972), 63–74.
13. W. SCHMIDT, Irregularities of distribution VIII, *Trans. Amer. Math. Soc.* **198** (1974), 1–22.
14. L. SHAPIRO, Irregularities of distribution in dynamical systems, *in* "Recent Advances in Topological Dynamics" (Anatole Beck, ed.), pp. 249–252, Springer-Verlag, Berlin and New York, 1973.

AMS (MOS) 1970 subject classifications: primary, 10K30; secondary, 10K05, 57E05.

STUDIES IN PROBABILITY AND ERGODIC THEORY
ADVANCES IN MATHEMATICS SUPPLEMENTARY STUDIES, VOL. 2

Strong Liftings on Topological Measured Spaces†

RICHARD J. MAHER

Department of Mathematics, Loyola University of Chicago, Chicago, Illinois

This paper extends the concepts of lifting and strong lifting to the setting of general topological measured spaces. Compatible and nearly compatible liftings are introduced as analogs of strong and almost strong liftings, and various theorems concerning the existence of compatible and nearly compatible liftings are given. The theorem of Bichteler is established for nearly compatible liftings in topological measured spaces. Finally, two examples of completely regular, nonlocally compact topological measured spaces possessing compatible, and therefore, strong, liftings are discussed.

1

Let X be a separated topological space and let μ be a Radon measure on X. We then say that the couple (X, μ) is a *topological measured space.* It is well known that much of the theory of integration valid for locally compact X can be extended to this more general setting [2, 4, 8, 13]. In this note we discuss some aspects of the theory of lifting for topological measured spaces (see [11]). In Section 2 we present the basic notations and results needed in the following sections. In Section 3 we introduce compatible and nearly compatible liftings as analogs, respectively, of strong and almost strong liftings. We also discuss other possible extensions. In Section 4 we show that the set of all measures μ on X such that $(X, |\mu|)$ admits a nearly compatible lifting is a band (in the set of all Radon measures on X). Finally, in Section 5 we give two examples of completely regular, nonmetrizable, nonlocally compact spaces which admit compatible (and therefore strong) liftings.

2

If X is a separated topological space, we denote by $\mathscr{M}(X)$ the set of all Radon measures on X and by $\mathscr{M}_+(X)$ the cone of all positive Radon measures

† Supported by the Loyola University Committee on Research.

ISBN 0-12-599102-9

on X. For (X, μ) a topological measured space, we write $M^\infty(X, \mu)$ (or M^∞, when no confusion is possible) for the space of all bounded, μ-measurable functions on X to R, and $C^b(X)$ for the set of all bounded, continuous functions on X to R. For a given topological measured space (X, μ), we write $\mathscr{B} = \{B : \phi_B \in M^\infty\}$ and $\mathscr{B}_0 = \{B : B \in \mathscr{B}, \mu(B) < \infty\}$. Let us recall that a family $(K_i)_{i \in I} \subseteq \mathscr{B}_0$ is called a *concassage* for (X, μ) if $(K_i)_{i \in I}$ is a locally denumerable family of pairwise disjoint compact subsets of X such that the set $X\text{-}\bigcup_{i \in I} K_i$ is locally μ-negligible. There always exists a concassage for (X, μ) [4, Section 1.8].

If (X, μ) is a topological measured space, then a mapping $\rho : M^\infty \to M^\infty$ is said to be a *lifting* of M^∞ if

(i) $\rho(f) \equiv f$
(ii) $f \equiv g$ implies $\rho(f) = \rho(g)$
(iii) $\rho(1) = 1$ and $\rho(0) = 0$
(iv) $f \geqslant 0$ implies $\rho(f) \geqslant 0$
(v) $\rho(af + bg) = a\rho(f) + b\rho(g)$
(vi) $\rho(fg) = \rho(f)\rho(g)$

Equivalently, a mapping $\rho : \mathscr{B} \to \mathscr{B}$ is said to be a *lifting* of $\mathscr{B}$ if

(i′) $\rho(B) \equiv B$
(ii′) $A \equiv B$ implies $\rho(A) = \rho(B)$
(iii′) $\rho(X) = X$ and $\rho(\varnothing) = \varnothing$
(iv′) $\rho(A \cap B) = \rho(A) \cap \rho(B)$
(v′) $\rho(A \cup B) = \rho(A) \cup \rho(B)$

In either of the above cases, we say that there exists a *lifting* of (X, μ).

The topological measured space (X, μ) is said to be *strictly localizable* [8] if there exists a partition $\mathscr{C}$ of X consisting of nonnegligible, integrable sets such that $\sup\{\tilde{K} : K \in \mathscr{C}\} = \tilde{X}$. The equivalence of the strict localizability of (X, μ) and the existence of a lifting of (X, μ) is well known [8, 12]. If $\mathscr{C} = (K_i)_{i \in I}$ is a concassage for (X, μ), then for any $B \in \mathscr{B}_0$, there exists a countable family $\mathscr{C}_B \subseteq \mathscr{C}$ such that $B\text{-}\bigcup_{K \in \mathscr{C}_B} K$ is locally negligible (those $K \in \mathscr{C}$ which intersect B, for example). By [8, Chapter 1, Proposition 3], we conclude that (X, μ) is strictly localizable and that every topological measured space possesses a lifting.

Let (X, μ) be a topological measured space with $\mu \in \mathscr{M}_+(X)$. A lifting ρ of $M^\infty(X, \mu)$ is said to be *strong* if

(vii) $\rho(f) = f$ for $f \in C^b(X)$

Note that we must have supp $\mu = X$ to discuss strong liftings. To deal with the case when supp $\mu \neq X$, we say that a lifting of M^∞ is *almost strong* if

(viii) $\rho(f)|\mathscr{C}_A = f|\mathscr{C}_A$ for $f \in C^B(X)$,
A a fixed, locally μ-negligible set

Many results concerning (almost) strong liftings are known in the case when X is locally compact [1, 8, 9, 11]. However, it is not known if every topological measured space (X, μ), with X locally compact, possesses an (almost) strong lifting.

Note. As a matter of convenience, we often say that a couple (X, μ) has the (almost) strong lifting property whenever there exists an (almost) strong lifting of $M^\infty(X, \mu)$.

3

If X is a locally compact space and if $\mu \in \mathscr{M}_+(X)$, then the existence of a strong lifting of $M^\infty(X, \mu)$ is equivalent to the existence of a lifting ρ of $\mathscr{B}$ such that $U \subseteq \rho(U)$ for all $U \subseteq X$, U open [8, Chapter 8, Theorem 1]. This equivalence leads to the following definition [7].

DEFINITION 1. Let (X, μ) be a topological measured space with $\mu \in \mathscr{M}_+(X)$. A lifting ρ of M^∞ is said to be *compatible* with the topology of X if $\phi_U \leqslant \phi_{\rho(U)}$ for all $U \subseteq X$, U open.

Clearly, we must have supp $\mu = X$ to discuss the concept of a lifting of $M^\infty(X, \mu)$ compatible with the topology of X.

Using the idea of a topology associated with a lifting and [8, Chapter 5, Theorem 1], it is readily seen that any lifting compatible with the topology of X is a strong lifting. The converse is valid *if* X is assumed *completely regular*; otherwise there may not be "enough" continuous functions [7].

Note. The preceding analog of strong liftings is not the only one possible. Schwartz [13] has introduced the concept of a *lifting of local type.* More specifically, if (X, μ) is a topological measured space and if $\mu \in \mathscr{M}_+(X)$, then a lifting ρ of M^∞ is said to be of local type if for any open set $U \subseteq X$ and any $f \in M^\infty$ with $f(x) \geqslant 0$ almost everywhere on U, we have $\rho(f)(x) \geqslant 0$ everywhere on U.

It is not difficult to see that the following two assertions are equivalent for a given topological measured space (X, μ), where $\mu \in \mathscr{M}_+(X)$ with $\operatorname{supp} \mu = X$:

(a) ρ is a lifting of M^∞ compatible with the topology of X;
(b) ρ is a lifting of local type.

(a) *implies* (b): If ρ is compatible with the topology of X and if $f \in M^\infty$, and $U \subseteq X$, U open, are such that $f \geqslant 0$ almost everywhere on U, then $f\phi_U \geqslant 0$ almost everywhere on X. Then $\rho(f\phi_U) \geqslant 0$ everywhere on X. If $x \in \rho(U)$, we conclude that $\rho(f\phi_U)(x) = \rho(f)\phi_{\rho(U)}(x) = \rho(f)(x) \geqslant 0$. Since $\phi_U \leqslant \phi_{\rho(U)}$, we conclude that $\rho(f) \geqslant 0$ everywhere on U.

(b) *implies* (a): Let ρ be a lifting of local type. Since $\phi_U(x) = 1$ for $x \in U$, we have $\phi_U(x) - c \geqslant 0$ for $x \in U$ and any $c \in (0, 1)$. Since ρ is of local type, we have $\rho(\phi_U - c)(x) \geqslant 0$ for all $x \in U$; i.e., $\phi_{\rho(U)}(x) - c \geqslant 0$ for all $x \in U$, so that $\phi_{\rho(U)}(x) \geqslant c > 0$ for $x \in U$. We conclude that $\phi_U \leqslant \phi_{\rho(U)}$, so that ρ is a lifting compatible with the topology of X.

To deal with the case when $\operatorname{supp} \mu \neq X$, we introduce the following definition:

DEFINITION 2. Let (X, μ) be a topological measured space, with $\mu \in \mathcal{M}_+(X)$. A lifting ρ of M^∞ is said to be *nearly compatible* with the topology of X if $\phi_{U|\mathscr{C}A} \leqslant \phi_{\rho(U)|\mathscr{C}A}$ (equivalently, $\phi_{U\cap\mathscr{C}A} \leqslant \phi_{\rho(U)\cap\mathscr{C}A}$) for all $U \subseteq X$ open and $A \subseteq X$ a fixed, locally μ-negligible set.

Note. To simplify the terminology in Definitions 1 and 2, we shall say that (X, μ) (or M^∞) possesses a (nearly) compatible lifting or that there exists a (nearly) compatible lifting for (X, μ) (or M^∞).

A useful relation between compatible and nearly compatible liftings is indicated in the following theorem.

THEOREM 1. *Let (X, μ) be a topological measured space, with $\mu \in \mathcal{M}_+(X)$ and* $\operatorname{supp} \mu = X_1 \subseteq X$. *Then the following assertions are equivalent*:

(1) *There exists a nearly compatible lifting for (X, μ).*
(2) *There exists a compatible lifting for (X_1, μ_{X_1}).*

Proof. (1) *implies* (2) Let ρ be a lifting of M^∞ such that $\phi_{U|\mathscr{C}A} \leqslant \phi_{\rho(U)|\mathscr{C}A}$ for $U \subseteq X$ open and $A \subseteq X$ a fixed, locally μ-negligible set. Let $f \in M^\infty(X_1, \mu_{X_1})$ and define $f' \in M^\infty(X, \mu)$ by $f'(x) = f(x)$ for $x \in X_1$, while $f'(x) = 0$ for $x \notin X_1$. For $f \in M^\infty(X_1, \mu_{X_1})$, define $\rho'(f)(x) = \rho(f')(x)$ for $x \in X_1 \cap \mathscr{C}A$; note that we thus define a lifting ρ^* of $(X_1 \cap \mathscr{C}A, \mu_{X_1\cap\mathscr{C}A})$. If $x \in X_1 \cap A$, let $\mathscr{U}_x$ be an ultrafilter on $X_1 \cap \mathscr{C}A$ converging to x. Define $\rho'(f)(x) = \lim_{\mathscr{U}_x} \rho^*(f)$, where ρ^* is defined above. It is clear that ρ' is a lifting of $M^\infty(X_1, \mu_{X_1})$ and that $\phi_U(x) \leqslant \phi_{\rho(U)}(x)$ for $x \in X_1 \cap \mathscr{C}A$ and $U \subseteq X_1$ open. If $x \in X_1 \cap A$, then $\phi_U(x) \leqslant \lim_{\mathscr{U}_x} \phi_{\rho(U)} \leqslant \lim_{\mathscr{U}_x} \rho^*(\phi_U) = \rho(\phi_U)(x) = \phi_{\rho(U)}(x)$. We conclude that $\phi_U(x) \leqslant \phi_{\rho(U)}(x)$ for all $x \in X_1$.

(2) *implies* (1) Let χ be any character of $L^\infty(X, \mu)$ and let ρ be a compatible lifting of $M^\infty(X_1, \mu_{X_1})$. Define $\rho' : M^\infty(X, \mu) \to M^\infty(X, \mu)$ by $\rho'(f)(x) = \rho(f|_{X_1})(x)$ for $x \in X_1$ and by $\rho'(f)(x) = \chi(f)$ for $x \in \mathscr{C}X_1$. It is clear that ρ'

is a lifting of $M^\infty(X, \mu)$ and that if $U \subseteq X$ is open, we have $\phi_{U|X_1} \leqslant \phi_{\rho'(U)|X_1}$. We conclude that ρ' is a nearly compatible lifting for (X, μ).

Remarks. (1) The proof that (1) implies (2) in Theorem 1 was suggested by a proof in [13].

(2) The existence of the ultrafilter $\mathscr{U}_x$ in Theorem 1 is not difficult to deduce. Suppose that $\mathscr{V}(x)$ is the neighborhood filter in X_1 of $x \in X_1 \cap A$. Since $\operatorname{supp} \mu_{X_1} = X_1$, $X_1 \cap \mathscr{C}A$ is dense in X_1, so $V \cap (X_1 \cap \mathscr{C}A) \neq \phi$ for all $V \in \mathscr{V}(x)$. Then $\{V \cap (X_1 \cap \mathscr{C}A) : V \in \mathscr{V}(x)\}$ is a filter basis on $X_1 \cap \mathscr{C}A$; any ultrafilter finer than this filter basis suffices.

(3) If $\operatorname{supp} \mu = X$, then the existence of a nearly compatible lifting for (X, μ) implies the existence of a compatible lifting for (X, μ).

The following theorem indicates that the question of the existence of a nearly compatible lifting for a topological measured space (X, μ) can be reduced to the case where X is compact (see also [5]).

THEOREM 2. *Let (X, μ) be a topological measured space, with $\mu \in \mathscr{M}_+(X)$.*

(1) *If $K \subseteq X$ is a compact set and if there exists a nearly compatible lifting for (X, μ), then there exists a nearly compatible lifting for (K, μ_K).*

(2) *If $(K_i)_{i \in I}$ is a concassage for (X, μ) and if (K_i, μ_{K_i}) possesses a nearly compatible lifting for each $i \in I$, then (X, μ) possesses a nearly compatible lifting.*

Proof. (1) Let $f \in M^\infty(K, \mu_K)$ and define $f' \in M^\infty(X, \mu)$ by the equality $f'(x) = f(x)$ for $x \in K$, while $f'(x) = 0$ if $x \notin K$. Let ρ be a nearly compatible lifting for (X, μ) with $A \subseteq X$ locally μ-negligible and $\phi_{U|\mathscr{C}A} \leqslant \phi_{\rho(U)|\mathscr{C}A}$ for all $U \subseteq X$, U open. Let χ be a character of $L^\infty(K, \mu_K)$. If $f \in M^\infty(K, \mu_K)$, define $\rho'(f)$ by $\rho'(f)(x) = \rho(f')(x)$ for $x \in K \cap \rho(K)$, while $\rho'(f)(x) = \chi(f)$ if $x \in K - \rho(K)$. It is immediate that ρ' is a lifting of $M^\infty(K, \mu_K)$. Furthermore, if $V \subseteq K$ is open, with $V = U \cap K$, where U is open in X, then $\phi_V(x) \leqslant \phi_{\rho'(V)}(x)$ for $x \notin K \cap (A \cup (K - \rho(K)))$, so that ρ' is a nearly compatible lifting for (K, μ_K).

(2) For each $i \in I$, let ρ_i be a nearly compatible lifting for (K_i, μ_{K_i}), with $\phi_{V_i|\mathscr{C}A_i} \leqslant \phi_{\rho_i(V_i)|\mathscr{C}A_i}$ for V_i open in K_i. Let χ be a character of $L^\infty(X, \mu)$. If $f \in M^\infty(X, \mu)$, define $\rho(f)(x) = \rho_i(f|_{K_i})(x)$ for $x \in K_i$, while $\rho(f)(x) = \chi(f)$ if $x \in X - \bigcup_{i \in I} K_i$. It is immediate that ρ is a lifting of $M^\infty(X, \mu)$. Furthermore, if $U \subseteq X$ is open, then $U_i = U \cap K_i$ is open in K_i and $\phi_{U_i|\mathscr{C}A_i} \leqslant \phi_{\rho(U_i)|\mathscr{C}A_i}$. We conclude that $\phi_{U|\mathscr{C}A} \leqslant \phi_{\rho(U)|\mathscr{C}A}$, where $A = (\bigcup_{i \in I} A_i) \cup (X - \bigcup_{i \in I} K_i)$, so that ρ is a nearly compatible lifting for (X, μ).

Note. The proof of part (1) of Theorem 2 remains valid if K is only assumed to be closed.

We can now establish the following useful theorem.

THEOREM 3. *Let (X, μ) be a topological measured space, with $\mu \in \mathcal{M}_+(X)$.*
(1) *Let ρ be a nearly compatible lifting of (X, μ). Then ρ is an almost strong lifting of (X, μ).*
(2) *Let ρ be an almost strong lifting of (X, μ) and assume that X is completely regular. Then ρ is a nearly compatible lifting of (X, μ).*

Proof. (1) Let ρ be a nearly compatible lifting of (X, μ), with $A \subseteq X$ a fixed, locally μ-negligible set such that $\phi_{U|\mathscr{C}A} \leqslant \phi_{\rho(U)|\mathscr{C}A}$ for all $U \subseteq X$, U open. Let $X_1 = \operatorname{supp} \mu$. Then (using the notation of Theorem 2, part (1)), there exists a nearly compatible lifting ρ' of (X_1, μ_{X_1}) such that if $f \in M^\infty(X_1, \mu_{X_1})$, then $\rho'(f)(x) = \rho(f')(x)$ for $x \in X_1$. Further, if $V \subseteq X_1$ is open in X_1, we have $\phi_V(x) \leqslant \phi_{\rho'(V)}(x)$ for $x \notin X_1 \cap A = A_1$. By Theorem 1, there exists a compatible lifting ρ'' of (X_1, μ_{X_1}) such that $\rho''(g)(x) = \rho'(g)(x)$ for $g \in M^\infty(X_1, \mu_{X_1})$ and $x \notin A_1$. By [8, Chapter 5, Theorem 1], we deduce that ρ'' is a strong lifting of (X_1, μ_{X_1}). By [8, Chapter 8, Proposition 12] (which does not use locally compact in the proof), there exists an almost strong lifting ρ^* for (X, μ) such that $\rho^*(f)|_{X_1} = \rho''(f|_{X_1})$. We conclude that strong lifting ρ^* for (X, μ) such that $\rho^*(f)|_{X_1} = \rho''(f|_{X_1})$. We conclude $\rho(f)(x) = \rho^*(f)(x)$ for $f \in M^\infty(X, \mu)$ and $x \notin A_1 \cup \mathscr{C}X_1$. Since $\mu(A_1 \cup \mathscr{C}X_1) = 0$ and since $\rho^*(f)(x) = f(x)$ for $f \in C^b(X)$ and $x \in \mathscr{C}X_1$, we find that $\rho(f)(x) = f(x)$ for $f \in C^b(X)$ and $x \notin A_1 \cup \mathscr{C}X_1$. We conclude that ρ is an almost strong lifting of (X, μ).

(2) Let ρ be an almost strong lifting for (X, μ), with $\rho(f)(x) = f(x)$ for $f \in C^b(X)$ and $x \in \mathscr{C}A$, where A is a fixed, locally μ-negligible set. Let $X_1 = \operatorname{supp} \mu$. By [9, Theorem 1], there exists an almost strong lifting ρ' of (X_1, μ_{X_1}) such that (again our notation is that of Theorem 2, part (1)) $\rho'(f)(x) = \rho(f')(x)$ for $x \in X_1$ and $\rho'(f)(x) = f(x)$ for $f \in C^b(X_1)$ and $x \notin X_1 \cap A = A_1$. By the remarks following [8, Chapter 8, Definition 5], there exists a strong lifting ρ'' for (X_1, μ_{X_1}) such that $\rho''(g)(x) = \rho'(g)(x)$ for $g \in M^\infty(X_1, \mu_{X_1})$ and $x \notin A_1$. Since X_1 is completely regular, ρ'' is a compatible lifting for (X_1, μ_{X_1}) [8, Chapter 5, Theorem 3]. By Theorem 1, there exists a nearly compatible lifting ρ^* of (X, μ) such that $\rho^*(f)|_{X_1} = \rho''(f|_{X_1})$. We conclude that $\rho(f)(x) = \rho^*(f)(x)$ for $f \in M^\infty(X, \mu)$ and $x \notin A_1 \cup \mathscr{C}X_1$. Since $\mu(A_1 \cup \mathscr{C}X_1) = 0$ and since $\phi_U(x) \leqslant \phi_{\rho^*(U)}(x)$ for $U \subseteq X$, U open, and $x \notin \mathscr{C}X_1$, we find that $\phi_U(x) \leqslant \phi_{\rho(U)}(x)$ for $U \subseteq X$, U open, and $x \notin A_1 \cup \mathscr{C}X_1$. We conclude that ρ is a nearly compatible lifting for (X, μ).

COROLLARY. *Let (X, μ) be a topological measured space, with $\mu \in \mathcal{M}_+(X)$. Then the following assertions are equivalent.*

(1) *ρ is a nearly compatible lifting of (X, μ).*
(2) *ρ is an almost strong lifting of (X, μ).*

Proof. By Theorem 3 we know that (1) implies (2). To see that (2) implies (1), let ρ be an almost strong lifting of (X, μ) and let $(K_i)_{i \in I}$ be a concassage for (X, μ). Let ρ_i be the lifting of (K_i, μ_{K_i}) constructed as in Theorem 2, part (1); it is immediate that ρ_i is an almost strong lifting of (K_i, μ_{K_i}). By Theorem 3, part (2), ρ_i is a nearly compatible lifting of (K_i, μ_{K_i}). Let ρ_1 denote the nearly compatible lifting of (X, μ) constructed in Theorem 2, part (2). If $f \in M^\infty(X, \mu)$, we have $\rho(f)(x) = \rho_1(f)(x)$ for $x \notin (X\text{-}\bigcup K_i) \cup (\bigcup (K_i - \rho(K_i)))$, so that ρ is also a nearly compatible lifting of (X, μ).

The preceding results enable us to establish the existence of a nearly compatible lifting in the following important case.

THEOREM 4. *Let (X, μ) be a topological measured space, with $\mu \in \mathscr{M}_+(X)$. Assume that there exists a concassage of metrizable sets for (X, μ). Then every lifting of (X, μ) is nearly compatible.*

Proof. Let ρ be a lifting of (X, μ) and let $(K_i)_{i \in I}$ be a concassage of metrizable sets for (X, μ). Let ρ_i denote the lifting of (K_i, μ_{K_i}) constructed as in Theorem 2, part (1). Since K_i is compact and metrizable, ρ_i is an almost strong lifting of (K_i, μ_{K_i}) [8, Chapter 8, Theorem 8] and therefore, by Theorem 3, part (2), ρ_i is a nearly compatible lifting of (K_i, μ_{K_i}). Let ρ_1 be the nearly compatible lifting of (X, μ) constructed in Theorem 2, part (2). Since for $f \in M^\infty(X, \mu)$ we have $\rho(f)(x) = \rho_1(f)(x)$ for $x \notin (X\text{-}\bigcup K_i) \cup (\bigcup (K_i - \rho(K_i)))$, we deduce that ρ is a nearly compatible lifting for (X, μ).

COROLLARY 1. *If (X, μ) is a metrizable topological measured space, then every lifting of (X, μ) is a nearly compatible lifting.*

COROLLARY 2. *If (X, μ) is a souslin topological measured space, then every lifting of (X, μ) is a nearly compatible lifting.*

We indicate yet another class of topological measured spaces for which every lifting is a nearly compatible lifting in the next theorem (see also [6]).

THEOREM 5. *Let (X, μ) be a topological measured space and assume that the topology of X is 2nd countable. Then every lifting of (X, μ) is a nearly compatible lifting.*

Proof. If X is regular, then the conclusion follows immediately from Theorem 4. If X is not regular, let $(U_n)_{n=1}^\infty$ be a countable basis for the topology of X and let ρ be any lifting of (X, μ). For each $n \in N$, define the set A_n by the equality $A_n = U_n \Delta \rho(U_n)$ and define $A = \bigcup_{n=1}^{\infty} A_n$. It is immediate that $U_n \cap \mathscr{C}A \subseteq \rho(U_n)$ for all n, so that $\phi_{U_n \cap \mathscr{C}A} \leqslant \phi_{\rho(U_n)} = \phi_{\rho(U_n \cap \mathscr{C}A)}$. Then, if $U \subseteq X$ is open in X, with $U = \bigcup_{j=1}^{\infty} U_{n_j}$, then $\phi_{U \cap \mathscr{C}A} \leqslant \phi_{\rho(U \cap \mathscr{C}A)} = \phi_{\rho(U)}$ [8, Chapter 3, Theorem 3]. Then $\phi_{U \cap \mathscr{C}A} \leqslant \phi_{\rho(U) \cap \mathscr{C}A}$, so we conclude that ρ is a nearly compatible lifting for (X, μ).

Notes. (1) Schwartz [13] has indicated that the concept of lifting of local type can be extended to the case when supp $\mu \neq X$. To do so, he modifies condition (iii) in the definition of lifting to condition

(iii*) $\rho(1) = 1$ on supp μ.

It is not difficult to see that a lifting of local type on (X, μ) induces a lifting of local type on (X_1, μ_{X_1}), where $X_1 =$ supp μ. The notes following Definition 1 and Theorem 1 then show that the existence of a lifting of local type implies the existence of a nearly compatible lifting. Conversely, the note following Theorem 2, Theorem 1, the note following Definition 1, and the analog to Theorem 1 for liftings of local type show that the existence of a nearly compatible lifting implies the existence of a lifting of local type.

(2) Bichteler [3] has introduced the concept of a lifting *strong on a set*. More specifically, if (X, μ) is a topological measured space and if $Y \subseteq X$, then a lifting ρ of (X, μ) is said to be strong on Y if $\rho(f)(x) = f(x)$ for $f \in C^b(X)$ and $x \in Y$. It is clear from the definition that we must have supp $\mu_Y = Y$; however, we need not have $\mu(\mathscr{C}Y) = 0$; in fact, if $\mu(\mathscr{C}Y) = 0$ we have the definition of an almost strong lifting. Using Theorems 1, 2, 3, and the note following Theorem 1, it is not difficult to show that the existence of a nearly compatible lifting for (X, μ) implies the existence of a lifting of (X, μ) strong on any set Y for which supp $\mu_Y = Y$. Converse implications are not readily available, since, as we have noted, $\mu(\mathscr{C}Y)$ need not be zero.

(3) If the existence of a strong lifting is established in the case of (X, μ) a compact topological measured space, then there exists a nearly compatible lifting for *any* topological measured space (X, μ) (Theorems 2 and 3, and the corollary to Theorem 3). However, even though a strong lifting need not be a compatible lifting, it will be a nearly compatible lifting for (X, μ). By Theorem 1 we can deduce that a compatible lifting for (X, μ) does exist.

4

In [1], Bichteler established that if X is a locally compact topological space, then the set of all $\mu \in \mathscr{M}(X)$ such that $(X, |\mu|)$ admits an almost strong lifting is a band in $\mathscr{M}(X)$. In [9], C. Ionescu Tulcea and the author established the same result by a technique different from that of Bichteler. In [5], Cohn extended a number of these results to completely regular and, in certain cases, to general separated spaces. In this section, we show that the set of all $\mu \in \mathcal{M}(X)$ such that $(X, |\mu|)$ admits a nearly compatible lifting forms a band in $\mathcal{M}(X)$.

We begin with the important theorem.

THEOREM 6. *Let X be a separated topological space and let μ and ν be two positive measures on X. Assume that ν is absolutely continuous with respect to μ. If (X, μ) possesses a nearly compatible lifting, then (X, ν) also possesses a nearly compatible lifting.*

Proof. It is not difficult to show that there exists a concassage $(K_i)_{i \in I}$ for (X, μ) such that μ_{K_i} and ν_{K_i} are equivalent for each $i \in I$ ([4, Section 1.8], and [9]). By Theorem 2, (K_I, μ_{K_i}) and therefore (K_i, ν_{K_i}) possess nearly compatible liftings. We conclude, again by Theorem 2, that (X, ν) admits a nearly compatible lifting.

THEOREM 7. *Let X be a separated topological space and let μ and ν be positive measures on X such that both (X, μ) and (X, ν) admit nearly compatible liftings. Then $(X, \mu + \nu)$ admits a nearly compatible lifting.*

Proof. Write $\mu = \mu_a + \mu_s$, where μ_a is absolutely continuous with respect to μ and μ_s is singular with respect to μ. Then $\mu + \nu = (\mu_a + \nu) + \mu_s$. By Theorem 6, $(X, \mu_a + \nu)$ admits a nearly compatible lifting. Likewise, we deduce the existence of a nearly compatible lifting for (X, μ_s).

Let X' and X'' be two universally measurable, disjoint sets with union X such that $(\mu_a + \nu)$ is concentrated on X' while μ_s is concentrated on X''. Let $(K_i)_{i \in I}$ be a concassage for $(X, \mu + \nu)$ such that for each $i \in I$, either $K_i \subseteq X'$ or $K_i \subseteq X''$, and define $I' = \{i : K_i \subseteq X'\}$ and $I'' = \{i : K_i \subseteq X''\}$. Since for $i \in I'$, $(\mu + \nu)_{K_i} = (\mu_a + \nu)_{K_i}$ while for $i \in I''$, $(\mu + \nu)_{K_i} = (\mu_s)_{K_i}$; we conclude by Theorem 6 that $(K_i, (\mu + \nu)_{K_i})$ admits a nearly compatible lifting for each $i \in I$. We conclude by Theorem 2 that $(X, (\mu + \nu))$ possesses a nearly compatible lifting.

COROLLARY. *Let X, μ, and ν, be as in the statement of Theorem 7. Then both $(X, \inf\{\mu, \nu\})$ and $(X, \sup\{\mu, \nu\})$ admit nearly compatible liftings.*

THEOREM 8. *Let X be a separated topological space and let A be a family of positive measures on X which is bounded above. Let $\lambda = \sup A$. If (X, μ) possesses a nearly compatible lifting for each $\mu \in A$, then (X, λ) admits a nearly compatible lifting.*

Proof. By the corollary to Theorem 7, we may assume that A is filtering. Furthermore, if $K \subseteq X$ is compact, then $\lambda_K = \sup\{\mu_K : \mu \in A\}$, so that by Theorem 2 it is enough to establish Theorem 8 in the case where X is compact. However, the theorem is true in this case [9, Theorem 4 and Theorem 1], so we conclude that (X, λ) admits a nearly compatible lifting.

We say that the topological measured space (X, μ) where $\mu \in \mathcal{M}(X)$, possesses a nearly compatible lifting if $(X, |\mu|)$ admits a nearly compatible lifting. We then obtain the following.

THEOREM 9. *The set of all measures $\mu \in \mathcal{M}(X)$ such that $(X, |\mu|)$ admits a nearly compatible lifting forms a band in $\mathcal{M}(X)$.*

Proof. The proof follows immediately from Theorems 6, 7, and 8.

5

In this section we give two examples of topological measured spaces (X, μ) which admit compatible liftings (thus we assume supp $\mu = X$). The topology of X will be completely regular, but neither metrizable nor locally compact. Recall that in this setting, the existence of a compatible lifting is equivalent with the existence of a strong lifting, so we will be also giving two examples of completely regular topological measured spaces which possess the strong lifting property.

Before discussing these two examples, the following preliminaries are needed:

(1) Let (X, μ) be a topological measured space and let $\mathcal{T}$ denote a completely regular topology on X. Let ρ be a lifting of (X, μ) and denote by $\mathcal{T}\rho$ the topology on X generated by taking sets of the form $\{\rho(B) : B \in \mathcal{B}\}$ for a basis. Then the following assertions are equivalent [8, Chapter 5, Theorem 3]:

(a) ρ is a compatible lifting for (X, μ);
(b) $\mathcal{T} \subseteq \mathcal{T}\rho$.

(2) Let R^m denote m dimensional Euclidean space endowed with the usual topology $\mathcal{T}_m$. Denote by μ_m the Lebesgue measure on R^m. Recall that a function $f : R^m \to R$ is *right-continuous* on R^m if

$$\lim_{x \to c \;\; x \geqslant c} f(x) = f(c).$$

If we denote by $C_+{}^b(R^m)$ the space of all bounded, right-continuous functions on R^m to R, then the following assertion is valid [10]:

(c) There exists a lifting ρ_m of $M^\infty(R^m, \mu_m)$ such that $\rho_m(f) = f$ for all $f \in C_+{}^b(R^m)$.

The lifting whose existence is asserted in (c) was used in [10] to construct examples of compact, nonmetrizable spaces possessing the strong lifting property. We now use this lifting and the equivalence of (a) and (b) to construct the above mentioned examples.

EXAMPLES. (1) Let $\mathscr{T}_1$ denote the usual topology on R and let $\mathscr{T}^1$ denote the topology on R generated by taking sets of the form $[a, b)$ as a basis. Then $\mathscr{T}_1 \subseteq \mathscr{T}^1$ and $\mathscr{T}^1$ is completely regular (in fact normal), nonmetrizable, and not locally compact. Let $\mathscr{T}\rho_1$ be the topology on R generated by the lifting in (c). Since $\rho_1([a, b)) = [a, b)$ (the characteristic function of $[a, b)$ is right-continuous), we have $\mathscr{T}^1 \subseteq \mathscr{T}\rho_1$. Since supp $\mu_1 = R$ when R has the topology $\mathscr{T}^1$, we conclude that the topological measured space (R, μ_1), where R is endowed with the topology $\mathscr{T}^1$, possesses a compatible (and therefore strong) lifting.

(2) Let $\mathscr{T}_2$ denote the usual topology on R^2 and let $\mathscr{T}^2$ denote the topology on R^2 generated by taking sets of the form $[a, b) \times [c, d)$ as a basis. Then $\mathscr{T}_2 \subseteq \mathscr{T}^2$ and $\mathscr{T}^2$ is completely regular, nonnormal, nonmetrizable, and not locally compact. Let $\mathscr{T}\rho_2$ be the topology on R^2 generated by the lifting in (c). Since $\rho_2([a, b) \times [c, d)) = [a, b) \times [c, d)$ (again, the characteristic function is right continuous), we have $\mathscr{T}^2 \subseteq \mathscr{T}\rho_2$. Since supp $\mu_2 = R^2$ when R^2 is given the topology $\mathscr{T}^2$, we conclude that the topological measured space (R^2, μ_2), where R^2 is endowed with the topology $\mathscr{T}^2$, possesses a compatible lifting. Therefore, there exists a strong lifting for (R^2, μ_2) in this setting.

REFERENCES

1. K. BICHTELER, On the strong lifting property, *Illinois J. Math.* **16** (1972), 370–380.
2. K. BICHTELER, Integration theory, "Lecture Notes in Mathematics," Vol. 315, Springer-Verlag, Berlin and New York, 1973.
3. K. BICHTELER, A weak existence theorem and weak permanence properties for strong liftings, *Manuscripta Math.* **8** (1973), 1–10.
4. N. BOURBAKI, "Integration," Ch. IX, Hermann, Paris, 1969.
5. D. COHN, Topics in liftings and stochastic processes, Ph.D. Thesis, Harvard Univ., 1975.
6. S. GRAFF, On the existence of strong liftings in second countable topological spaces, *Pacific J. Math.* **58**, No. 2 (1975), 419–426.
7. A. IONESCU TULCEA, Liftings compatible with topologies, *Bull. Soc. Math. Grèce*, **8** (1967), 116–126.

8. A. Ionescu Tulcea and C. Ionescu Tulcea, Topics in the theory of lifting, "Ergebnisse der Mathematik und ihrer Grenzgebiete," Vol. 48, Springer-Verlag, Berlin and New York, 1969.
9. C. Ionescu Tulcea and R. Maher, A note on almost strong liftings, *Ann. Inst. Fourier* (*Grenoble*), **21**, No. 2 (1971), 35–41.
10. R. Maher, A note on strong liftings, *J. Math. Anal. Appl.* **29** (1970), 633–639.
11. R. Maher, Strong liftings and borel liftings, *Advances in Math.* **13**, No. 1 (1974), 55–72.
12. R. Ryan, Representative sets and direct sums, *Proc. Amer. Math. Soc.* **15** (1964), 387–390.
13. L. Schwartz, "Radon Measures," Oxford University Press, London and New York, 1973.

AMS (MOS) 1970 subject classification: 46G15.

STUDIES IN PROBABILITY AND ERGODIC THEORY
ADVANCES IN MATHEMATICS SUPPLEMENTARY STUDIES, VOL. 2

Mixing Transformations in an Infinite Measure Space†

NATHANIEL A. FRIEDMAN

Department of Mathematics, State University of New York, Albany, New York

1. INTRODUCTION

During the past nine years a great deal of progress has been made on the isomorphism problem based on the fundamental work of D. S. Ornstein. In particular, it follows from [5, 10] that mixing Markov shifts in a probability space are isomorphic if they have the same finite or infinite entropy.

Mixing Markov shifts in a probability space can be constructed from mixing Markov chains with a finite invariant measure or by the stacking construction as in [3, 4, 12]. The purpose of this paper is to consider the isomorphism problem for analogous examples of mixing transformations in an infinite measure space, as formulated by Krickeberg [9] and Papengelou [11]. The definition of mixing below will also imply the mixing condition of Krengel and Sucheston [8].

The general result that entropy characterizes a mixing Markov shift in the case of a probability space does not carry over to the infinite measure case. We will show below that a convenient invariant referred to as the spreading rate can be utilized to distinguish transformations with the same entropy that are not isomorphic in the sense of Krickeberg (referred to as strong isomorphism below). We only know that the spreading rate is a strong isomorphism invariant. It is an open question whether the spreading rate is actually an isomorphism invariant, where isomorphism is defined in the general sense (see Section 2). However, results in [1] imply that the spreading rate is a general isomorphism invariant in the special case of mixing transformations generated by random walks.

In a subsequent paper [2] it will be shown that the spreading rate can be utilized to construct another invariant that is a certain subclass $W(T)$ of the class of weakly wandering sequences for T [6]. The class $W(T)$ is a

† Research partially supported by National Science Foundation Grant MPS71–02757 AO3.

ISBN 0–12–599102–9

general isomorphism invariant. In [2] we utilize the invariant $W(T)$ to show that certain transformations with the same entropy are not isomorphic.

Preliminary definitions and notation are given in Section 2. The stacking construction for mixing transformations on the real line is presented in Section 3. The stacking construction is particularly well-suited for slowing down the spreading rate. In Section 4 we discuss transformations generated by Markov chains with infinite invariant measure.

2. Preliminaries

Let $(X, \mathscr{B}, \mu)$ be a measure space with $\mu(X) = \infty$ and let $\mathscr{R}$ be a ring of sets of finite measure such that $\mathscr{R}$ generates $\mathscr{B}$. Let T be an invertible measure preserving transformation on X such that $T\mathscr{R} = \mathscr{R}$. T is said to be *mixing with respect to* $\mathscr{R}$ (*$\mathscr{R}$-mixing*) if there exists a sequence of positive numbers ρ_n such that

$$\lim_{n\to\infty} \rho_n \mu(T^n A \cap B) = \mu(A)\mu(B) \tag{2.1}$$

for A and B in $\mathscr{R}$. Since ρ_n does not depend on A and B, we see that (2.1) implies

$$\lim_{n\to\infty} \frac{\mu(T^n A \cap B)}{\mu(T^n C \cap D)} = \frac{\mu(A)\mu(B)}{\mu(C)\mu(D)}, \tag{2.2}$$

for A, B, C, and D in $\mathscr{R}$.

In [9, 11] transformations were obtained from Markov chains with infinite invariant measure. The space X was also a topological space, and the ring $\mathscr{R}$ consisted of a certain class of sets of finite measure that had boundary measure zero. The transformations T were required to be continuous a.e. We will extend the stacking construction in [3, 4, 12] in order to obtain mixing transformations on $X = (-\infty, \infty)$ with Lebesgue measure μ. The transformations constructed by stacking will be piecewise linear and hence continuous a.e. A set A will be in $\mathscr{R}$ if A is contained in a finite interval and has boundary measure zero. In order to verify that T is $\mathscr{R}$-mixing, it will suffice to show that T is mixing for sets in a class $\mathscr{C}$ of finite unions of finite intervals that generate $\mathscr{B}$ [9]. The appropriate class $\mathscr{C}$ will arise from the stacking construction.

Since $T\mathscr{R} = \mathscr{R}$, it is not unnatural to require that isomorphisms preserve the ring structure, as in [9]. In general, given measure spaces $(X_i, \mathscr{B}_i, \mu_i)$, $i = 1, 2$, we say that S mapping X_1 into X_2 is an *isomorphism* if S is measure preserving and invertible a.e. If T_i is $\mathscr{R}_i$-mixing, $i = 1, 2$, then we will say T_1 and T_2 are *strongly isomorphic* if $T_1 = S^{-1}T_2S$ and $\mathscr{R}_2 = S\mathscr{R}_1$ for some isomorphism S. T_1 and T_2 are *isomorphic* if $T_1 = S^{-1}T_2S$ for some isomorphism S.

A strong isomorphism invariant referred to as the spreading rate is defined as follows. Let T be $\mathscr{R}$-mixing and let C be in $\mathscr{R}$, with $\mu(C) > 0$. A sequence $\rho(T, C)$ is defined by

$$\rho(T, C)_n = \mu(C)^2/\mu(T^nC \cap C). \tag{2.3}$$

Equations (2.2) and (2.3) imply

$$\lim_{n\to\infty} \rho(T, C)_n\mu(T^nA \cap B) = \mu(A)\mu(B), \tag{2.4}$$

for A and B in $\mathscr{R}$. We assume there exist sets B in $\mathscr{R}$ with arbitrarily large measure; hence (2.4) implies

$$\lim_{n\to\infty} \rho(T, C)_n = \infty, \qquad C \text{ in } \mathscr{R}. \tag{2.5}$$

Sequences ρ and σ are *equivalent* ($\rho \sim \sigma$) if $\lim_{n\to\infty} \rho_n/\sigma_n = 1$. It follows that $\rho(T, C) \sim \rho(T, D)$ for C and D in $\mathscr{R}$. Let $\rho(T, \mathscr{R})$ denote the equivalence class of sequences $\rho(T, C)$ for C in $\mathscr{R}$. We refer to $\rho(T, \mathscr{R})$ as the *spreading rate of T with respect to $\mathscr{R}$.*

If T_1 and T_2 are strongly isomorphic, then $\rho(T_1, \mathscr{R}_1) = \rho(T_2, \mathscr{R}_2)$. Actually the spreading rates are equal as long as $\mathscr{R}_2 \cap S\mathscr{R}_1$ contains one set of positive measure.

We will now present some preliminary definitions for the stacking construction. All intervals will be assumed left-closed and right-open. A *column* C is a finite *ordered* set of disjoint intervals of the same length. A column C with h intervals is denoted by

$$C = (I_0, I_1, I_2, \ldots, I_{h-1}) = (I_i : 0 \leqslant i < h). \tag{2.6}$$

C in (2.6) has *height* $h(C) = h$, *base* $\underline{C} = I_0$, *top* $\bar{C} = I_{h-1}$, and *width* $w(C) = \mu(\underline{C})$. We refer to I_i as the ith *level* in C, $0 \leqslant i < h$. We also let C denote the union of the levels in C and therefore $\mu(C) = hw(C)$.

A column of height h can be pictured as h rungs in the ladder, where the base is the bottom interval and the top is the top interval.

Let T_C denote the transformation that maps I_i linearly onto I_{i+1}, $0 \leqslant i < h - 2$. The domain of T_C is $C - \bar{C}$ and the range of T_C is $C - \underline{C}$. We can denote $\underline{C}$ as

$$C = (T_C{}^i\underline{C} : 0 \leqslant i < h).$$

Let I be a left-closed right-open subinterval of $\underline{C}$. A subcolumn C_I of C is defined by

$$C_I = (T_C{}^iI : 0 \leqslant i < h).$$

Given r, $0 < r < 1$, we let rC denote a subcolumn of C with $w(rC) = rw(C)$.

Columns C_1 and C_2 are *disjoint* columns if C_1 and C_2 are disjoint as sets. If C_1 and C_2 are disjoint and have the same width, then we can define C_1C_2 to be the single column consisting of the intervals in C_1 followed by the intervals in C_2. Intuitively, C_1C_2 is obtained by placing C_2 above C_1.

If $C = C_1C_2$, then $\underline{C} = \underline{C}_1$, $\bar{C} = \bar{C}_2$, and T_C extends T_{C_1} to map $\bar{C}_1$ linearly onto $\underline{C}_2$.

A *tower* G is an *ordered* set of disjoint columns, which may have different heights and widths. The *base* of G is $\underline{G} = \bigcup \underline{C}$, the *top* of G is $\bar{G} = \bigcup \bar{C}$, and the *width* of G is $w(G) = \mu(\underline{G})$, where unions extend over the columns C in G.

The transformation T_G is defined as T_C on $C - \bar{C}$ for C in G. Thus T_G is defined on $G - \bar{G}$ and T_G^{-1} is defined on $G - \underline{G}$.

Towers G_1 and G_2 are *disjoint* towers if G_1 and G_2 are disjoint as sets. Suppose G_1 and G_2 are disjoint towers with the same number of columns, where corresponding columns have the same width. Define the single tower G_1G_2 to be the tower obtained by placing G_2 above G_1, column by column. That is, if

$$G_i = (C_{i,j}:1 \leqslant j \leqslant k), \qquad i = 1, 2,$$

then

$$G_1G_2 = (C_{1,j}C_{2,j}:1 \leqslant j \leqslant k).$$

If $G = G_1G_2$, then $\underline{G} = \underline{G}_1$, $\bar{G} = \bar{G}_2$, and T_G extends T_{G_1} to map $\bar{G}_1$ onto $\underline{G}_2$.

If G_1 and G_2 are disjoint towers, then we can also define the single tower (G_1, G_2) consisting of the columns in G_1 followed by the columns in G_2. We can view (G_1, G_2) as the tower obtained by placing G_2 to the right of G_1. If $G = (G_1, G_2)$, then $\underline{G} = \underline{G}_1 \cup \underline{G}_2$ and $\bar{G} = \bar{G}_1 \cup \bar{G}_2$.

Let $G = (C_j:1 \leqslant j \leqslant k)$ and $0 < r < 1$. A tower rG is an *r-copy of G* if

$$rG = (rC_j:1 \leqslant j \leqslant k).$$

Let C be a column and let G be a tower such that C and G are disjoint and $w(C) = w(G)$. Let G have k columns and define $p_j = w(C_j)/w(G)$, $1 \leqslant j \leqslant k$; hence $\sum_{j=1}^{k} p_j = 1$. We can now cut C into disjoint subcolumns p_jC, $1 \leqslant j \leqslant k$, and form the tower $G(C)$ defined by

$$G(C) = (p_jC:1 \leqslant j \leqslant k).$$

Since C and G have the same width, we can define $G(C)G$, which consists of of G placed above $G(C)$. We denote $G(C)G$ simply by CG.

In general, let G_1 and G_2 be disjoint towers with $w(G_1) = w(G_2)$. Let G_1 have q columns $C_{1,j}$, $1 \leqslant j \leqslant q$, and let G_2 have k columns. Let $s_j = w(C_{1,j})/(w(G_1)$, $1 \leqslant j \leqslant q$; hence $\sum_{j=1}^{q} s_j = 1$. We can now form disjoint copies s_jG_2, $1 \leqslant j \leqslant q$, of G_2. Note that

$$w(s_jG_2) = s_jw(G_2) = \frac{w(C_{1,j})}{w(G_1)}\, w(G_2) = w(C_{1,j}).$$

We now define

$$G_1 * G_2 = (C_{1,1}s_1G_2, C_{1,2}s_2G_2, \ldots, C_{1,k}s_kG_2).$$

Thus $G_1 * G_2$ consists of a copy s_jG_2 above the jth column $C_{1,j}$ of G_1, $1 \leqslant j \leqslant q$. In this case we say that G_2 is *stacked independently above* G_1.

Let G be a tower and let us decompose G into two disjoint copies $\frac{1}{2}G$ and $\frac{1}{2}G$. We now define

$$S(G) = \tfrac{1}{2}G * \tfrac{1}{2}G.$$

Thus $S(G)$ consists of a half-size copy of G and a copy of G above each column in the half-size copy. An example is shown below.

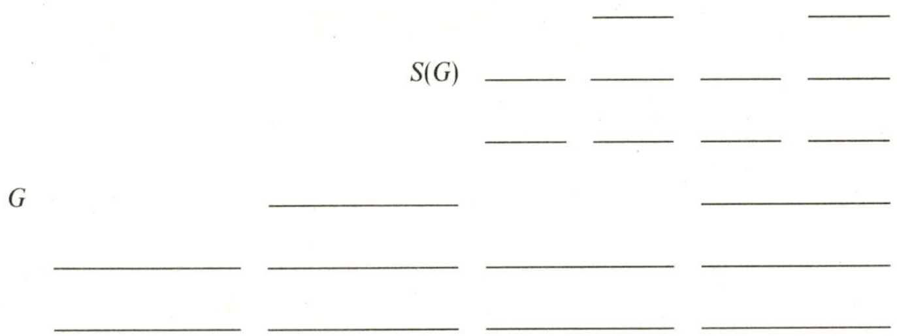

In general, if G has k columns, then $S(G)$ has k^2 columns. $T_{S(G)}$ extends T_G to map a subset of $\bar{G}$ of measure $w(G)/2$ onto a subset of G of measure $w(G)/2$.

We next consider a tower G and a disjoint interval I. The interval I will be cut into subintervals to form a tower $G(I)$. Let G have k columns C_j, $1 \leqslant j \leqslant k$, with heights h_j and widths $w_j = w(C_j)$, $1 \leqslant j \leqslant k$. Let α be defined by

$$\alpha = m(I) \Big/ \sum_{j=1}^{k} h_j w_j$$

Let us decompose I into intervals $I_{j,i}$, $1 \leqslant j \leqslant k$, $1 \leqslant i \leqslant h_j$ such that $m(I_{j,i}) = \alpha w_j$, $1 \leqslant i \leqslant h_j$, $1 \leqslant j \leqslant k$. Denote $C_j' = (I_{j,i}: 1 \leqslant i \leqslant h_j)$ and $G(I) = (C_j': 1 \leqslant j \leqslant k)$.

Towers G_1 and G_2 are *similar* if they have the same number of columns with corresponding columns having the same heights and the ratio of the widths of corresponding columns is a constant α. In this case we denote $G_1 \sim G_2$. Note that for a copy αG of G, we have $\alpha G \sim G$. Also $G(I) \sim G$ for $G(I)$ as defined above.

3. Stacking Construction for Mixing Transformations

In this section we will utilize the stacking construction in order to define mixing transformations on the real line with Lebesgue measure μ. A set will be in the ring $\mathscr{R}$ if it is contained in a finite interval and has boundary measure zero. The stacking construction is particularly well-suited to control the divergence of sequences in $\rho(T, \mathscr{R})$ to ∞. In particular, if X_k denotes the set (of finite measure) being mixed at the kth stage in the construction and I is the unit interval, then one can obtain $\rho(T, I)_n$ essentially equal to $\mu(X_k)$ for arbitrarily large n. Thus the divergence of $\rho(T, I)_n$ can be slowed down as much as desired. A sequence of transformations T_i, $i = 1, 2, 3, \ldots,$ will be obtained such that for each i,

$$\lim_{n\to\infty} \rho(T_i, I)_n/\rho(T_{i+1}, I)_n = \infty.$$

Thus T_i cannot be strongly isomorphic to T_j for $i \neq j$. The transformations also have property that if $T_{i,n}$ denotes the transformation induced by T_i on $[-n, n]$, then $T_{i,n}$ is isomorphic to a Bernoulli shift with infinite entropy. Thus $T_{i,n}$ is isomorphic to $T_{j,n}$ for all n [10]. However, in [2] it will be shown that the sequence of transformations can be constructed so that T_i is not isomorphic to T_j for $i \neq j$.

The construction will proceed by mixing a finite part of the space at each stage. Then a small amount of measure will be added and the larger set will be mixed during the next stage. Thus gradually more and more of the space is mixed, eventually obtaining mixing on the whole line.

In general, let G_1 be a tower, let $u = (u_k)$ be a sequence of positive integers, and let $\eta = (\eta_k)$ be a sequence of positive numbers such that $\sum_{k=1}^{\infty} \eta_k = \infty$. A transformation $T = T(G_1, u, \eta)$ will be defined corresponding to the tower G_1 and the sequences u and η. For appropriate G_1, u, and η the transformation T will be $\mathscr{R}$-mixing.

Let I_k, $k \leqslant 1$, be a sequence of disjoint intervals that are also disjoint from G_1 and $\mu(I_k) = \eta_k$. Define $X = G_1 \cup \bigcup_{k=1}^{\infty} I_k$. Since $\sum_{k=1}^{\infty} \eta_k = \infty$ we have $\mu(X) = \infty$; hence X can be regarded as the real line. A sequence of towers is defined inductively as follows:

$$g_k = G_k(I_k),\ k \geqslant 1, \qquad G_{k+1} = S^{u_k}(G_k, g_k). \tag{3.1}$$

Equation (3.1) implies $X = \bigcup_{k=1}^{\infty} G_k$. Also (3.1) implies that $T_{G_{k+1}}$ extends T_{G_k} to at least one-half of $\bar{G}_k$. Let T be defined by

$$T = \lim_{k\to\infty} T_{G_k}. \tag{3.2}$$

Since we are considering left-closed right-open intervals, it follows from (3.2) that T will be defined for all x in X.

Given a tower G, a transformation $T(G)$ is defined on G by

$$T(G) = \lim_{n\to\infty} T_{S^nG} \tag{3.3}$$

An *M-tower* is a tower that has at least two columns whose heights differ by one. It is shown in [4, 12] that if G is an M-tower, then $T(G)$ is isomorphic to a Bernoulli shift with respect to normalized measure on G.

Some results for transformations of the type $T(G)$ will now be obtained. Let $\mathscr{R}(G)$ denote the ring consisting of unions of levels in columns in G.

LEMMA 3.4. *Let G be a tower and let g be disjoint from G with $g \sim G$. Let $T_1 = T(G)$ and $T_2 = T(G, g)$. For A and B in $\mathscr{R}(G)$ we have*

(a) $$\mu(T_2{}^nA \cap B) = \mu(T_1{}^nA \cap B)\frac{\mu(G)}{\mu(G)+\mu(g)}, \qquad n > h(G).$$

(b) $$\frac{\mu(G)}{\mu(G)+\mu(g)}\mu(T_1{}^nA \cap B) \leqslant \mu(T_2{}^nA \cap B) \leqslant \mu(T_1{}^nA \cap B),$$

$$0 \leqslant n \leqslant h(G).$$

Proof. We have $T_1 = \lim_{u\to\infty} T_{S^uG}$ and $T_2 = \lim_{u\to\infty} T_{S^u(G,g)}$. Fix n and let $\epsilon > 0$. Choose u so large that

$$w((G, g))/2^u < \epsilon/n. \tag{1}$$

Denote $G_1 = S^uG$ and $G_2 = S^u(G, g)$. The set $T_1{}^nA \cap B$ consists of subsets of B that occur as levels L_1 in copies of G in G_1 and a subset E_1 contained in $\bigcup_{i=0}^{n-1} T_1{}^iG_1$. Let B_1 be the union of the L_1 levels; hence

$$T_1{}^nA \cap B = B_1 \cup E_1. \tag{2}$$

The set $T_2{}^nA \cap B$ consists of subsets of B that occur as levels L_2 in copies of G in G_2 and a subset E_2 contained in $\bigcup_{i=0}^{n-1} T_2{}^iG_2$. Let B_2 be the union of the levels L_2; hence

$$T_2{}^nA \cap B = B_2 \cup E_2. \tag{3}$$

Now (1) implies

$$\mu(E_i) < \epsilon, \qquad i = 1, 2. \tag{4}$$

If $n > h(G)$, then all the levels L_2 are subsets of $T_2{}^nA$ in a copy of (G, g) that is not the bottom copy of (G, g) in G_2. We can match L_2 with a level L_1 that is a subset of $T_1{}^nA$ in the corresponding copy of G in G_1. The definition of the S operator and the fact that $g \sim G$ imply

$$\mu(L_2) = \mu(L_1)\frac{\mu(G)}{\mu(G)+\mu(g)}. \tag{5}$$

Adding (5) over the L_1 and L_2 levels yields

$$\mu(B_2) = \mu(B_1)\frac{\mu(G)}{\mu(G) + \mu(g)}. \tag{6}$$

Since $\epsilon > 0$ is arbitrary, the conclusion (a) follows from (2), (3), (4), and (6).

Now suppose $0 \leqslant n \leqslant h(G)$. In this case there will be a certain subset F of A such that $T_1{}^nF = T_2{}^nF$, where $T_1{}^nF$ does not pass through $\bar{G}$. Let B_1 and E_1 and B_2 and E_2 be defined as before with respect to $T_1{}^n(A - F)$ and $T_2{}^n(A - F)$, respectively. Thus we have

$$T_1{}^nA \cap B = T_1{}^nF \cup B_1 \cup E_1 \tag{7}$$

$$T_2{}^nA \cap B = T_2{}^nF \cup B_2 \cup E_2. \tag{8}$$

Also B_1 and B_2 satisfy (6). Therefore (8) implies

$$\begin{aligned}\mu(T_2{}^nA \cap B) &= \mu(T_2{}^nF) + \mu(B_2) + \mu(E_2)\\ &= \mu(T_1{}^nF) + \mu(B_1)\frac{\mu(G)}{\mu(G) + \mu(g)} + \mu(E_2). \end{aligned}\tag{9}$$

Equation (9) implies

$$\mu(T_2{}^nA \cap B) \leqslant \mu(T_1{}^nF) + \mu(B_1) + \mu(E_2). \tag{10}$$

$$\mu(T_2{}^nA \cap B) \geqslant (\mu(T_1{}^nF) + \mu(B_1))\frac{\mu(G)}{\mu(G) + \mu(g)} + \mu(E_2). \tag{11}$$

Part (b) follows from (10) and (11) since $\epsilon > 0$ is arbitrary.

LEMMA 3.5. *Let G, g, T_1, and T_2 be as in Lemma 3.4 and let A, B, C, and D be in $\mathscr{R}(G)$.*

(a) *If $n > h(G)$, then*

$$\frac{\mu(T_2{}^nA \cap B)}{\mu(T_2{}^nC \cap D)} = \frac{\mu(T_1{}^nA \cap B)}{\mu(T_1{}^nC \cap D)}.$$

(b) *If $0 \leqslant n \leqslant h(G)$, then*

$$\frac{\mu(G)}{\mu(G) + \mu(g)}\frac{\mu(T_1{}^nA \cap B)}{\mu(T_1{}^nC \cap D)} \leqslant \frac{\mu(T_2{}^nA \cap B)}{\mu(T_2{}^nC \cap D)} \leqslant \frac{\mu(T_1{}^nA \cap B)}{\mu(T_1{}^nC \cap D)}\frac{\mu(G) + \mu(g)}{\mu(G)}$$

Proof. Parts (a) and (b) follow from Lemma 3.4(a) and (b), respectively.

As started above, $T(G)$ will be isomorphic to a Bernoulli shift if G is an M-tower. In particular, $T(G)$ will be mixing on G with respect to the measure

μ_G defined by

$$\mu_G(B) = \mu(B)/\mu(G), \qquad B \subset G.$$

Hence for A and B contained in G we have

$$\lim_{n\to\infty} \mu(T(G)^n A \cap B) = \frac{\mu(A)\mu(B)}{\mu(G)}. \tag{3.6}$$

For any transformation T and finite ring $\mathscr{R}$, let us denote

$$\mu(n, \mathscr{R}, T) = \max_{\mathscr{R}} \left| \frac{\mu(T^n A \cap B)}{\mu(T^n C \cap D)} - \frac{\mu(A)\mu(B)}{\mu(C)\mu(D)} \right|, \tag{3.7}$$

where the maximum is taken over the finite number of sets A, B, C, and D in $\mathscr{R}$. In particular, we denote $\mu(n, G, T) = \mu(n, \mathscr{R}(G), T)$.

If G is an M-tower, then (3.6) implies that for $\epsilon > 0$ there exists a positive integer $N(G, \epsilon)$ such that

$$\mu(n, G, T(G)) < \epsilon, \qquad n \geqslant N(G, \epsilon). \tag{3.8}$$

If a transformation T extends a transformation T_G, then we will simply say that T extends G.

LEMMA 3.9. *Let G be an M-tower and let g be disjoint from G and $g \sim G$. Let $\epsilon > 0$ and $N > N(G, \epsilon)$. There exists a positive integer $u = u(G, N, \epsilon)$ such that if T extends $S^u(G, g)$, then*

$$\mu(n, G, T) < 2\epsilon, \qquad N(G, \epsilon) \leqslant n \leqslant N.$$

Proof. Let T_1 and T_2 be as in Lemma 3.4. Note that $T_2 = T(S^k(G, g))$. Since G is an M-tower, (3.8) implies

$$\mu(n, G, T_1) < \epsilon, \qquad n \geqslant N(G, \epsilon). \tag{1}$$

We can also assume $n > h(G)$ since it will always be the case that $N(G, \epsilon) > h(G)$. Therefore (1) and Lemma 3.5(a) imply

$$\mu(n, G, T_2) < \epsilon, \qquad n \geqslant N(G, \epsilon). \tag{2}$$

Since $N > N(G, \epsilon)$ is fixed, we can guarantee that T^n will agree with $T_2{}^n$ on $G \cup g$, except on a set of arbitrarily small measure, for the specified range of n by choosing u sufficiently large. Thus (2) implies $\mu(n, G, T) < 2\epsilon$ if u is sufficiently large, for the specified range of n.

The next lemma says that if there is a certain amount of mixing in a tower (G_1, g_1), then we can add a set g_2 to the space without perturbing the mixing very much if $\mu(g_2)$ is small.

LEMMA 3.10. *Let G_1 be an M-tower and let g_1 be disjoint from G_1 with $g_1 \sim G_1$. Let $\epsilon_1 > 0$ and let $N_1 > N(G_1, \epsilon_1)$. Assume u_1 is chosen so that if T extends $G_2 = S^{u_1}(G_1, g_1)$, then*

$$\text{(a)} \qquad \mu(n, G_1, T) < 2\epsilon_1, \qquad N(G_1, \epsilon_1) \leqslant n \leqslant N_1,$$

Let g_2 be disjoint from G_2 with $g_2 \sim G_2$. Let $\epsilon_2 > 0$ and let $N_2 > N(G_2, \epsilon_2)$. There exists u_2 such that if T extends $G_3 = S^{u_2}(G_2, g_2)$, then

$$\text{(b)} \qquad \mu(n, G_2, T) < 2\epsilon_2, \qquad N(G_2, \epsilon_1) \leqslant n \leqslant N_2.$$

Let $\mathscr{R} \subset \mathscr{R}(G_1)$ and let $\gamma_1 > 0$. There exists $\delta = \delta(\mathscr{R}, \gamma_1)$ such that if $\mu(g_2) < \delta$ and u_2 is sufficiently large, then

$$\text{(c)} \qquad \mu(n, \mathscr{R}, T) < \epsilon_1 + \epsilon_2 + \gamma_1, \qquad N_1 \leqslant n \leqslant N_2.$$

Proof. Part (b) follows from Lemma 3.9 with $G = G_2$, $\epsilon = \epsilon_2$, and $N = N_2$. For (c) we let $T_i = T(G_i)$, $i = 1, 2, 3$. Note that

$$T_2 = T(G_2) = T(S^{k_1}(G_1, g_1)) = T(G_1, g_1). \tag{1}$$

$$T_3 = T(G_3) = T(S^{k_2}(G_2, g_2)) = T(G_2, g_2). \tag{2}$$

For A, B, C, and D in $\mathscr{R}$ we have

$$\left|\frac{\mu(T^nA \cap B)}{\mu(T^nC \cap D)} - \frac{\mu(A)m(B)}{\mu(C)m(D)}\right| \leqslant \left|\frac{\mu(T^nA \cap B)}{\mu(T^nC \cap D)} - \frac{\mu(T_3{}^nA \cap B)}{\mu(T_3{}^nC \cap D)}\right|$$
$$+ \left|\frac{\mu(T_3{}^nA \cap B)}{\mu(T_3{}^nC \cap D)} - \frac{\mu(T_2{}^nA \cap B)}{\mu(T_2{}^nC \cap D)}\right| + \mu(n, G_1, T_2). \tag{3}$$

We can assume $N(G_1, \epsilon_1) \geqslant h(G_1)$. Lemma 3.5(a) with $G = G_1$ and $g = g_1$ implies

$$\mu(n, G_1, T_2) < \epsilon_1, \qquad n \geqslant N(G_1, \epsilon_1). \tag{4}$$

Note that since $\mathscr{R}(G_1) \subset \mathscr{R}(G_2)$, (b) implies that we need only consider $N_1 \leqslant n < N(G_2, \epsilon_2)$ in (c). Since T extends T_3 we can also choose u_2 sufficiently large to guarantee that T^n will agree with $T_3{}^n$ on $G_2 \cup g_2$ for the specified range of n, except on a set of small measure. Hence u_2 can be chosen so that

$$\left|\frac{\mu(T^nA \cap B)}{\mu(T^nC \cap D)} - \frac{\mu(T_3{}^nA \cap B)}{\mu(T_3{}^nC \cap D)}\right| < \epsilon_2, \qquad N_1 \leqslant n < N(G_2, \epsilon_2). \tag{5}$$

Now $T_2 = T(G_2)$ and $T_3 = T(G_2, g_2)$ by (2). Lemma 3.5(b) with G replaced by G_2, T_1 replaced by T_2, and T_2 replaced by T_3 implies

$$\left|\frac{\mu(T_3{}^nA \cap B)}{\mu(T_3{}^nC \cap D)} - \frac{\mu(T_2{}^nA \cap B)}{\mu(T_2{}^nC \cap D)}\right| \leqslant \frac{\mu(T_2{}^nA \cap B)}{\mu(T_2{}^nC \cap D)} \frac{\mu(g_2)}{\mu(G_2)}. \tag{6}$$

By (6) we can choose $\delta = \delta(\gamma_1, \mathscr{R})$ so that the right side of (6) is less than γ_1 if $\mu(g_2) < \delta$. Therefore (c) follows from (3), (4), (5), and (6).

THEOREM 3.11. *Let G_k, $k \geqslant 1$, be defined as in* (3.1) *and let T be defined as in* (3.2). *If G_1 is an M-tower, then u_k and η_k, $k \geqslant 1$, can be chosen so that T is $\mathscr{R}$-mixing.*

Proof. Let ϵ_k be a sequence of positive numbers converging to 0. Let $N_1 > N(G_1, \epsilon_1)$. By Lemma 3.9 with $G = G_1$, $\epsilon = \epsilon_1$, and $N = N_1$ we obtain u_1 such that if T extends G_2, then

$$\mu(n, G_1, T) < 2\epsilon_1, \qquad N(G_1, \epsilon_1) \leqslant n \leqslant N_1. \tag{1}$$

Let $\mathscr{R} = \mathscr{R}(G_1)$ and let $\gamma_1 > 0$ in Lemma 3.10. Lemma 3.10 implies there exists $\delta_2 = \delta$ and u_2 such that if $\mu(g_2) < \delta_2$ and T extends G_3, then

$$\mu(n, G_2, T) < 2\epsilon_2, \qquad N(G_2, \epsilon_2) \leqslant n \leqslant N_2. \tag{2}$$

$$\mu(n, G_1, T) < \epsilon_1 + \epsilon_2 + \gamma_1, \qquad N_1 \leqslant n \leqslant N_2. \tag{3}$$

We now apply Lemma 3.10 with (a) replaced by (2). Hence $G_1, G_2, \epsilon_1, \epsilon_2, N_1$, and N_2 are replaced by $G_2, G_3, \epsilon_2, \epsilon_3, N_2$, and $N_3 > N(G_3, \epsilon_3)$, respectively. However, we still let $\mathscr{R} = \mathscr{R}(G_1)$ and $\gamma_2 = \gamma_1$ so that $\delta_3 = \delta(\mathscr{R}, \gamma_2) = \delta_2$. Thus we obtain u_3 so that if T extends G_4, then

$$\mu(n, G_3, T) < 2\epsilon_3, \qquad N(G_3, \epsilon_3) \leqslant n \leqslant N_3. \tag{4}$$

Also if $\mu(g_3) > \delta_3$ and u_3 is sufficiently large, then

$$\mu(n, G_1, T) < \epsilon_2 + \epsilon_3 + \gamma_1, \qquad N_2 \leqslant n \leqslant N_3. \tag{5}$$

We now proceed inductively. Suppose we have

$$\mu(n, G_k, T) < 2\epsilon_k, \qquad N(G_k, \epsilon_k) \leqslant n \leqslant N_k. \tag{6}$$

$$\mu(n, G_1, T) < \epsilon_{k-1} + \epsilon_k + \gamma_1, \qquad N_{k-1} \leqslant n \leqslant N_k. \tag{7}$$

We now apply Lemma 3.10 with (a) replaced by (6). Hence $G_1, G_2, \epsilon_1, \epsilon_2, N_1$, and N_2 are replaced by G_k, G_{k+1}, ϵ_k, ϵ_{k+1}, N_k, and $N_{k+1} > N(G_k, \epsilon_k)$, respectively. However, let $\mathscr{R} = \mathscr{R}(G_1)$ and $\gamma_k = \gamma_1$ so that $\delta_{k+1} = \delta(\mathscr{R}, \gamma_k) = \delta_2$. Thus we obtain u_{k+1} so that if T extends G_{k+2}, then

$$\mu(n, G_{k+1}, T) < 2\epsilon_{k+1}, \qquad N(G_{k+1}, \epsilon_{k+1}) \leqslant n \leqslant N_{k+1}. \tag{8}$$

$$\mu(n, G_1, T) < \epsilon_k + \epsilon_{k+1} + \gamma_1, \qquad N_k \leqslant n \leqslant N_{k+1}. \tag{9}$$

By induction we can choose k_1 arbitrarily large so that for $1 \leqslant i \leqslant k_1$ we have

$$\mu(n, G_i, T) < 2\epsilon_i, \qquad N(G_i, \epsilon_i) \leqslant n \leqslant N_i. \tag{10}$$

$$\mu(n, G_1, T) < \epsilon_{i-1} + \epsilon_i + \gamma_1, \qquad N_{i-1} \leqslant n \leqslant N_i. \tag{11}$$

Since $\gamma_i = \gamma_1$, $1 \leqslant i \leqslant k_1$, we have $\delta_i = \delta_2$, $1 \leqslant i \leqslant k_1$. Thus we can choose $\mu(g_i) = \eta_i = \delta_2/2$, $1 \leqslant i \leqslant k_1$. Hence we have added measure $k_1\eta_{k_1}$ to G_1 to obtain G_{k_1}.

We can now consider $i > k_1$ and repeat the above steps with G_1 replaced by G_{k_1} and γ_1 replaced by $\gamma_2 > 0$. We choose $k_2 > k_1$ and for $k_1 < i \leqslant k_2$ obtain (10) and

$$\mu(n, G_{k_1}, T) < \epsilon_{i-1} + \epsilon_i + \gamma_2, \qquad N_{i-1} \leqslant n \leqslant N_i. \tag{12}$$

Let $\mu(g_i) = \eta_i = \delta(\mathscr{R}(G_{k_1}), \gamma_2)/2$ for $k_1 < i \leqslant k_2$. Thus a set of measure $(k_2 - k_1)\eta_{k_2}$ is added to G_{k_1} to obtain G_{k_2}.

We now proceed inductively as above to choose an increasing sequence k_l and a decreasing sequence γ_l such that $\lim_{l\to\infty} \gamma_l = 0$. Let $N_i > N(G_i, \epsilon_i)$ and let $\gamma_i = \gamma_{l+1}$ for $k_l < i < k_{l+1}$. We apply Lemma 3.10 to obtain (10) for all i. Also for $k_l < i \leqslant k_{l+1}$ we obtain

$$\mu(n, G_{k_l}, T) < \epsilon_{i-1} + \epsilon_i + \gamma_{l+1}, \qquad N_{i-1} \leqslant n \leqslant N_i. \tag{13}$$

The total amount of measure added is

$$\sum_{i=1}^{\infty} \mu(g_i) = \sum_{l=1}^{\infty} (k_l - k_{l-1})\eta_{k_l}, \tag{14}$$

where $k_0 = 0$. (14) implies that k_l can be chosen so that $\mu(X) = \infty$.

Now $\lim_{i\to\infty} \epsilon_i = 0$ and $\lim_{l\to\infty} \gamma_l = 0$. Therefore (10) and (13) guarantee that T will be mixing on $\mathscr{C} = \bigcup_{k=1}^{\infty} \mathscr{R}(G_k)$. Since $\mathscr{C}$ generates $\mathscr{B}$, we conclude that T is $\mathscr{R}$-mixing [9, p. 435].

The stacking construction in Theorem 3.11 implies that T is ergodic and measure preserving with respect to Lebesgue measure μ on $X = (-\infty, \infty)$. The entropy $h(T)$ of an ergodic measure preserving transformation in an infinite measure space is defined as $h(T) = h(T_E)$, where E is any set of positive measure and T_E is the transformation induced by T on E [7]. The entropy $h(T_E)$ is computed with respect to the nonnormalized measure μ/E, where $\mu/E(B) = \mu(E \cap B) = \mu_E(B)\mu(E)$. If $\bar{h}(T_E)$ denotes the entropy of T_E computed with respect to normalized measure μ_E on E, then $\bar{h}(T_E)\mu(E) = h(T_E)$. In order to compute $h(T)$, we first need some preliminary results for transformations of the type $T(G)$. The following result is contained in [4].

LEMMA 3.12. *Let G consist of k columns with widths w_j, $1 \leqslant j \leqslant k$, and let $w = w(G)$. The entropy of $T(G)$ is given by*

$$h(T(G)) = -\sum_{j=1}^{k} w_j \log(w_j/w)$$

LEMMA 3.13. *Let G be as in Lemma 3.12 and let g be disjoint from G with $g \sim G$. Let $\alpha = \mu(g)/\mu(G)$. The entropy of $T(G, g)$ is given by $h(T(G, g)) = (1 + \alpha)h(T(G)) + w[\log(1 + \alpha) + \alpha \log(1 + 1/\alpha)]$.*

Proof. Lemma 3.12 applied to $T(G, g)$ implies

$$\begin{aligned} h(T(G, g)) &= -\sum_{j=1}^{k} w_j \log w_j/(1 + \alpha)w \\ &\quad - \sum_{j=1}^{k} \alpha w_j \log \alpha w_j/(1 + \alpha)w \\ &= -\sum_{j=1}^{k} w_j \log w_j/w + w \log(1 + \alpha) \\ &\quad -\alpha \sum_{j=1}^{k} w_j \log w_j/w + \alpha w \log(1 + 1/\alpha) \\ &= (1 + \alpha)h(T(G)) + w[\log(1 + \alpha) + \alpha \log(1 + 1/\alpha)]. \end{aligned}$$

In order to compute the entropy $h(T)$, we first specify η_k, $k \geqslant 1$ in Theorem 3.11. Let r_l be a sufficiently large integer, $l \geqslant 1$, so that we can take $\eta_k = 1/r_l$, $k_{l-1} < k \leqslant k_l$ ($k_0 = 0$). Also assume $k_l = k_{l-1} + r_l$, $l \geqslant 1$; hence

$$(k_l - k_{l-1})\eta_{k_l} = r_l/r_l = 1, \qquad l \geqslant 1. \tag{3.14}$$

Let us take $G_1 = [0, 1]$. Equation (3.14) implies $\mu(G_{k_l}) = \mu(G_{k_{l-1}}) + 1$; hence

$$\mu(G_{k_l}) = l, \qquad l \geqslant 1. \tag{3.15}$$

In particular, we can assume $G_{k_{2n}} = [-n, n]$.

LEMMA 3.16. *The sequence r_l can be chosen so that $h(T) = \infty$.*

Proof. Let k satisfy $k_{l-1} < k \leqslant k_l$. Hence (3.15) implies

$$\begin{aligned} \alpha_k &= \mu(g_k)/\mu(G_k) \\ &> \mu(g_k)/\mu(G_{k_l}) = 1/lr_l. \end{aligned} \tag{1}$$

Lemma 3.13 and (1) imply

$$\begin{aligned} h(T(G_{k_l})) &> \prod_{k_{l-1}}^{k_l - 1} (1 + \alpha_k)h(T(G_k)) \\ &> (1 + 1/lr_l)^{r_l} h(T(G_{k_{l-1}})). \end{aligned} \tag{2}$$

Equation (2) implies that r_l can be chosen sufficiently large so that

$$h(T(G_{k_l})) \geqslant e^{1/l} h(T(G_{k_{l-1}})). \tag{3}$$

Equation (3) implies

$$h(T(G_{k_l})) \geqslant \left(\exp \sum_{i=1}^{l} 1/i\right) h(T(G_1)). \tag{4}$$

From (4) we conclude

$$\lim_{l\to\infty} h(T(G_{k_l})) = \infty. \tag{5}$$

Let P_k denote the partition consisting of the levels in the columns in G_k. Let T_k denote the transformation induced by T on G_k. The stacking construction implies that if T_k/P_k is the factor of T_k with respect to the partition P_k, then

$$T_k/P_k = T(G_k). \tag{6}$$

Since $h(T_k) \geqslant h(T_k/P_k)$, (5) and (6) imply that $h(T) = \infty$.

In the following theorem we show that T can be constructed so that $\rho(T, I)$ diverges to ∞ arbitrarily slowly, where $I = [0, 1]$. Also the sets of finite measure on which the induced transformation is isomorphic to a Bernoulli shift are dense in the class of sets of finite measure.

THEOREM 3.17. *Let s_n be a diverging sequence of positive numbers. There exists an $\mathscr{R}$-mixing transformation T such that:*

(a) $\lim_{n\to\infty} s_n/\rho(T, I)_n = \infty$.
(b) $h(T) = \infty$.
(c) *If $\mu(B) < \infty$ and $\epsilon > 0$, then there exists A such that $\mu(A \triangle B) < \epsilon$ and the induced transformation T_A is isomorphic to a Bernoulli shift with infinite entropy.*

Proof. Let T be as in Lemma 3.16; hence T is $\mathscr{R}$-mixing and $h(T) = \infty$. It will next be shown that u_k, $k \geqslant 1$, can be chosen sufficiently large, independent of $\mu(g_k)$, to guarantee (a). Let N_k be an increasing sequence such that

$$s_n \geqslant k^3, \qquad n \geqslant N_k. \tag{1}$$

Since $T(G_k)$ is mixing, we have

$$\lim_{n\to\infty} \mu(T(G_k)^n I \cap I) = 1/\mu(G_k). \tag{2}$$

By (3.15) it follows that $\mu(G_k) \leqslant k$ since $l \leqslant k$, $l \geqslant 1$. Thus (2) implies

$$\lim_{n\to\infty} \mu(T(G_k)^n I \cap I) \geqslant 1/k. \tag{3}$$

It follows from (3) and (3.1) that u_k can be chosen sufficiently large to guarantee

$$\mu(T^n I \cap I) \geqslant 1/k, \qquad N_k \leqslant n \leqslant N_{k+1}. \tag{4}$$

Equations (1) and (4) imply

$$s_n/\rho(T, I_n) \geqslant k^2, \qquad N_k \leqslant n \leqslant N_{k+1}. \tag{5}$$

Equation (5) implies (a).

For (c) we apply an extension of the method of proof in [3, 4, 12]. Since $\mu(B) < \infty$ there exists a tower G_k such that $\mu(B \cap (X - G_k)) < \epsilon/2$. Since $\lim_{k\to\infty} w(G_k) = 0$, the maximum length of an interval that is a level in a column in G_k will converge to 0. It follows that k can be chosen so large that there exists a set A that is a union of levels in columns in G_k such that $\mu(A \triangle B) < \epsilon/2$; hence $\mu(A \triangle B) < \epsilon$. We can also assume that A contains G_k, so that A contains a level in each column in G_k. Let $(G_k)_A$ denote the tower G_k restricted to A and let $(P_k)_A$ denote the partition consisting of the levels in columns in $(G_k)_A$. The stacking construction implies

$$T_A/(P_k)_A = T((G_k)_A). \tag{6}$$

As in [4], we can also assume $(G_k)_A$ is an M-tower, or else modify A by a sufficiently small set consisting of levels in two columns in G_k. Therefore (6) implies that the factor $T_A/(P_k)_A$ is isomorphic to a Bernoulli shift.

Also A appears as a union of levels in G_l for $l > k$. The stacking construction implies

$$T_A/(P_l)_A = T((G_l)_A), \qquad l > k. \tag{7}$$

The sequence of sigma algebras generated by the partitions $(P_l)_A$ converge to $\mathscr{B}$ restricted to A. Hence by (7) and [10] we conclude that T_A is isomorphic to a Bernoulli shift. Also $h(T_A) = \infty$ since $h(T) = \infty$.

COROLLARY 3.18. *There exists a sequence of $\mathscr{R}$-mixing transformations T_i, $i \geqslant 1$, such that*

(a) $\lim_{n\to\infty} \rho(T_i, I)_n/\rho(T_{i+1}, I)_n = \infty$, $i \geqslant 1$.
(b) $h(T_i) = \infty$.
(c) *Each transformation T_i, $i \geqslant 1$, induces a Bernoulli shift with infinite entropy on $[-n, n]$, $n \geqslant 1$.*

Proof. Let T_1 be constructed as in Theorem 3.17 with respect to $s_n = n$. Therefore $h(T_1) = \infty$ and (c) follows for T_1 if we let $A = B = G_{k_{2n}} = [-n, n]$. Now assume T_i, $1 \leqslant i \leqslant j$, have been constructed. Let s_n be defined by

$$s_n = \rho(T_j, I)_n. \tag{1}$$

We apply Theorem 3.17 to obtain $T_{j+1} = T$. Therefore (a) is satisfied for $i = j$ and $h(T_{j+1}) = \infty$. Also T_{j+1} satisfies (c), as verified above for T_1. Thus the sequence of transformations T_i is defined for $i \geqslant 1$ by induction.

By (a) we can only conclude that T_i is not strongly isomorphic to T_j for $i \neq j$. If it can be shown that the spreading rate is a general isomorphism invariant, then (a) would imply that T_i is not isomorphic to T_j for $i \neq j$. In [2] it will be shown that the spreading rate can be utilized to construct another invariant that can be applied to verify that T_i is not isomorphic to T_j for $i \neq j$.

4. Mixing Markov Shifts

We will now give a brief discussion of mixing Markov shifts with infinite invariant measure. A more detailed discussion is given in [9]. Let Z denote the set of all integers i and let $X = \prod_{i=-\infty}^{\infty} Z_i$, where $Z_i = Z$ for all i. Let $\Pi = (\Pi(i, j): i, j \in Z)$ be a stochastic matrix, where $\Pi(i, j) \geqslant 0$ and $\sum_j \Pi(i, j) = 1$. A vector $\lambda = (\lambda(j): j \in Z)$ such that $\lambda(j) \geqslant 0$, $\sum_j \lambda(j) = \infty$, and $\lambda = \lambda\, \Pi$ is an *infinite invariant measure* for Π. We fix one such λ.

An *elementary cylinder* is a set of the form

$$A = \{x : x_{n_j} = i_j,\ 1 \leqslant j \leqslant k\}, \tag{4.1}$$

where $n_1 < n_2 < \cdots < n_k$ and $i_j \in Z$, $1 \leqslant j \leqslant k$. Let $\mathscr{C}$ denote the class of all elementary cylinders.

Let $\Pi^n = (\Pi^n(i, j): i, j \in Z)$ be the matrix of n-step transition probabilities. A measure μ is defined for A in (4.1) as

$$\mu(A) = \lambda(i_1)\, \Pi^{n_2 - n_1}(i_1, i_2) \cdots \Pi^{n_k - n_{k-1}}(i_{k-1}, i_k). \tag{4.2}$$

Let $\mathscr{B}$ be the sigma algebra generated by $\mathscr{C}$. The measure μ can be extended to a sigma finite measure on $\mathscr{B}$ with $\mu(X) = \sum_i \lambda(i)$. We can assume $\lambda(i) > 0$ for each i [9]. The *shift* T defined by $T(x)_i = x_{i+1}$, $i \in Z$ is an invertible measure preserving transformation on X. If (2.2) holds for A, B, C, and D in $\mathscr{C}$, then T is a mixing Markov shift with infinite invariant measure.

We can also consider X as a topological space as in [9]. A topology in X is defined as the product topology of the discrete topologies in the factor spaces Z_i. The class $\mathscr{C}$ is a basis of clopen sets for the topology. In this case the shift T is a measure preserving homeomorphism. If (2.2) holds for A, B, C, and D in $\mathscr{C}$, then (2.2) holds for the ring $\mathscr{R}$ generated by $\mathscr{C}$ since $\mathscr{C}$ is closed under finite intersections. Thus T is $\mathscr{R}$-mixing if T is referred to as a mixing Markov shift with infinite invariant measure.

Let $\mathscr{M}$ denote the class of mixing Markov shifts with infinite invariant measure. Conditions for T corresponding to (Π, λ) to be in $\mathscr{M}$ are given in

[9]. The entropy $h(T)$ corresponding to T in $\mathscr{M}$ is given by [7] as follows.

$$h(T) = -\sum_i \lambda_i \sum_j \Pi(i,j) \log \Pi(i,j). \tag{4.3}$$

A *random walk* corresponds to the case where $\Pi(i,j) = \Pi(i-j)$ is a function of $i-j$. We fix the invariant measure $\lambda(i) = 1$ for all i for the case of a random walk. Equation (4.3) implies that $h(T) = \infty$ for a random walk T in $\mathscr{M}$.

In particular, let r be a positive integer and consider the random walk defined by $\pi_r(u) = 1/(2u+1)$, $-r \leqslant u \leqslant r$. The corresponding shift is in $\mathscr{M}$ and will be denoted by T_r.

Let $I = \{x : x_0 = 0\}$; hence $\mu(I) = \lambda(0) = 1$. The spreading rate of T_r is given by applying Proposition 9 of Spitzer ([13], p. 75) as follows:

$$\rho(T_r, I)_n \sim [(r^3 + 3r^2 + 2r^3)(2\pi n)/3]^{1/2}. \tag{4.4}$$

In particular, (4, 4) implies T_r is not strongly isomorphic to T_s, $r \neq s$. Furthermore, in [1] it is shown that the spreading rate is a general isomorphism invariant for random walks. Therefore T_r is not isomorphic to T_s, $r \neq s$.

A class of mixing Markov shifts with finite entropy can be defined by first choosing positive numbers $\epsilon_i < \frac{1}{2}$, $i \geqslant 0$, such that

$$-\sum_{i=0}^{\infty} (\epsilon_i \log \epsilon_i + (1-\epsilon_i)\log(1-\epsilon_i)) < \infty. \tag{4.5}$$

A corresponding doubly stochastic matrix $\Pi(i,j)$ with invariant measure $\chi(i) = 1$, all i, is defined as follows:

$$\begin{aligned}
&\Pi(0, i) = \epsilon_0, \qquad |i| = 1.\\
&\Pi(0, 0) = 1 - 2\epsilon_0.\\
i \geq 1: &\Pi(i, i-1) = \epsilon_i,\\
&\Pi(i, i+1) = \epsilon_i,\\
&\Pi(i, i) = 1 - \epsilon_{i-1} - \epsilon_i.\\
i \leq 1: &\Pi(i, i+1) = \epsilon_{-i-1},\\
&\Pi(i, i-1) = \epsilon_{-i},\\
&\Pi(i, i) = 1 - \epsilon_{-i-1} - \epsilon_{-i}.
\end{aligned} \tag{4.6}$$

It is not difficult to choose the ϵ_i in (4.6) in two different ways so as to obtain transformations with the same finite entropy but different spreading rates; hence the transformations are not strongly isomorphic. One can also choose the ϵ_i so that the spreading rates guarantee that the corresponding transformations are not isomorphic. These examples will be discussed further in [2].

The author wishes to thank Martin Ellis and Arshag Hahian for their helpful discussions.
Related results are proved by S. M. Rudolter in, Some metric invariants for Markov shifts, *Z. Wahrscheinlichkeitstheorie und Verw. Gebiete* **15** (1970), 202–207.

References

1. J. Aaronson, Ratio limit properties and a metric invariant for Markow shifts, *Israel J. Math.* **27** (1977), 93–123.
2. M. Ellis and N. A. Friedman, On eventually weakly wandering sequences, this volume.
3. N. A. Friedman, On mixing, entropy, and generators, *J. Math. Anal. Appl.* **26** (1969), 512–528.
4. N. A. Friedman, Bernoulli shifts induce Bernoulli shifts, *Advances in Math.* **10** (1973), 39–48.
5. N. A. Friedman and D. S. Ornstein, On isomorphism of weak Bernoulli transformations, *Advances in Math.* **5** (1970), 365–394.
6. A. B. Hajian and S. Kakatani, Weakly wandering sets and invariant measures, *Trans. Amer. Math. Soc.* **110** (1964), 136–151.
7. U. Krengel, Entropy and conservative transformations, *Z. Wahrscheinlichkeitstheorie und Verw. Gebiete* **7** (1967), 161–181.
8. U. Krengel and L. Sucheston, On mixing in infinite measure spaces, *Z Wahrscheinlichkeitstheorie und Verw. Gebiete* **13** (1969), 150–164.
9. K. Krickeberg, Strong mixing properties of Markov chains with infinite invariant measure, *Proc. 5th Berkeley Symp., Math. Stat. Prob.* **2**, Part II (1965), 431–445.
10. D. S. Ornstein, Bernoulli shifts with infinite entropy are isomorphic, *Advance in Math.* **5** (1970), 339–348.
11. F. Papengelou, Strong ratio limits, R-recurrence and mixing properties of discrete parameter Markov processes, *Z. Wahrscheinlichkeitstheorie und Verw. Gebiete* **8** (1967), 259–297.
12. P. Shields, Cutting and independent stacking of intervals, *Math. Systems Theory* **7**, (1973), 1–4.
13. F. Spitzer, "Principles of Random Walk," Van Nostrand–Reinhold, New York, 1964.

STUDIES IN PROBABILITY AND ERGODIC THEORY
ADVANCES IN MATHEMATICS SUPPLEMENTARY STUDIES, VOL. 2

On Eventually Weakly Wandering Sequences†

MARTIN H. ELLIS AND NATHANIEL A. FRIEDMAN

Department of Mathematics, State University of New York, Albany, New York

1. INTRODUCTION

In [5] Hajian and Kakutani proved that a transformation admits an equivalent finite invariant measure if and only if the transformation does not admit a weakly wandering sequence (see Section 2 for definitions). If an ergodic transformation has an equivalent infinite invariant measure, then an equivalent finite invariant measure cannot exist. Thus the transformation admits weakly wandering sequences and these sequences are isomorphism invariants for the transformation.

In this paper we will introduce a special type of wandering sequence called an eventually weakly wandering sequence. Briefly, a sequence is eventually weakly wandering for a transformation if a set of arbitrarily small measure can be removed from any set and the remaining set is weakly wandering on the sequence after an intial finite segment of the sequence is removed. That is, the remaining set is eventually weakly wandering on the sequence.

The class of eventually weakly wandering sequences for a transformation will be applied as an isomorphism invariant for transformations that are mixing with respect to an infinite invariant measure [4].

Preliminary definitions will be introduced in Section 2. In Section 3 it will be proven that the existence of weakly wandering sequences for an ergodic transformation implies the existence of eventually weakly wandering sequences for the transformation. In Section 4 it will be shown that the spreading rate introduced in [4] can be utilized to give an alternative construction of eventually weakly wandering sequences. For example, it will be shown that certain mixing transformations corresponding to random walks on the integers admit eventually weakly wandering sequences of the form $n_k \sim \alpha^k$, where $\alpha > 1$, and $n_k \sim k^\beta$, where $\beta > 4$. In Section 5 we will also discuss examples of mixing transformations with the same entropy that are not isomorphic.

† Research partially supported by National Science Foundation Grant MPS71–02858 AO3.

ISBN 0-12-599102-9

2. Definitions

Let (X, B, m) be a measure space with $m(X) = 1$. Let T be a one-to-one point transformation mapping X into X such that the domain and range of T are sets of measure one. T is *measurable* if $B \in \mathscr{B}$ implies $TB \in \mathscr{B}$ and $T^1 B \in \mathscr{B}$. A measurable transformation T is *nonsingular* if $m(B) = 0$ implies $m(TB) = 0$ and $m(T^{-1}B) = 0$. Hereafter T will be assumed nonsingular. A measure μ defined on $\mathscr{B}$ is *equivalent to* $m(\mu \sim m)$ if μ and m have the same sets of measure zero. A measure μ is an *invariant measure for* T if $\mu(TB) = \mu(B)$, $B \in \mathscr{B}$. T admits a *weakly wandering* sequence (n_k) if there exists a set W of positive measure such that the images $T^{n_k}W$, $k = 1, 2, 3, \ldots$, are pairwise disjoint. In this case we also say that W is *weakly wandering on* (n_k). T admits a finite invariant measure $\mu \sim m$ if and only if T does not admit a weakly wandering set [5].

We will say that T admits an *eventually weakly wandering sequence* (n_k) if for each $\epsilon > 0$ there exists A_ϵ and K_ϵ such that $m(A_\epsilon) < \epsilon$ and $X - A_\epsilon$ is weakly wandering on $(n_k : k \geqslant K_\epsilon)$. In this case $B = B \cap A_\epsilon$ satisfies $m(B_\epsilon) < \epsilon$ and $B - B_\epsilon$ is weakly wandering on $(n_k : k \geqslant K_\epsilon)$, $B \in \mathscr{B}$.

If T admits an infinite sigma-finite invariant measure $\mu \sim m$, then it is not difficult to show that a sequence (n_k) is eventually weakly wandering for T if and only if for each $B \in \mathscr{B}$ with $\mu(B) < \infty$ and $\epsilon > 0$, there exists B_ϵ and K_ϵ such that $\mu(B_\epsilon) < \epsilon$ and $B - B_\epsilon$ is weakly wandering on $(n_k : k \geqslant K_\epsilon)$.

The class of eventually weakly wandering sequences for a transformation T will be denoted by $W(T)$. An explicit construction of a transformation admitting the sequence $(4^k : k = 1, 2, \ldots)$ as an eventually weakly wandering sequence is given in [3, p. 85]. Henceforth it will be assumed that eventually weakly wandering sequences are increasing.

A transformation T is *ergodic* if $TA = A$ implies $m(A) = 0$ or 1. If m is nonatomic, then T is ergodic if and only if $m(A) > 0$ implies $m(\bigcup_{n=0}^{\infty} T^n A) = 1$. Henceforth we assume m is nonatomic.

Let $(X_i, \mathscr{B}_i, \mu_i)$, $i = 1, 2$, be measure spaces and let S be a one-to-one mapping of X_1 into X_2 such that $\mu_1(B_1) = \mu_2(SB_1)$, $B \in \mathscr{B}_1$, and $\mu_1(S^{-1}B_2) = \mu_2(B_2)$, $B_2 \in \mathscr{B}_2$. In this case we refer to S as an *isomorphism*. Let T_i be measure preserving with respect to μ_i, $i = 1, 2$. T_1 and T_2 are *isomorphic* if there exists an isomorphism S such that $T_1 = S^{-1} T_2 S$.

Let $(X, \mathscr{B}, \mu)$ be a measure space with $\mu(X) = \infty$ and let $\mathscr{R}$ be a ring of sets of finite measure such that $\mathscr{R}$ generates $\mathscr{B}$. Let μ be an invariant measure for T and assume $T\mathscr{R} = \mathscr{R}$. T is *mixing with respect to* $\mathscr{R}$ ($\mathscr{R}$-mixing) if there exists a sequence of positive numbers ρ_n such that A and B in $\mathscr{R}$ imply

$$\lim_{n \to \infty} \rho_n \mu(T^n A \cap B) = \mu(A)\mu(B). \tag{2.1}$$

Since ρ_n does not depend on A and B, (2.1) is equivalent to

$$\lim_{n\to\infty} \frac{\mu(T^n A \cap B)}{\mu(T^n C \cap D)} = \frac{\mu(A)\mu(B)}{\mu(C)\mu(D)} \tag{2.2}$$

for A, B, C, and D in $\mathscr{R}$.

Let T be $\mathscr{R}$-mixing and let $A \in \mathscr{R}$. A sequence $\rho(T, A)$ is defined by

$$\rho(T, A)_n = \mu(A)^2/\mu(T^n A \cap A). \tag{2.3}$$

Sequences ρ and σ are *equivalent* ($\rho \sim \sigma$) if $\lim_{n\to\infty} \rho_n/\sigma_n = 1$. It follows that $\rho(T, A) \sim \rho(T, B)$ for A and B in $\mathscr{R}$. Let $\rho(T, \mathscr{R})$ denote the equivalence class of sequences $\rho(T, A)$ for $A \in \mathscr{R}$. We refer to $\rho(T, \mathscr{R})$ as the *spreading rate of T with respect to* $\mathscr{R}$ [4].

If T_i is $\mathscr{R}_i$-mixing, $i = 1, 2$, then we will say that T_1 and T_2 are *strongly isomorphic* if $T_1 = S^{-1}T_2S$ and $\mathscr{R}_2 = S\mathscr{R}_1$ for some isomorphism S. If T_1 and T_2 are strongly isomorphic, then $\rho(T_1, \mathscr{R}_1) = \rho(T_2, \mathscr{R}_2)$. Actually the spreading rates are equal as long as $\mathscr{R}_2 \cap S\mathscr{R}_1$ contains one set of positive measure.

In [4] the spreading rate was utilized in order to show that certain mixing transformations were not strongly isomorphic. We will utilize eventually weakly wandering sequences in order to show that certain mixing transformations are not isomorphic.

3. Existence of Eventually Weakly Wandering Sequences

Methods in [2] [5] will be utilized to prove the following result.

THEOREM 3.1. *Let T be ergodic. If there exists a sequence (n_k) and a set B of positive measure such that $\lim_{k\to\infty} m(T^{n_k}B) = 0$, then there exists a subsequence of (n_k) that is an eventually weakly wandering sequence for T.*

Proof. Let $\epsilon_i = 2^{-i}$, $i = 1, 2, 3, \ldots$. Since T is ergodic we can choose a positive integer p_1 such that $B_1 = \bigcup_{i=0}^{p_1} T^i B$ satisfies

$$m(B_1) > 1 - \epsilon_1. \tag{1}$$

Since T is nonsingular, the hypothesis implies

$$\lim_{k\to\infty} m(T^{n_k} B_1) = 0. \tag{2}$$

Choose k_1 such that $q_1 = n_{k_1}$ satisfies

$$m(T^{q_1} B_1) < \epsilon_1. \tag{3}$$

Let $A_1 = B_1 - T^{q_1}B_1$. Hence (1) and (3) imply $m(A_1) > 1 - 2\epsilon_1$ and $A_1 \cap T^{q_1}A_1 = \varnothing$.

Let $n_0 = 0$. Assume we have an increasing sequence $q_i = n_{k_i}$, $1 \leqslant i \leqslant j$, and a set A_j such that

$$m(A_j) > 1 - 2\epsilon_j \quad \text{and} \quad T^{q_i}A_j \cap T^{q_j}A_j = \varnothing, \qquad 0 \leqslant i < j. \tag{4}$$

We now choose p_{j+1} such that $B_{j+1} = \bigcup_{i=0}^{p_{j+1}} T^iB$ satisfies

$$m(B_{j+1}) > 1 - \epsilon_{j+1}. \tag{5}$$

The hypothesis implies

$$\lim_{k\to\infty} m(T^{n_k}B_{j+1}) = 0. \tag{6}$$

Choose k_{j+1} such that $q_{j+1} = n_{k_{j+1}} > q_j$ and

$$\sum_{i=0}^{j} m(T^{q_{j+1}-q_i}B_{j+1}) < \epsilon_{j+1}. \tag{7}$$

Let $A_{j+1} = B_{j+1} - \bigcup_{i=0}^{j} T^{q_{j+1}-q_i}B_{j+1}$. Equations (5) and (7) imply (4) holds with j replaced by $j+1$. Thus by induction we have an increasing sequence (q_j) that is a subsequence of (n_k) and A_j such that (4) holds for all j.

It will now be shown that (q_j) is an eventually weakly wandering sequence for T. Let $\epsilon > 0$ and choose K_ϵ such that

$$\sum_{k=K_\epsilon}^{\infty} 2^{-k+1} < \epsilon. \tag{8}$$

Let $W_\epsilon = \bigcap_{k=K_\epsilon}^{\infty} A_k$. Equations (4) and (8) imply $m(W_\epsilon) > 1 - \epsilon$. Consider $q_j > q_i$ for $j \geqslant K_\epsilon$. Therefore $W_\epsilon \subset A_j$ and (4) implies $T^{q_i}W_\epsilon \cap T^{q_j}W_\epsilon = \varnothing$ Thus W_ϵ is weakly wandering on $(q_j : j \geqslant K)$. It follows that (q_j) is an eventually weakly wandering sequence for T.

COROLLARY 3.2. *If T is ergodic, then every weakly wandering sequence admits an eventually weakly wandering subsequence.*

Proof. If B is weakly wandering on (n_k) then B satisfies the hypothesis of Theorem 3.1. Hence (n_k) admits an eventually weakly wandering subsequence.

4. SPREADING RATES AND EVENTUALLY WEAKLY WANDERING SEQUENCES

In this section we will show that if T is mixing with respect to an infinite sigma-finite invariant measure $\mu \sim m$ then we can use the spreading rate to give a criterion for finding eventually weakly wandering sequences.

Let $\mathcal{R}$ be a ring of sets of finite measure that generates B and let T be $\mathcal{R}$-mixing. Since $\rho(T, A) \sim \rho(T, B)$ for A and B in $\mathcal{R}$ it follows that

$$\lim_{n\to\infty} \mu(T^n A \cap A) = 0, \qquad A \in \mathcal{R}. \tag{4.1}$$

Let $0 = n_0 < n_1 < n_2 < \cdots$ be an increasing sequence of positive integers and denote

$$r_k(A) = \sum_{i=0}^{k-1} \mu(T^{n_k - n_i} A \cap A). \tag{4.2}$$

It will be shown that (4.1) implies that (n_k) can be choosen so that

$$\lim_{k\to\infty} (n_{k+1} - n_k) = \infty \quad \text{and} \quad \sum_{k=1}^{\infty} r_k(A) < \infty. \tag{4.3}$$

By (4.1) we can choose n_1 such that

$$r_1(A) = \mu(T^{n_1} A \cap A) < \tfrac{1}{2}. \tag{1}$$

By (4.1) we can now choose $n_2 > n_1 + 2$ such that

$$r_2(A) = \mu(T^{n_2} A \cap A) + \mu(T^{n_2 - n_1} A \cap A) < \tfrac{1}{4}. \tag{2}$$

In general, given n_{k-1}, (4.1) implies that we can choose $n_k > n_{k-1} + k$ such that $r_k(A) < (\frac{1}{2})^k$. Hence we obtain $\sum_{k=1}^{\infty} r_k(A) < 1$.

It will now be proven that the sequences satisfying (4.3) are eventually weakly wandering sequences for T.

THEOREM 4.4. *Let (n_k) be an increasing sequence of positive integers and fix $A \in \mathcal{R}$. Let $r_k(A)$ be defined as in* (4.2) *and assume* (4.3) *is satisfied. Then (n_k) is an eventually weakly wandering sequence for T.*

Proof. Let $\epsilon > 0$. Choose K_ϵ such that

$$\sum_{k=K_\epsilon}^{\infty} r_k(A) < \epsilon. \tag{1}$$

Let A_ϵ be defined by

$$A_\epsilon = \bigcup_{k=K_\epsilon}^{\infty} \bigcup_{i=0}^{k-1} (T^{n_k - n_i} A \cap A) \tag{2}$$

Equations (1) and (2) imply $\mu(A_\epsilon) < \epsilon$. Let $W = A - A_\epsilon$ and let $k > i \geqslant K_\epsilon$. Equation (2) implies

$$T^{n_k - n_i} W \cap W \subset (T^{n_k - n_i} A) \cap (A - T^{n_k - n_i} A) = \varnothing.$$

Thus W is weakly wandering on $(n_k : k \geqslant K_\epsilon)$.

Now let $B \in \mathscr{R}$ and $r = \mu(B)^2/\mu(A)^2$. Hence $\rho(T, A) \sim \rho(T, B)$ implies

$$\lim_{n\to\infty} \frac{\mu(T^n B \cap B)}{\mu(T^n A \cap A)} = r. \tag{3}$$

Equation (3) implies there exists N such that

$$\mu(T^n B \cap B) < (r+1)\mu(T^n A \cap A), \qquad n \geqslant N. \tag{4}$$

Choose L such that $k > L$ implies $n_{k+1} - n_k \geqslant N$. Hence (4) implies

$$\sum_{k=1}^{\infty} r_k(B) \leqslant \sum_{k=1}^{L} r_k(B) + \sum_{k=L+1}^{\infty} (r+1) r_k(A) < \infty. \tag{5}$$

Equation (5) implies we can now choose K_ϵ and B_ϵ such that $\mu(B_\epsilon) < \epsilon$ and $B - B_\epsilon$ is weakly wandering on $(n_k : k \geqslant K_\epsilon)$.

Lastly, let $E \in \mathscr{B}$ with $\mu(E) < \infty$ and let $\epsilon > 0$. Since $\mathscr{R}$ generates $\mathscr{B}$ there exists $B \in \mathscr{R}$ such that $\mu(E \triangle B) < \epsilon/2$. Let $B_{\epsilon/2}$ and $K_{\epsilon/2}$ be defined as above and let $E_\epsilon = (E - B) \cup B_{\epsilon/2}$. Then $\mu(E_\epsilon) < \epsilon$ and $E - E_\epsilon$ is weakly wandering on $(n_k : k \geqslant K_{\epsilon/2})$.

5. Examples

We will first consider the case where $X = (-\infty, \infty)$ with Lebesgue measure μ. Let $\mathscr{R}$ be the ring of sets that are contained in a finite interval and have boundary measure zero. A stacking construction for $\mathscr{R}$-mixing transformations is given in [4] and this construction will be referred to in the proof below.

THEOREM 5.1. *Let (s_n) be an increasing sequence of positive integers. There exists an $\mathscr{R}$-mixing transformation T on $(-\infty, \infty)$ and K such that $k \geqslant K$ implies*

(a) $\mu(\bigcup_{i=1}^{k} T^{s_i}[0, 1]) \leqslant k/3$.

(b) $h(T) = \infty$.

(c) *If $\mu(B) < \infty$ and $\epsilon > 0$, then there exists A such that $\mu(A \triangle B) < \epsilon$ and the induced transformation T_A is isomorphic to a Bernoulli shift with infinite entropy.*

Proof. Let T be as in Theorem 3.17 [4]. Therefore T is $\mathscr{R}$-mixing and $h(T) = \infty$. It will be shown that the sequence u_k, $k \geqslant 1$, in the construction in [4] can be chosen so that (a) is satisfied. We can take $G_1 = [0, 1]$ and there ore

$$[0, 1] \subset G_{k+1} = S^{u_k}(G_k, g_k), \qquad k \geqslant 1. \tag{1}$$

We can also assume $\mu(g_k) < 1\ 10$ for $k \geqslant 1$. Therefore u_k can be chosen so large that

$$\mu\left(\bigcup_{i=1}^{k} T^{s_i}[0, 1]\right) < 2\mu(G_{k+1}) < 2(1 + k/10). \tag{2}$$

Part (a) follows from (2) if $k \geqslant K = 15$. The proof of (c) is as in Theorem 3.17 [4].

COROLLARY 5.2. *There exists a sequence of $\mathscr{R}$-mixing transformations T_i, $i \geqslant 1$, on $(-\infty, \infty)$ such that*

(a) $W(T_i) \neq W(T_j), i \neq j$.
(b) $h(T_i) = \infty, i \geqslant 1$.
(c) *Each transformation T_i, $i \geqslant 1$, induces a Bernoulli shift with infinite entropy on $[-n, n]$, $n \geqslant 1$.*

Proof. Let T_1 be constructed as in Theorem 5.1 with respect to $s_n = n$. Therefore (b) holds and (c) follows from (c) of Theorem 5.1 with $A = B = [-n, n]$ and the construction of T_1 in [4].

Assume T_i, $1 \leqslant i \leqslant l$, have been obtained satisfying (a)–(c). Choose an eventually weakly wandering sequence $\sigma_i = (s_{i,n}) \in W(T_i)$, $1 \leqslant i \leqslant l$. We now form a single increasing sequence (s_n) by taking blocks of consecutive terms in the sequences σ_i, where the jth block has length 2^j. Let $s_1 = s_{1,1}$ and $s_2 = s_{1,2}$. In general, suppose s_n has been defined for $n \leqslant n_u = \sum_{i=1}^{u} 2^i$. Suppose $s_{n_u} \in \sigma_p$, where $p < l$. Let r be the smallest positive integer such that $s_{p+1,r} > s_{n_u}$. Define $s_n = s_{p+1,r+n}$, $n_u < n \leqslant n_{u+1}$. If $p = l$, then $s_n = s_{1,r+n}$, $n_u < n \leqslant n_{u+1}$. Thus (s_n) is defined for all n by induction.

By Theorem 5.1 we obtain T_{l+1} such that for all sufficiently large k we have

$$\mu\left(\bigcup_{i=1}^{k} T_{l+1}^{s_i}[0, 1]\right) \leqslant k/3. \tag{1}$$

Now if $\sigma_j \in W(T_{l+1})$ for some j, $1 \leqslant j \leqslant l$, then for $\epsilon > 0$ and all k sufficiently large we have

$$\mu\left(\bigcup_{i=1}^{k} T_{l+1}^{s_i}[0, 1]\right)\Big/ k > 1 - \epsilon. \tag{2}$$

Now the construction of (s_n) above implies that (s_n) coincides with $(s_{j,r+n})$ for $n_u < n \leqslant n_{u+1}$ for infinitely many u. Hence (2) and $n_{u+1} = 2n_u + 2$ contradicts (1) for sufficiently large k. Therefore $\sigma_j \notin W(T_{l+1})$, $1 \leqslant j \leqslant l$, which implies part (a). Parts (b) and (c) follows for T_{l+1} as verified for T_1 above. Therefore the sequence T_i, $i \geqslant 1$, is defined by induction.

Note that Corollary 5.2 (c) implies that if $T_{i,n}$ is the transformation induced by T_i on $[-n, n]$ then $T_{i,n}$ is isomorphic to $T_{j,n}$ for all n [6]. However, (a) implies that T_i is not isomorphic to T_j.

We will now consider examples of transformations generated by random walks on the integers, as described in [4]. The following result follows from Spitzer [7, p. 75].

THEOREM 5.3. *Let T correspond to a strongly aperiodic random walk with mean zero and finite second moment. There exists a constant c such that for every set I of measure one in the ring generated by cylinder sets,* $\rho(T, I)_n \sim c\sqrt{n}$.

COROLLARY 5.4. *Let T be as in Theorem 5.3. The following conditions are each sufficient for a sequence* (n_k) *to be eventually weakly wandering for T.*

(a) $n_k \sim \alpha^k$, $\alpha > 1$.
(b) $n_k \sim k^\beta$, $\beta > 5$.

Proof. It will be shown that the criterion of Theorem 4.4 is satisfied. The condition of Theorem 5.3 implies there exists a constant c_1 such that

$$\begin{aligned} r_k(I) &= \sum_{i=0}^{k-1} \mu(T^{n_k - n_i} I \cap I) \\ &\leqslant c_1 \sum_{i=0}^{k-1} \frac{1}{\sqrt{n_k - n_i}} \end{aligned} \tag{1}$$

Equation (1) and (a) imply there exists a constant c_2 such that

$$\begin{aligned} r_k(I) &\leqslant c_2 \sum_{i=0}^{k-1} \frac{1}{\sqrt{\alpha^k - \alpha^i}} \\ &\leqslant c_2 \frac{k}{\sqrt{\alpha^k - \alpha^{k-1}}} \\ &= c_3 \frac{k}{\alpha^{k/2}}, \end{aligned} \tag{2}$$

where $c_3 = c_2/\sqrt{1 - 1/\alpha}$. Since $\alpha > 1$, (2) implies $\sum_{k=1}^{\infty} r_k(I) < \infty$.

For (b) there exists a constant c_4 such that for $k > 1$

$$\begin{aligned} r_k(I) &\leqslant c_4 \sum_{i=0}^{k-1} \frac{1}{\sqrt{k^\beta - i^\beta}} \leqslant c_4 \frac{k}{\sqrt{k^\beta - (k-1)^\beta}} \\ &\leqslant c_4 \frac{k}{\sqrt{\beta (k-1)^{\beta - 1}}} \leqslant \frac{2c_4}{\sqrt{\beta}\, k^{\gamma}}, \end{aligned} \tag{3}$$

where $\gamma = \beta/2 - \frac{3}{2}$. If $\beta > 5$, then $\gamma > 1$ and (3) imply $\sum_{k=1}^{\infty} r_k(I) < \infty$.

A more detailed argument shows that condition (b) implies $(n_k) \in W(T)$ if $\beta > 4$.

Mixing transformations generated by random walks have infinite entropy. Examples were given in [3] of mixing transformations generated by random walks with different spreading rates; hence the transformations cannot be strongly isomorphic. Furthermore, results in [1] imply that the spreading rate is a general isomorphism invariant for random walks as in Theorem 5.3; hence such random walks with different spreading rates are not isomorphic.

Using eventually weakly wandering sequences, we will construct examples of mixing Markov shifts with the same finite entropy that are not isomorphic. Let ϵ_i, $i \geqslant 0$, satisfy $0 < \epsilon_i < \frac{1}{2}$ and

$$-\sum_{i=0}^{\infty} (\epsilon_i \log \epsilon_i + (1 - \epsilon_i)\log(1 - \epsilon_i)) < \infty \tag{5.5}$$

A corresponding doubly stochastic matrix $\pi(i, j)$ with invariant measure $\lambda(i) = 1$ for all i is defined as follows.

$$\begin{aligned}
i = 0: \quad & \pi(0, i) = \epsilon_0, \quad |i| = 1. \\
& \pi(0, 0) = 1 - 2\epsilon_0. \\
i \geqslant 1: \quad & \pi(i, i-1) = \epsilon_{i-1}, \\
& \pi(i, i+1) = \epsilon_i, \\
& \pi(i, i) = 1 - \epsilon_{i-1} - \epsilon_i. \\
i \leqslant -1: \quad & \pi(i, i+1) = \epsilon_{-i-1}, \\
& \pi(i, i-1) = \epsilon_{-i}, \\
& \pi(i, i) = 1 - \epsilon_{-i-1} - \epsilon_{-i}.
\end{aligned} \tag{5.6}$$

We can now let T_1 correspond to $\epsilon_i = 1/2^{i+2}$, $i \geqslant 0$. In this case $h(T_1) < \infty$ since (5.5) is satisfied. Let $\sigma \in W(T_1)$. It is not difficult to choose ϵ_i, $i \geqslant 0$, so that if T_2 is the corresponding mixing Markov shift, then $h(T_1) = h(T_2)$ and $\sigma \notin W(T_2)$. Briefly, the ϵ_i for large i are chosen very small to guarantee $\sigma \notin W(T_2)$. The ϵ_i for small i are chosen sufficiently large to guarantee $h(T_2) = h(T_1)$. Thus T_1 and T_2 have the same entropy but are not isomorphic since $W(T_1) \neq W(T_2)$.

We will now give simple examples of isomorphic mixing transformations on an infinite measure space that are obtained from isomorphic transformations on a finite measure space. Let S_1 and S_2 be isomorphic mixing transformations on the unit interval; for example, two Bernoulli shifts with the same entropy. Let T be any mixing transformation on $(-\infty, \infty)$ and let $T_i = T \times S_i$, $i = 1, 2$, be the corresponding product transformation on $(-\infty, \infty) \times [0, 1]$. Since T is mixing and S_1 and S_2 are mixing, it follows that T_1 and T_2 are mixing. Furthermore, since S_1 and S_2 are isomorphic, it follows that T_1 and T_2 are isomorphic.

It is not hard to show that any ergodic transformation that admits weakly wandering sequences also admits weakly wandering sequences that are not eventually weakly wandering. In a subsequent paper we will explore this and other properties of eventually weakly wandering sequences.

Lastly, note that if the definition of an isomorphism S is weakened to require only the S and S^{-1} preserve sets of positive measure, then $W(T)$ still remains an isomorphism invariant. In particular, this is the case if S and S^{-1} are each measure preserving up to multiplication by a positive constant.

Acknowledgments

The authors wish to thank Arshag Hajian for his very helpful suggestions.

References

1. J. Aaronson, Ratio limit properties and a metric invariant for Markov shifts, *Israel J. Math.* **27** (1977), 93–123.
2. N. A. Friedman, On transformations with weakly wandering sets, *Acta Math. Acad. Sci. Hung.* **17** (1966), 451–455.
3. N. A. Friedman, "Introduction to Ergodic Theory," Van Nostrand Reinhold, New York, 1970.
4. N. A. Friedman, Mixing transformations in an infinite measure space, this volume.
5. A. B. Hajian and S. Kakutani, Weakly wandering sets and invariant measures, *Trans. Amer. Math. Soc.* **110** (1964), 136–151.
6. D. S. Ornstein, Bernoulli shifts with infinite entropy are isomorphic, *Advances in Math.* **5** (1970), 339–348.
7. F. Spitzer, "Principles of Random Walk," Van Nostrand Reinhold, New York, 1964.

STUDIES IN PROBABILITY AND ERGODIC THEORY
ADVANCES IN MATHEMATICS SUPPLEMENTARY STUDIES, VOL. 2

Gap Sequences and Eventually Weakly Wandering Sequences[†]

MARTIN H. ELLIS AND NATHANIEL A. FRIEDMAN

Department of Mathematics, State University of New York, Albany, New York

1. INTRODUCTION[1]

In this paper we shall examine under what conditions an ergodic transformation T admits a sequence of positive integers $G = (g_i)$ such that if $W = (w_i)$ satisfies $w_{i+1} - w_i \geqslant g_i$, $i = 1, 2, \ldots$, then W is e.w.w. for T. In this case G is a *gap sequence for* T. Every w.w. sequence admits e.w.w. subsequences [1]. However, not all ergodic transformations admitting e.w.w. sequences admit gap sequences. In particular, a gap sequence cannot exist if a transformation admits an increasing sequence with no w.w. subsequence, such as a recurrent sequence [4]. In Section 3 we shall show that gap sequences exist for transformations that are mixing with respect to an infinite invariant measure, which includes the case of mixing Markov shifts with infinite σ-finite invariant measure. Examples of gap sequences will be given for mixing transformations generated by random walks.

If an ergodic transformation has an infinite σ-finite invariant measure, then a necessary and sufficient condition for the existence of a gap sequence is that the transformation be of type zero as defined in [5]. The definition of type zero will also be extended to transformations that may not have a σ-finite invariant measure and it will be shown that type zero is sufficient for the existence of a gap sequence in this case. We do not know if type zero is also a necessary condition.

In Section 4 it will be shown that any finite union of translates of an e.w.w. sequence is w.w. but not e.w.w. This provides many examples of w.w. sequences that are not e.w.w. It is an open problem whether all finite unions of translates of a w.w. sequence are w.w. However, it will be shown that finite unions of certain translates of a w.w. sequence are w.w.

[†] Research partially supported by National Science Foundation Grant MCS 7606735.

[1] For a general review of w.w. (weakly wandering) and e.w.w. (eventually weakly wandering) sequences, see [1, Introduction].

ISBN 0-12-599102-9

2. DEFINITIONS

Let $(X, \mathscr{B}, m)$ be a measure space with $m(X) = 1$. Let T be a one-to-one point transformation mapping X into X such that the domain and range of T are sets of measure one. T is *measurable* if $B \in \mathscr{B}$ implies $TB \in \mathscr{B}$ and $T^{-1}B \in \mathscr{B}$. A measurable transformation T is *nonsingular* if $m(B) = 0$ implies $m(TB) = m(T^{-1}B) = 0$. Hereafter T will be assumed nonsingular. A sigma-finite measure μ defined on $\mathscr{B}$ is *equivalent to* $m(\mu \sim m)$ if μ and m have the same sets of measure zero. A measure μ is an *invariant measure for* T if $\mu(TB) = \mu(B)$, $B \in \mathscr{B}$.

A sequence of integers $W = (w_i : i = 1, 2, \ldots)$ is *weakly wandering* (w.w.) *for* T if there exists a set B of positive measure such that the images $T^{w_i}B$, $i = 1, 2, \ldots,$ are pairwise disjoint. We also say that B is a w.w. set *for* T and B is w.w. *on* W. A transformation admits a finite invariant measure (f.i.m.) $\mu \sim m$ if and only if the transformation does not admit a w.w. set [5].

A sequence of integers $W = (w_i : i = 1, 2, \ldots)$ is *eventually weakly wandering* (e.w.w.) *for* T if for each $\epsilon > 0$ there exists a set A_ϵ and a positive integer i_ϵ such that $m(A_\epsilon) < \epsilon$ and $X - A_\epsilon$ is w.w. on $(w_i : i \geqslant i_\epsilon)$. In this case $B = B \cap A_\epsilon$ satisfies $m(B_\epsilon) < \epsilon$ and $B - B_\epsilon$ is w.w. on $(w_i : i \geqslant i_\epsilon)$, $B \in \mathscr{B}$.

If T admits an infinite sigma-finite invariant measure (σ-f.i.m.) $\mu \sim m$, then it is not difficult to show that W is e.w.w. for T if and only if for each $B \in \mathscr{B}$ with $\mu(B) < \infty$ and $\epsilon > 0$, there exist B_ϵ and i_ϵ such that $\mu(B_\epsilon) < \epsilon$ and $B - B_\epsilon$ is w.w. on $(w_i : i \geqslant i_\epsilon)$.

An ergodic transformation T is of *type zero* if there exists a set B of positive measure such that $\lim_{n \to \infty} m(T^n B) = 0$. This condition will be shown to be equivalent to the condition in [4] when T admits an infinite σ-f.i.m. $\mu \sim m$.

Let T be ergodic and admit an infinite σ-f.i.m. $\mu \sim m$. Let $\mathscr{R}$ be a ring of sets of finite measure that generates $\mathscr{B}$ and assume $T\mathscr{R} = \mathscr{R}$. T is *mixing with respect to* $\mathscr{R}$ ($\mathscr{R}$-mixing) if there exists a sequence of positive numbers (ρ_n) such that A and B in $\mathscr{R}$ imply

$$\lim_{n \to \infty} \rho_n \mu(T^n A \cap B) = \mu(A)\mu(B). \tag{2.1}$$

Since (ρ_n) does not depend on A and B, (2.1) is equivalent to

$$\lim_{n \to \infty} \frac{\mu(T^n A \cap B)}{\mu(T^n C \cap D)} = \frac{\mu(A)\mu(B)}{\mu(C)\mu(D)} \tag{2.2}$$

for A, B, C, and D in $\mathscr{R}$. This definition of mixing was suggested by [6] and implies the mixing condition in [7]. Mixing transformations were constructed by the stacking method in [3].

If T is $\mathscr{R}$-mixing, the fact that $\mu(X) = \infty$ and (2.2) imply

$$\lim_{n \to \infty} \mu(T^n A \cap B) = 0, \qquad A, B \in \mathscr{R}. \tag{2.3}$$

Since μ is σ-finite and $\mu \sim m$, (2.3) implies

$$\lim_{n\to\infty} m(T^n A) = 0, \qquad A \in \mathscr{R}. \tag{2.4}$$

In particular, (2.4) implies T is type zero. Since μ is an invariant measure for T, (2.2) holds with T^{-1} replacing T. Hence we also have

$$\lim_{n\to\infty} m(T^{-n} A) = 0, \qquad A \in \mathscr{R}. \tag{2.5}$$

3. Gap Sequences

We shall first verify that the definition of type zero transformations coincides with [5] when an invariant measure exists.

LEMMA 3.1. *Let T be ergodic and admit a σ-f.i.m. $\mu \sim m$. Then T is of type zero if and only if $\mu(A) < \infty$ implies*

$$\lim_{n\to\infty} \mu(T^n A \cap A) = 0. \tag{0}$$

Proof. Suppose T is of type zero; hence there exists a set B of positive measure such that

$$\lim_{n\to\infty} m(T^n B) = 0. \tag{1}$$

Let $\epsilon > 0$. Since T is ergodic and μ is σ-finite, we can choose k so large that

$$E = \left(\bigcup_{i=0}^{k} T^i B\right) \cap A \qquad \text{satisfies} \quad \mu(A - E) < \epsilon. \tag{2}$$

Since T is nonsingular, (1) implies

$$\lim_{n\to\infty} m(T^n E) = 0. \tag{3}$$

Since $\mu(A) < \infty$ and $\mu \sim m$, (3) implies

$$\lim_{n\to\infty} \mu(T^n E \cap A) = 0. \tag{4}$$

Since μ is invariant, (2) implies that for all integers n

$$\mu(T^n(A - E)) < \epsilon. \tag{5}$$

Since ϵ is arbitrary, (4) and (5) imply (0).

Now suppose (0) holds. Since μ is σ-finite we can choose A_k such that $m(A_k) > 1 - 3^{-k}$ and $\mu(A_k) < \infty$. Now (0) implies we can choose n_k such that

$$m(T^n A_k \cap A_k) < 3^{-k}, \qquad n \geqslant n_k. \tag{6}$$

Now (6) and the choice of A_k implies

$$m(T^n A_k) < 2 \cdot 3^{-k}, \qquad n \geqslant n_k. \tag{7}$$

Let $B = \bigcup_{k=1}^{\infty} A_k$; hence

$$m(B) > 1 - \sum_{k=1}^{\infty} 3^{-k} = \tfrac{1}{2}. \tag{8}$$

Equation (7) implies $\lim_{n\to\infty} m(T^n B) = 0$.

THEOREM 3.2. *If T is ergodic and type zero, then T admits gap sequences. If T also has a σ-f.i.m. $\mu \sim m$ and admits gap sequences, then T is type zero.*

Proof. Let T be type zero and B a set of positive measure such that $\lim_{n\to\infty} m(T^n B) = 0$. Since T is ergodic we can choose positive integers p_k such that $B_k = \bigcup_{i=0}^{p_k} T^i B$ satisfies

$$m(B_k) > 1 - 2^{-k}, \qquad k = 1, 2, 3, \ldots. \tag{1}$$

Since T is nonsingular, there exist positive integers g_k such that $d_k \geqslant g_k$ implies

$$m(T^{d_k} B_k) < 2^{-k}/k, \qquad k = 1, 2, 3, \ldots. \tag{2}$$

Let $n_k - n_{k-1} \geqslant g_k$, where $n_0 = 0$, and define A_k by

$$A_k = B_k - \bigcup_{i=0}^{k-1} T^{n_k - n_i} B_k. \tag{3}$$

Since $n_k - n_i \geqslant n_k - n_{k-1} \geqslant g_k$, $0 \leqslant i \leqslant k-1$, (1), (2), and (3) imply

$$m(A_k) \geqslant m(B_k) - k 2^{-k}/k > 1 - 2^{-k+1}. \tag{4}$$

Let $\epsilon > 0$ and choose L such that

$$\sum_{k=L}^{\infty} 2^{-k+1} < \epsilon. \tag{5}$$

Let $W = \bigcap_{k=L}^{\infty} A_k$. Equations (4) and (5) imply $m(W) > 1 - \epsilon$. Then for $k > i \geqslant L$, $W \subset A_k$ and (3) imply $T^{n_k} W \cap T^{n_i} W = \varnothing$. Thus W is w.w. on $(n_k : k \geqslant L)$. It follows that (n_k) is e.w.w.

Let us now assume T admits a σ-f.i.m. $\mu \sim m$. We shall show that if T is not type zero, then T cannot admit gap sequences. Thus we assume

$$m(B) > 0 \quad \text{implies} \quad \limsup_{n\to\infty} m(T^n B) > 0. \tag{6}$$

Fix B with $\infty > \mu(B) > 0$ and choose $\delta > 0$ such that

$$\limsup_{n\to\infty} m(T^n B) \geqslant 2\delta. \tag{7}$$

Choose k such that

$$m\left(\bigcup_{i=0}^{k} T^i B\right) > 1 - \delta. \tag{8}$$

Now (7) and (8) imply

$$\limsup_{n\to\infty} m\left(T^n B \cap \bigcup_{i=0}^{k} T^i B\right) \geqslant \delta \tag{9}$$

and (9) implies there exists r, $0 \leqslant r \leqslant k$, such that

$$\limsup_{n\to\infty} m(T^n B \cap T^r B) \geqslant \delta/(k+1). \tag{10}$$

Since T is nonsingular, (10) implies there exists $\eta > 0$ such that

$$\limsup_{n\to\infty} m(T^n B \cap B) \geqslant \eta. \tag{11}$$

Now (11) and Theorem 2 [4] imply there exists an increasing sequence which does not admit a w.w. subsequence for T. Hence gap sequences cannot exist since otherwise all increasing sequences would admit e.w.w. subsequences.

COROLLARY 3.2. *Let T be ergodic and type zero. Then every increasing sequence admits an e.w.w. subsequence.*

COROLLARY 3.3. *Let T be mixing with respect to an infinite invariant measure $\mu \sim m$. Then T admits gap sequences.*

Proof. T is type zero by (2.4).

It is not difficult to construct ergodic transformations that admit a σ-f.i.m. $\mu \sim m$ but are not type zero. Hence they cannot admit gap sequences although they admit e.w.w. sequences by Corollary 3.2 [1]. One can also construct transformations with no σ-f.i.m. $\mu \sim m$ that admit increasing sequences with no w.w. subsequences. Hence they cannot admit gap sequences. A sequences of positive integers $I = (i_n : n = 1, 2, \ldots)$ will be called an *identity sequence for T* if

$$\lim_{n\to\infty} m(T^{i_n} B \triangle B) = 0, \qquad B \in \mathscr{B}.$$

The stacking construction in Example 6.7 [2] can be combined with the stacking construction in Theorem 6.3 [2] to obtain a transformation with no σ-f.i.m. $\mu \sim m$ that still admits identity sequences. An identity sequence cannot have an e.w.w. subsequence; hence gap sequences cannot exist for this transformation.

The following result gives a criterion for a sequence to be a gap sequence for a mixing transformation.

THEOREM 3.4. *Let T be $\mathscr{R}$-mixing and let $G = (g_k : k = 1, 2, 3, \ldots)$ be an increasing sequence of positive integers. If there exists $A \in \mathscr{R}$ and $c > 0$ such that*

(a) $\sum_{k=1}^{\infty} k\mu(T^{g_k}A \cap A) < \infty$,

and

(b) $d \geqslant g_k$ *implies* $\mu(T^d A \cap A) \leqslant c\mu(T^{g_k}A \cap A)$, $k = 1, 2, \ldots,$

then G is a gap sequence for T.

Proof. Suppose $W = (w_k : k = 0, 1, 2, \ldots)$ satisfies $w_k - w_{k-1} \geqslant g_k$, $k = 1, 2, \ldots$; hence

$$w_k - w_i \geqslant w_k - w_{k-1} \geqslant g_k, \qquad 0 \leqslant i < k. \tag{1}$$

Then (1) and (b) imply

$$\sum_{k=l}^{\infty} \sum_{i=0}^{k-1} \mu(T^{w_k - w_i}A \cap A)$$
$$\leqslant \sum_{k=l}^{\infty} \sum_{i=0}^{k-1} c\mu(T^{g_k}A \cap A) = c \sum_{k=l}^{\infty} k\mu(T^{g_k}A \cap A). \tag{2}$$

The left side of (2) is therefore finite by (a). Hence Theorem 4.4 [1] implies W is e.w.w. for T.

The following theorem is similar to Theorem 3.4 and has a similar proof. Theorem 3.5 gives sharper results since condition (a) of Theorem 3.5 is easier to satisfy then condition (a) of Theorem 3.4. For example, see the proof of Corollary 3.7 below.

THEOREM 3.5. *Let T be $\mathscr{R}$-mixing, $G = (g_k : k = 1, 2, 3, \ldots)$ be a sequence of positive integers with $\lim_{n \to \infty} g_k = \infty$, $n_k = \sum_{i=1}^{k} g_k$ for $k \geqslant 1$, and $n_0 = 0$. Suppose there exists an $A \in \mathscr{R}$, $c > 0$, and a positive integer L such that*

(a) $\sum_{k=1}^{\infty} \sum_{i=0}^{k-1} \mu(T^{n_k - n_i}A \cap A) < \infty$

and

(b) *for all integers i and k with $0 \leqslant i < k$ and $k \geqslant L$, if $d \geqslant n_k - n_i$ then $\mu(T^d A \cap A) \leqslant c\mu(T^{n_k - n_i}A \cap A)$.*

Then G is a gap sequence for T.

Proof. Suppose $W = (w_k : k = 0, 1, 2, \ldots)$ satisfies $w_k - w_{k-1} \geqslant g_k$. Then $\lim_{n\to\infty} w_{k+1} - w_k = \infty$, and

$$w_k - w_i \geqslant n_k - n_i, \qquad 0 \leqslant i < k. \tag{1}$$

Now (1) and (b) imply

$$\sum_{k=1}^{\infty} \sum_{i=0}^{k-1} \mu(T^{w_k - w_i} A \cap A)$$

$$\leqslant c \sum_{k=L}^{\infty} \sum_{i=0}^{k-1} \mu(T^{n_k - n_i} A \cap A) + \frac{(L)(L-1)}{2} \mu(A). \tag{2}$$

The left side of (2) is therefore finite by (a). Hence Theorem 4.4 [1] implies W is e.w.w. for T.

COROLLARY 3.6. *Let T correspond to a strongly aperiodic random walk on the integers with mean zero and finite variance. Then $G = (g_k : k = 1, 2, \ldots)$ is a gap sequence for T in the following cases*:

(a) $g_k = \alpha^k$, $\alpha > 1$;
(b) $g_k = k^\beta$, $\beta > 3$.

Proof. The proof of Corollary 5.4 [1] shows that the hypotheses of Theorems 3.4 and 3.5 are satisfied for $\alpha > 1$ and for $\beta > 4$.

A more detailed argument would show that the hypotheses of Theorem 3.5 are satisfied for $\beta > 3$; such an argument will be used to prove the following corollary.

COROLLARY 3.7. *Let T correspond to a strongly aperiodic random walk on $Z \times Z$ with mean zero and finite variance. Then for $\gamma > 1$ $(k^\gamma : k = 1, 2, \ldots)$ is a gap sequence for T.*

Proof. Let A be some fixed set in the ring generated by the cylinder sets. Then

$$\lim_{n\to\infty} \frac{n\mu(T^n A \cap A)}{\mu(A)^2} = c \tag{1}$$

where c is a constant depending upon the variance. Hence condition (b) of Theorem 3.5 is satisfied for large L. To satisfy condition (a) of Theorem 3.5 it suffices to show that for $\beta > 2$

$$\sum_{k=1}^{\infty} \sum_{i=0}^{k-1} \mu(T^{k^\beta - i^\beta} A \cap A) < \infty \tag{2}$$

and from (1), (2) will be satisfied if

$$\sum_{k=1}^{\infty} \sum_{i=0}^{k-1} (k^\beta - i^\beta)^{-1} < \infty. \tag{3}$$

For every positive integer n

$$\begin{aligned}
&\sum_{i=0}^{k-1} (k^\beta - i^\beta)^{-1} \\
&\quad = \sum_{i=0}^{[k-k^{n/(n+1)}]} (k^\beta - i^\beta)^{-1} + \sum_{i=[k-k^{n/(n+1)}]+1}^{[k-k^{(n-1)/(n+1)}]} (k^\beta - i^\beta)^{-1} \\
&\quad\quad + \cdots + \sum_{i=[k-k^{1/(n+1)}]}^{[k-1]} (k^\beta - i^\beta)^{-1} \\
&\quad \leqslant k(k^\beta - (k - k^{n/(n+1)})^\beta)^{-1} + k^{n/(n+1)}(k^\beta - (k - k^{(n-1)/(n+1)})^\beta)^{-1} \\
&\quad\quad + \cdots + k^{1/(n+1)}(k^\beta - (k-1)^\beta)^{-1} \\
&\quad = k^{1-\beta}(1 - (1 - k^{-1/(n+1)})^\beta)^{-1} + k^{n/(n+1)-\beta}(1 - (1 - k^{-2/(n+1)})^\beta)^{-1} \\
&\quad\quad + \cdots + k^{1/(n+1)-\beta}(1 - (1 - k^{-1})^\beta)^{-1} \\
&\quad \leqslant k^{1-\beta+1/(n+1)} + k^{n/(n+1)-\beta+2/(n+1)} + \cdots + k^{1/(n+1)-\beta+1} \\
&\quad = (n+1)k^{1+1/(n+1)-\beta}.
\end{aligned}$$

Hence for every positive integer n

$$\sum_{k=1}^{\infty} \sum_{i=0}^{k-1} (k^\beta - i^\beta)^{-1} \leqslant (n+1) \sum_{k=1}^{\infty} k^{1+1/(n+1)-\beta}. \tag{4}$$

If $\beta > 2$, then for large n, $\beta - 1 - (n+1)^{-1} > 1$, whence the right side of (4) is finite. Hence (3) is satisfied for $\beta > 2$ and the proof is complete.

Note that Corollary 3.7 cannot be extended to $\gamma = 1$ since a set which is weakly wandering on a sequence with gaps $(1, 2, 3, \ldots)$ must be a wandering set, whereas T has no wandering sets.

4. E.W.W. Sequences and W.W. Sequences

It will be shown here that any finite union of translates of an e.w.w. sequence is w.w. but not e.w.w. Some preliminary results will be obtained first.

LEMMA 4.1. *If $W = (w_k)$ is e.w.w. for T, then* $\lim_{k\to\infty} (w_{k+1} - w_k) = \infty$.

Proof. Suppose there exists a positive integer u and k_l, $l = 1, 2, \ldots$, such that

$$w_{k_l+1} - w_{k_l} = u, \qquad l = 1, 2, \ldots. \tag{1}$$

Since T is nonsingular, there exists $\delta > 0$ such that

$$m(A) > 1 - \delta \quad \text{implies} \quad m(T^u A \cap A) > 0. \tag{2}$$

However, since W is e.w.w. for T, (1) implies $T^u A \cap A = \varnothing$ for sets A of measure arbitrarily close to 1, contradicting (2). Thus $\lim_{k\to\infty}(w_{k+1} - w_k) = \infty$.

LEMMA 4.2. *If* $m(E) > 0$ *and* $\lim_{n\to\infty}\inf m(T^n E) = 0$, *then* $E \not\subset \bigcup_{i=1}^{n} T^i E$, $n = 1, 2, \ldots.$

Proof. If $E \subset \bigcup_{i=1}^{k} T^i E$, then

$$E \subset \bigcup_{i=1}^{k} T^i E \subset \bigcup_{i=2}^{k+1} T^i E \subset \cdots \subset \bigcup_{i=n+1}^{n+k} T^i E. \tag{1}$$

However, since T is nonsingular, the hypothesis implies

$$\liminf_{n\to\infty} \sum_{i=n+1}^{n+k} m(T^i E) = 0. \tag{2}$$

Equation (2) contradicts (1) since $m(E) > 0$.

The following result states that a positive integer can be inserted in a w.w. sequence and the new sequence will still be w.w.

LEMMA 4.3. *If* $W = (w_i : i = 1, 2, \ldots)$ *is w.w. for* T, *then for each positive integer* k, $W \cup \{k\}$ *is w.w. for* T.

Proof. Let A be a set of positive measure that is w.w. on W. If $A \not\subset \bigcup_{i=1}^{\infty} T^{w_i - k} A$, then $B = A - \bigcup_{i=1}^{\infty} T^{w_i - k} A$ is w.w. on $W \cup \{k\}$. If $A \subset \bigcup_{i=1}^{\infty} T^{w_i - k} A$, then fix j such that $E = A \cap T^{w_j - k} A$ has positive measure. Let $B = E - T^{w_j - k} E$. Note that $E \subset A$ implies E is w.w. for T; hence E is also w.w. for T^{-1}. Therefore Lemma 4.2 holds for T and T^{-1}. In particular, B has positive measure. Furthermore,

$$\begin{aligned} T^{w_i} B \cap T^{w_l} B &\subset T^{w_i} A \cap T^{w_l} A = \varnothing, \qquad i \neq l, \\ T^{w_i} B \cap T^{k} B &\subset T^{w_i} A \cap T^{w_j} A = \varnothing, \qquad i \neq j, \\ T^{w_j} B \cap T^{k} B &\subset T^{w_j} E \cap (T^{k} E - T^{w_j} E) = \varnothing. \end{aligned}$$

Therefore B is w.w. on $W \cup \{k\}$.

THEOREM 4.4. *If* (w_i) *is e.w.w. for* T, *then* $\bigcup_{j=0}^{l}(w_i + j)$ *is w.w. for* T *but not e.w.w.*

Proof. There exist sets A of measure arbitrarily close to 1 that are e.w.w. on (w_i). Let $E = \bigcap_{i=0}^{l} T^{-i}A$. Since T is nonsingular, E has positive measure if A has measure sufficiently close to 1. Fix A so that E has positive measure and A is w.w. on $(w_i : i \geqslant r)$. Let $B = E - \bigcup_{i=1}^{l} T^i E$. Since $E \subset A$, Lemma 4.2 implies B has positive measure. Furthermore, $B \subset E$ implies that for $i, j \geqslant r$ and $0 \leqslant u \leqslant v \leqslant l$ we have

$$T^{w_i+u}B \cap T^{w_j+v}B \subset T^{w_i}(T^u E) \cap T^{w_j}(T^v E) \subset T^{w_i}A \cap T^{w_j}A = \varnothing, \qquad i \neq j, \tag{1}$$

and

$$T^{w_i+u}B \cap T^{w_i+v}B \subset T^{w_i+u}(B \cap T^{v-u}B) \subset T^{w_i+u}(B \cap T^{v-u}E) = \varnothing, \quad u \neq v. \tag{2}$$

From (1) and (2) it follows that B is w.w. on $\bigcup_{j=0}^{l}(w_i + j : i \geqslant r)$. Lemma 4.3 now implies $\bigcup_{j=0}^{l}(w_i + j : i \geqslant 1)$ is w.w. for T. Lemma 4.1 implies that a union of translates is not e.w.w.

We do not know if every finite union of translates of a w.w. sequence is w.w. However, the method of proof in Theorem 4.4 can be applied to obtain the following result.

THEOREM 4.5. *Let* A *be w.w. on* (w_i) *and* $0 = k_0 < k_1 < k_2 < \cdots < k_l$. *Let* $E = \bigcap_{i=0}^{l} T^{-k_i}A$ *and* $B = E - \bigcup_{0 \leqslant i < j \leqslant l} T^{k_j - k_i}E$. *If* $m(E) > 0$, *then* $m(B) > 0$ *and* B *is w.w. on* $\bigcup_{j=0}^{l}(w_i + k_j)$.

EXAMPLE 4.6. If A is w.w. for T, then A is w.w. for T^{-1} and Lemma 4.2 implies $A \not\subset \bigcup_{i=1}^{n} T^i A$ and $A \not\subset \bigcup_{i=1}^{n} T^{-i}A$. However, an example will be constructed where A is w.w. and $A \subset TA \cup T^{-1}A$. Consider an e.w.w. sequence (w_i) where $w_{i+1} - w_i \geqslant 2$ (by Lemma 4.1 any e.w.w. sequence (w_i) will satisfy $w_{i+1} - w_i \geqslant 2$ after removing finitely many (terms).

Theorem 4.4 implies $(w_i) \cup (w_i + 1)$ is a w.w. sequence. Let B be a w.w. set on this sequence. Then $A = B \cup TB$ is w.w. on (w_i) and $A \subset TA \cup T^{-1}A$.

REFERENCES

1. M. H. ELLIS AND N. A. FRIEDMAN, On eventually weakly wandering sequences, this volume.
2. N. A. FRIEDMAN, "Introduction to Ergodic Theory," Van Nostrand Reinhold, New York, 1970.
3. N. A. FRIEDMAN, Mixing transformations in an infinite measure space, this volume.

4. A. B. Hajian and Y. Ito, Weakly wandering and related sequences, *Z. Wahrscheinlichkeitstheorie und Verw. Gebiete* **8** (1969), 150–164.
5. A. B. Hajian and S. Kakutani, Weakly wandering sets and invariant measures, *Trans. Amer. Math. Soc.* **110** (1964), 136–151.
6. U. Krengel and L. Sucheston, On mixing in infinite measure spaces, *Z. Wahrscheinlichkeitstheorie und Verw. Gebiete* **13** (1969), 150–164.
7. K. Krickenberg, Strong mixing properties of Markov chains with infinite invariant measure, *Proc. 5th Berkeley Symp., Math. Stat. Prob.* **2**, Part II (1965), 431–445.

STUDIES IN PROBABILITY AND ERGODIC THEORY
ADVANCES IN MATHEMATICS SUPPLEMENTARY STUDIES, VOL. 2

The Breakdown of Automorphisms of Compact Topological Groups

G. MILES AND R. K. THOMAS

Department of Mathematics, Birkbeck College, London, England

1. INTRODUCTION

This paper is one of three which together prove that an ergodic group automorphism of a compact separable topological group is Bernoullian (i.e., isomorphic to a Bernoulli shift—see [1, 2] for terminology, etc.). The result is being published in separate papers because each is self-contained and has wider interest than this particular result. The present paper concerns the breakdown of a group automorphism into special types. This has been attempted before—(see [3, 4], [5, 6])—but the present result is much simpler and more powerful. It is hoped that our result will lead to a greater understanding of compact group automorphisms and perhaps eventually to some sort of classification of them.

2. STATEMENT OF RESULT

We shall prove the following:

THEOREM A. *Let T be an automorphism of a compact separable topological group G. Then G contains a sequence G_i of closed normal subgroups such that $G = G_0 \supset G_1 \supset G_2 \supset \cdots$, $TG_i = G_i$ for all i and $\bigcap_{i=0}^{\infty} G_i = e$ and for each i there exists a finite sequence $G = G_i^0 \supset G_i^1 \supset G_i^2 \supset \cdots \supset G_i^{k(i)} = G_i$ of closed normal subgroups satisfying $TG_i^j = G_i^j$ for all j such that:*

(i) *the automorphism T_{G/G_i} induced on G/G_i is a Markov process with a Lie group state space;*

(ii) *the automorphism $T_{G_i^j/G_i^{j+1}}$ induced on G_i^j/G_i^{j+1} is a group Bernoulli automorphism for $j \geqslant 2$;*

(iii) *the automorphism $T_{G_i^1/G_i^2}$ is a generalized torus automorphism;*

(iv) *the automorphism T_{G/G_i^1} is an automorphism of the nontoroidal Lie group G/G_i^1.*

ISBN 0-12-599102-9

The exact meaning of (i) will be made clear in Lemma 4. Generalized tori and their automorphisms will be considered in Section 3. Now we explain (ii).

A Bernoulli automorphism (or shift) is a Markov process with complete independence between terms, i.e., it sends a sequence $\{x_i\}_{i=-\infty}^{\infty}$, $x_i \in X$ (the state space), to $\{y_i\}$ where $y_i = x_{i+1}$ (shifts to the left). X is a Lebesgue space with normalized measure and the sequence space is given the product measure which is of course preserved by the shift.

A Bernoulli automorphism is called a *group Bernoulli automorphism* when the state space is a group with Haar measure. It is a group automorphism in the ordinary sense (with component-wise multiplication of sequences).

The automorphisms (i)–(iv) are all allowed to be trivial or degenerate. The state space of the Bernoulli automorphism can be one point—the group identity; the generalized torus can be a torus or the identity group; G/G_i^1 can be any Lie group (including the identity group) which is not a torus (and does not have a normal subgroup that is a nontrivial torus).

In proving that an ergodic group automorphism is Bernoullian, the automorphisms in (ii) and (iv) can be quickly disposed with. If T is ergodic on G (with respect to Haar measure), then the induced automorphism on G/G_i^1 must be ergodic for each i. But there are no ergodic group automorphisms of nontrivial Lie groups other than tori [7]. So in this case $G_i^1 = G$ for all i, disposing with (iv). The following lemma deals with (ii).

LEMMA 1. *If T is an automorphism of a group G and H is a closed normal subgroup of G such that $TH = H$ and T_H (T restricted to H) is a group Bernoulli automorphism, then T is measure-theoretically* (*but not algebraically in general*) *isomorphic to $T_{G/H} \otimes T_H$.*

Proof. We put $S = T_{G/H}$ and $\sigma = T_H$. Using a cross-section for G/H we can express T (measure-theoretically) as a skew-product transformation on $G/H \otimes H: T(x, y) = (Sx, \varphi(x)\sigma(y))$, where φ is some measureable map of G/H into H. σ is group Bernoulli, therefore $\varphi(x)$ is a sequence $\{\varphi_i(x)\}$. Let F be given by $F(x, y) = (x, f(x)y)$, where $f = \{f_i\}$ is some measureable map of G/H into H. Then we have

$$F^{-1}TF(x, y) = (Sx, [f(Sx)]^{-1}\varphi(x)\sigma(f(x))\sigma(y))$$

and so if

$$[f(Sx)]^{-1}\varphi(x)\sigma(f(x)) = e \tag{*}$$

for all $x \in G/H$, then $F^{-1}TF = S \otimes \sigma$, i.e., T is measure isomorphic to $S \otimes \sigma$. So to complete the proof we must find an f that satisfies (*). But (*) is equivalent to the set of equations:

$$\varphi_i(x)f_{i+1}(x) = f_i(Sx)$$

which has a solution:

$$f_0(x) = e;$$
$$f_1(x) = [\varphi_0(x)]^{-1};$$
$$f_i(x) = [\varphi_0(S^{i-1}x)\varphi_1(S^{i-2}x)\cdots\varphi_{i-1}(x)]^{-1}, \qquad i > 1;$$
$$f_i(x) = \varphi_{-1}(S^i x)\varphi_{-2}(S^{i+1}x)\cdots\varphi_i(S^{-1}x), \qquad i < 0.$$

This completes the proof.

We also need

LEMMA 2. *If the measure algebra for a measure-preserving transformation T is generated by an increasing sequence of T-invariant algebras $\{\mathscr{A}_i\}_{i=1}^{\infty}$ such that T restricted to each $\mathscr{A}_i$ is Bernoullian, then T is Bernoullian.*

The proof of this is given by Ornstein [2].

So for an ergodic group automorphism T, if we know that ergodic generalized torus automorphisms are Bernoullian, then, using Theorem A, we can apply Lemma 1 repeatedly to show that the automorphism induced on G/G_i is Bernoullian for every i (since products of Bernoullian transformations are Bernoullian). Application of Lemma 2 then shows that T is Bernoullian. So Theorem A together with these two lemmas reduces the problem of a general ergodic group automorphism to that of an ergodic generalized torus automorphism—the remaining two papers [8] and [9] are concerned with this case only.

3. PROOF OF THEOREM A

We recall the well known fact that a compact separable topological group G is a direct limit of Lie group representations, i.e., there exists a decreasing sequence of closed normal subgroups $H_1 \supset H_2 \supset \cdots$ such that $\bigcap_{i=1}^{\infty} H_i = e$ and $G/H_i = L_i$ is a Lie group for all i [10]. We shall consider the effect of the automorphism T on these representations. G_i in Theorem A will be put equal to $\bigcap_{n=-\infty}^{\infty} T^n H_i$. We now have a lemma which shows that the H_i can be "normalized" with respect to T in a very important way.

LEMMA 3. *Let $L' = G/H'$ be a Lie group representation of G and T be an automorphism. Then there exists a closed normal subgroup $H \subset H'$ such that G/H is a Lie group, $\bigcap_{n=-\infty}^{\infty} T^n H = \bigcap_{n=-\infty}^{\infty} T^n H'$ and*

(i) $p_H(\bigcap_{i=1}^{m} T^i H) = p_H(TH)[= D]$ *for all $m > 1$;*
(ii) $p_H(\bigcap_{i=1}^{n} T^{-i} H) = p_H(T^{-1}H)[= C]$ *for all $n > 1$.*

p_H denotes the natural projection of G onto $G/H = L$.

Proof. Observe that $L^1 \rhd p_{H'}(TH') \rhd p_{H'}(TH' \cap T^2H') \rhd \cdots$ is a descending chain of closed normal subgroups and so, since L^1 is compact Lie, there exists a number M and a subgroup D' of L' such that $p_{H'}(\bigcap_{i=1}^k T^iH') = D'$ for all $k \geqslant M$. Similarly there exist N and C' such that $p_{H'}(\bigcap_{i=1}^k T^{-i}H') = C'$ for all $k \geqslant N$.

We put $q = \max\{M-1, N-1\}$ and take H to be $\bigcap_{i=0}^q T^iH'$. Now $p_H(\bigcap_{i=1}^m T^iH)$ is isomorphic to $p_{H'}(\bigcap_{i=1}^{m+q} T^iH') = D'$ for all $m \geqslant 1$ and $p_H(\bigcap_{i=1}^n T^{-i}H)$ is isomorphic to $p_{T^qH'}(\bigcap_{i=-n}^{q-1} T^iH')$ which is in turn isomorphic (through T^{-q}) to $p_{H'}(\bigcap_{i=1}^{q+n} T^{-i}H') = C'$ for all $n \geqslant 1$. Now G/H_1 Lie and G/H_2 Lie implies $G/H_1 \cap H_2$ is Lie [10]. Hence G/H is Lie and so H satisfies all the requirements.

DEFINITION. Let G, T, H, L and p_H be as in Lemma 3. With each element $x \in G$ we associate a sequence $\{x_i\}_{-\infty}^{\infty}$, putting $x_i = p_H(T^ix)$. The map $x \mapsto \{x_i\}$ is a homomorphism h of G onto a closed subgroup S_H of the two-way product of copies of $L = G/H$; the kernel of h is $\bigcap_{i=-\infty}^{\infty} T^iH = K$. S_H is given the induced topology and normalized Haar measure so that h is continuous and measure-preserving. We shall call S_H *the sequence space of T on G associated with H*.

LEMMA 4. *Using the notation of Lemma 3 and the definition, T induces an automorphism $T_{G/K}$ on G/K which is isomorphic to the (left) shift operator on S_H. There is a homomorphism τ from L onto L/D such that S_H consists of all sequences $\{x_i\}$ that satisfy $x_{i+1} \in \tau(x_i)$; the kernel of τ is C. The shift on S_H is a Markov process with stationary probability given by the Haar measure on L and transitional probability determined by τ and the Haar measure on D.*

Proof. The first sentence is clear from the definition of S_H.

Let $A_j(a) = \{x \in G \mid x_j = a\}$. We have $T^jA_j(a) = \alpha H$ (α some element of $p_H^{-1}(a)$). The set $\{x_{j+1} \mid x \in A_j(a)\}$ is equal to $p_H(T^{j+1}A_j(a)) = p_H(T\alpha TH)$ which is some coset $\tau(a)$ of D. $\tau = p_H T p_H^{-1}$ is a homomorphism of L onto L/D. The kernel of τ is $p_H(HT^{-1}H) = C$. The conditional measure induced on $\tau(a)$ is just the Haar measure on D (translated). τ is clearly independent of j. Note that $\{x_{j-1} \mid x \in A_j(a)\} = \tau^{-1}(aD)$ (which is a coset of C). So we have determined the one term dependence in S_H—we show now that the many term dependence is the same as the one term, i.e., that we have a Markov process.

Let $A_j(a_1, \ldots, a_k) = \{x \in G \mid x_j = a_1, \ldots, x_{j+k-1} = a_k\}$. From the previous paragraph we see that this set is empty unless $a_i \in \tau(a_{i-1})$ for $i = 2, \ldots, k$. Let $p_H^{-1}(a_i) = \alpha_i H$. We assume now that $a_i \in \tau(a_{i-1})$, $i=2, \ldots, k$. This means that α_i can be taken to be any element of $T\alpha_{i-1}TH$ for each $i > 1$; we shall show that we can take $\alpha_i \in T\alpha_{i-1}(TH \cap \cdots \cap T^{i-1}H)$. Assuming this to be true for i we prove it for $i+1$: we have that $\alpha_i H \cap T\alpha_{i-1}(TH \cap \cdots \cap T^{i-1}H)$ is nonempty and equal to $\alpha_i(H \cap \cdots \cap T^{i-1}H) \subset \alpha_i H$—so we can take α_{i+1} to be any element of $T\alpha_i(TH \cap \cdots \cap T^i H)$. The result is trivially true for $i = 2$ and so it is true for all $i = 2, \ldots, k$. With this choice of α_i it is easy to see that $T^{j+k-1}A_j(a_1, \ldots, a_k) = T^{k-1}\alpha_1 T^{k-1}H \cap T^{k-2}\alpha_2 T^{k-2}H \cap \cdots \cap \alpha_k H = \alpha_k(T^{k-1}H \cap T^{k-2}H \cap \cdots \cap H)$. $\{x_{j+k} \mid x \in A_j(a_1, \ldots, a_k)\} = p_H(T^{j+k}A_j(a_1, \ldots, a_k)) = p_H(T\alpha_k)$. $p_H(\bigcap_{i=1}^k T^i H) = \tau(a_k)$ since $p_H(\bigcap_{i=1}^k T^i H) = D$ by Lemma 3. The conditional measure is once again Haar measure on $\tau(a_k)$—note in this context that τ is measure-preserving. We extend the result to an infinite number of fixed terms by taking limits. The proof of the lemma is complete.

Note. We do not rule out the possibility in Lemmas 3 and 4 that C or D or both may be the identity. If C and D are both the identity, then S_H consists of orbits of τ which is then an automorphism of the Lie group L; it follows that $T_{G/K}$ is isomorphic to τ. If D is trivial but not C, then τ is an endomorphism and $T_{G/K}$ is its natural (automorphism) extension; C trivial but not D gives $T_{G/K}$ as the inverse of a natural extension. $C = D = L$ is also possible. We point out that L/C must be isomorphic to L/D.

Lemma 4 gives us statement (i) of Theorem A (with $G_i = K$). Interesting though this process is, the general Lie group state space is too difficult to handle as it stands—so we break it down giving us the G_i^j's.

LEMMA 5. *Using the notation of Lemmas 3 and 4, let $C \cap D = E$ which we assume to be larger than the identity of L. Then G contains a closed normal subgroup $G' \triangleright H$ such that $TG' = G'$ and*

(i) *T induces a group Bernoulli automorphism on G'/K with state space E;*
(ii) *putting $H' = p_H^{-1}(E)$, $p_{H'}$ the natural projection onto G/H',*

$$p_{H'}(TH') = p_{H'}\left(\bigcap_{i=1}^{m} T^i H'\right)[=D'] \qquad \textit{for all} \quad m > 1$$

and

$$p_{H'}(T^{-1}H') = p_{H'}\left(\bigcap_{i=1}^{n} T^i H'\right)[=C'] \qquad \textit{for all} \quad n > 1;$$

(iii) *the induced automorphism $T_{G/G'}$ on G/G' is a Markov process with Lie group state space G/H' $[=L']$;*

(iv) *C' is isomorphic to C/E and D' to D/E and $C' \cap D'$ is the identity element of L':*

Proof. (i) Let F be the set of all sequences $\{f_i\}$ in S_H with $f_i \in E$ for all i and let $G' = h^{-1}(F)$, h being the homomorphism of G onto S_H. $G' = \bigcap_{i=-\infty}^{\infty} T^i(p_H^{-1}E)$ and so $TG' = G'$. It is clear that G'/K is isomorphic to F and that the induced automorphism $T_{G'/K}$ is isomorphic to the shift on F which is Bernoulli since τ maps E (contained in C) onto D which contains E.

(ii) We prove only the first part of (ii), the proof of the second part being similar. As $p_{H'}(\bigcap_{i=1}^{j} T^iH'), j = 1, 2, \ldots$, is decreasing, it suffices to show that for every $x \in TH'$ there exists an $x' \in \bigcap_{i=1}^{m} T^iH'$ such that $xH' = x'H'$ (m fixed).

Let $x \in TH'$. Now $H' = p_H^{-1}E$ is contained in $p_H^{-1}D = H(TH \cap \cdots \cap T^mH)$ (see Lemma 3). So we can put $x = TaTh$ for some $h \in H$ and $a \in (TH \cap \cdots \cap T^mH)$. Then $xH = TaThH$. But since $p_H^{-1}D = HTH = H(TH \cdots T^mH)$, $ThH = yH$ for some $y \in (TH \cap \cdots \cap T^mH)$. We put $x' = (Ta)y$. Now $x'H = xH$ implying that $x'H' = xH'$ as required. Also we have $y \in T^iH \subset T^iH'$ for $i = 1, \ldots. m$, $Ta \in T^iH \subset T^iH'$ for $i = 2, \ldots, m$ and $a \in H'$ implies $Ta \in TH'$. Hence $x' \in \bigcap_{i=1}^{m} T^iH'$ as required.

(iii) Part (ii) means that H' is normalized with respect to T so that Lemma 4 can be applied (with H' for H). The kernel of the homomorphism h' of G onto $S_{H'}$ is $\bigcap_{i=-\infty}^{\infty} T^iH' = G'$.

(iv) With each element $x \in G$ there is associated a sequence $\{x_i\} \in S_H$ and a sequence $\{x_i'\} \in S_{H'}$ (the latter obtained by (iii)). It is clear that $x_i' = p_E(x_i)$, where p_E is the natural projection of G/H onto G/H'. Thus $C' = p_E(C)$, $D' = p_E(D)$ and $C' \cap D' = p_E(E)$ proving (iv).

LEMMA 6. *Using the notation of Lemma* 3, *if* $C \cap D =$ *the identity element of* L, *then, putting* $\hat{H} = p_H^{-1}(D)$;

(i) $p_{\hat{H}}(T\hat{H}) = p_{\hat{H}}(\bigcap_{i=1}^{m} T^i\hat{H})$ $[=\hat{D}]$ *for all* $m > 1$ *and* $p_{\hat{H}}(T^{-1}\hat{H}) = p_{\hat{H}}(\bigcap_{i=1}^{n} T^{-i}\hat{H})$ $[=\hat{C}]$ *for all* $n > 1$;

(ii) $K = \bigcap_{n=-\infty}^{\infty} T^nH = \bigcap_{n=-\infty}^{\infty} T^n\hat{H}$;

(iii) $\hat{C}$ *is isomorphic to* C *and* $\hat{D}$ *is isomorphic to* D. (i), (ii), *and* (iii) *are also true for* $\hat{H} = p_H^{-1}(C)$.

Proof. We give the proof only for $\hat{H} = p_H^{-1}(D)$. The proof of (i) is essentially the same as the proof of part (ii) of Lemma 5—we omit the details. (i) tells us that $\hat{H}$ is normalized with respect to T and so we can apply Lemma 3. The point $x \in G$ is associated with the sequence $\{\hat{x}_i\}$ in $S_{\hat{H}}$, where $\hat{x}_{i+1} \in \hat{\tau}\hat{x}_i$; x is of course also associated with the sequence $\{x_i\}$ in S_H as before. It is clear that $\hat{x}_i = p_D(x_i)$ for all i, here p_D denotes the natural projection of G/H onto $G/\hat{H}$—(iii) is easily proved from this observation. Now we prove (ii).

Suppose that points x and y in G have the same sequence $\{\hat{x}_i\}$ in $S_{\hat{H}}$ but different sequences $\{x_i\}$ and $\{y_i\}$ in S_H. Then we must have $x_i \in y_i D$ for all i. For a fixed k we have $x_{k+1}D = y_{k+1}D$. But $x_{k+1}D = \tau(x_k)$. Hence $\tau(x_k) = \tau(y_k)$. But the kernel of τ is C. Hence $x_k \in y_k C \cap y_k D$ which is a single point. So $x_k = y_k$. This is true for all k and so we have a contradiction. Thus the kernel of the homomorphism of G onto $S_{\hat{H}}$ is contained in the kernel of the homomorphism of G onto S_H, i.e., $\bigcap_{n=-\infty}^{\infty} T^n\hat{H} \subset \bigcap_{n=-\infty}^{\infty} T^n H$. The inclusion the other way follows from $H \subset \hat{H}$ and so (ii) is proved.

Now we are in a position to break down the automorphism $T_{G/K}$ of Lemma 4. We start with H, C, D, and K as in Lemma 4 and put these equal to H^1, C^1, D^1, and $G^{(1)}$, respectively. We apply Lemma 5 to obtain $(H^2, C^2, D^2, G^{(2)})$—if $C^1 \cap D^1$ is the identity of G/H_1, then we put $(H^2, C^2, D^2, G^{(2)}) = (H^1, C^1, D^1, G^{(1)})$. Then we apply Lemma 6 to $(H^2, C^2, D^2, G^{(2)})$ to obtain $(\hat{H}^2, \hat{C}^2, \hat{D}^2, G^{(2)})$. Now we apply Lemma 5 again to obtain $(H^3, C^3, D^3, G^{(3)})$—equal to $(\hat{H}^2, \hat{C}^2, \hat{D}^2, G^{(2)})$ if $\hat{C}^2 \cap \hat{D}^2$ is the identity of $G/\hat{H}^2$. We continue in this way, applying Lemmas 5 and 6 alternately, to obtain a sequence $\{(H^i, C^i, D^i, G^{(i)})\}$ (with corresponding $\hat{H}^i$'s, $\hat{C}^i$'s and $\hat{D}^i$'s). A (possibly trivial) Bernoulli automorphism is produced at each step. Note that in applying Lemma 6 it does not matter whether we use C^i or D^i in forming $\hat{H}^i$—we use whichever is the larger. So unless C^i and D^i are both the identity of G/H^i, H^{i+1} is strictly greater than H^i $(i \neq 1)$ though it may be equal to $\hat{H}^i$.

Now we apply an important result on the structure of compact topological groups [10]: if the connected component of the center of a Lie group A contains only the identity, then A has no infinite ascending chain of closed normal subgroups—this result is trivial for finite groups so it is essentially a result about semi-simple Lie groups. A connected abelian Lie group, i.e., a finite-dimensional torus, does contain infinite ascending chains of closed subgroups but in any such chain after a finite number of terms each term must be a finite extension of the previous term—an extension by a nonfinite group increases the dimension. Now $p_H(H^i)$ is an ascending chain of closed normal subgroups in $L = G/H$. Combining the above with the observation that C^{i+1} and D^{i+1} are isomorphic to $C^i/C^i \cap D^i$ and $D^i/C^i \cap D^i$, respectively, we deduce that after a finite number of terms the C^i's and D^i's are all finite. Eventually they both reach their minimum size C^{k-2} and D^{k-2} at the $(k-2)$th stage. C^{k-2} and D^{k-2} must both be in the connected component of the center of G/H^{k-2} (which is a finite-dimensional torus). We cannot reduce C^{k-2} or D^{k-2} any further, but we can take one more step in the breakdown. For this we need the following lemma:

LEMMA 7. *Using the notation of Lemmas* 3 *and* 4, *if* C *and* D *are both finite subgroups of the connected component of the center* L_0 *of* $L = G/H$ *and*

$C \cap D$ is the identity of L_0, then G contains a closed normal subgroup G' such that $TG' = G'$ and

(i) *the automorphism $T_{G'/K}$ induced on G'/K is a generalized torus automorphism;*

(ii) *the automorphism $T_{G/G'}$ induced on G/G' is an automorphism of a nontoroidal Lie group.*

Proof. We put $H' = p_H^{-1}(L_0)$ and $G' = \bigcap_{i=-\infty}^{\infty} T^i H'$. Clearly $TG' = G'$. It is clear that the homomorphism h mapping G onto S_H maps G'/K isomorphically onto the set of all sequences $\{x_i\}$ for which $x_i \in L_0$ for all i. Since L_0 is a torus, $C \cap D = e$ and C and D are both finite and contained in L_0, the shift on these sequences is a generalized torus automorphism (see the next section). This proves (i).

Now $p_H(TH') = L_0$ and $p_H(H') = L_0$; thus $p_{H'}(TH')$ is the identity of G/H' as is $p_{H'}(T^{-1}H')$. Hence $T_{G/G'}$ is isomorphic to an automorphism of the Lie group G/H' (of course G' must equal H')—see the note following the proof of Lemma 4. G/H' is clearly not a torus (unless it be the identity)—so we have (ii).

We apply Lemma 7 to $(H^{k-2}, C^{k-2}, D^{k-2}, G^{(k-2)})$ to obtain $G^{(k-1)}$. We change the superscripts on the $G^{(i)}$'s to fit in with the notation used in the statement of Theorem A—we put $G^j = G^{(k-j)}$.

The proof of Theorem A is now complete. We give the G^j's the subscripts i corresponding to the representations $G \to G/H_i$ and $G_i = \bigcap_{n=-\infty}^{\infty} T^n H_i$.

4. Generalized Torus Automorphisms

Generalized torus automorphism is the name we give to the Markov process obtained in Lemma 4 when the state space G/H is a finite-dimensional torus, C and D are both finite, and $C \cap D$ is the identity of G/H—this is the situation described in the proof of Lemma 7. To examine these automorphisms we assume that K is the identity of G; G is then a "generalized torus." We shall give some examples and discuss some of the properties of this important class of groups.

Generalized torus automorphisms include automorphisms of tori—these occur when H, C, and D are all equal to the identity of G (see the note following the proof of Lemma 4). In the torus case the sequences of S_H are just the orbits of T. We wish to consider the generalized torus case in the same way. The elements of S_H are in general not orbits of T but, since D is discrete, we can regard them *locally* as orbits of T in $L = G/H$, i.e., we consider T as a relation on L taking a point to a coset of D. This relation is locally euclidean and so is locally represented by a matrix. We now find this matrix.

We represent the n-dimensional tori L and L/D in the usual way as R^n factored by the integer lattice (the unit cube with addition mod 1). The natural projection $L \to L/D$ is then represented by some integer matrix P. The map $\tau: L \to L/D$ is a torus endomorphism and so represented by a matrix V. The local behavior of T is given by the matrix $U = P^{-1}V$; U is clearly a rational matrix.

U in fact represents the global behavior of T if we interpret the situation correctly: we pull L back to R^n, apply U and then reduce mod 1. D is then the image of the integer lattice in R^n. So we are to regard T as a rational matrix acting on a torus as opposed to an integer matrix acting on a torus—hence the term "generalized torus automorphism." We call U the matrix associated with T. We note that U is not unique—it is dependent on the representations of L and L/D. These are subject to changes of basis and so U is unique only up to similarity.

It will be useful to find the dual group Γ and the dual automorphism. We will use the convention of writing the induced automorphism on the right so that $(\gamma T)(x) = \gamma(Tx)$ for all $x \in G$ and $\gamma \in \Gamma$. We start by taking the annihilator of H to be $Z^n \subset Q^n$ (Z = integers, Q = rationals). The annihilator of $T^{-1}H$ consists of the characters of Z^n composed with T; it is natural to take this group to be $(Z^n)U \subset Q^n$—this is readily justified. Similarly we take the annihilator of $T^{-i}H$ to be $(Z^n)U^i$ for $i = -1, \pm 2, \pm 3, \ldots$. Γ is then the smallest subgroup of Q^n containing Z^n and invariant under U—we say that Γ is generated by Z^n and U (or T). The dual automorphism is represented by U (acting on the right).

It is often easier to describe T in terms of its dual. The dual group is only unique up to isomorphism and so one has to find a suitable subgroup, isomorphic to Z^n, to be the annihilator of H. We can now give the examples.

Example 1. Let Γ be the dyadic rationals, i.e., all rationals of the form $p/2^m$, p and m integers. Let T be multiplication by 2. We take H to be the annihilator of $Z \subset \Gamma$. G/H is then the circle (the reals mod 1). Now TH is annihilated by $ZT^{-1} = \frac{1}{2}Z$; thus $TH \subset H$ and $p_H(TH) =$ the identity in $G/H = L$; so D is the identity. $T^{-1}H$ is annihilated by $ZT = 2Z$, $T^{-2}H$ is annihilated by $4Z$, etc. Hence $p_H(T^{-1}H) = p_H(\bigcap_{i=1}^{m} T^{-i}H) = C = \{0, \frac{1}{2}\}$. It is clear that Γ is generated by Z and T so that $\bigcap_{i=-\infty}^{\infty} T^iH = e$. Thus G is isomorphic to S_H which consists of all sequences $\{x_i\}$ satisfying $2x_i = x_{i+1}$ for all i. G is usually called the solenoidal group. T is the natural extension of the endomorphism on the circle sending x to $2x$ mod 1.

Example 2. Let Γ be the 6-adic rationals and let T be multiplication by $\frac{2}{3}$. We take H to be the annihilator of Z as before. Using the same techniques as in Example 1, we get that $C = \{0, \frac{1}{2}\}$ and $D = \{0, \frac{1}{3}, \frac{2}{3}\}$; $\tau: L \to L/D$ is the map $x \to 2x$. Γ is again generated by Z and T and so G is isomorphic to S_H which consists of all sequences $\{x_i\}$ satisfying $2x_i = 3x_{i+1}$ for all i. This is the simplest possible example with both C and D nontrivial (remember that $C \cap D = e$).

EXAMPLE 3. Finally we give a two-dimensional example. Let the matrix associated with T be

$$\begin{pmatrix} 0 & \frac{2}{3} \\ 1 & \frac{1}{2} \end{pmatrix}$$

(this is U in the notation used above). Γ is the group generated by Z^2 and this matrix. H is the annihilator of Z^2 and $L = G/H$ is the two-dimensional torus which we represent in the usual way as the unit square mod 1. It is easy to see that D is the subgroup $\{(0, 0), (0, \frac{1}{2}), (\frac{1}{3}, 0), (\frac{1}{3}, \frac{1}{2}), (\frac{2}{3}, 0), (\frac{2}{3}, \frac{1}{2})\}$ and C is the subgroup $\{(0, 0), (\frac{1}{2}, 0), (\frac{1}{4}, \frac{1}{2}), (\frac{3}{4}, \frac{1}{2})\}$. If the natural projection $L \to L/D$ is represented by the matrix

$$P = \begin{pmatrix} 3 & 0 \\ 0 & 2 \end{pmatrix},$$

then $\tau: L \to L/D$ is represented by

$$V = \begin{pmatrix} 0 & 2 \\ 2 & 1 \end{pmatrix}.$$

Now we list some of the properties of generalized tori and their automorphisms.

A generalized torus is a connected finite-dimensional abelian group. It is not however locally connected (unless it is a torus)—this follows from Lemma 4 since the image under τ (or τ^{-1}) of a sufficiently small neighborhood in L consists of a union of disjoint open sets (D (or C) being discrete).

The ergodicity condition is the same as for a torus: a generalized torus automorphism is ergodic if and only if the associated matrix U has no eigenvalues that are roots of unity. As with the torus, eigenvalues on the unit circle are possible. Unlike the torus however, it is possible for all the eigenvalues to lie on the unit circle

$$U = \begin{pmatrix} \frac{1}{2} & 1 \\ 1 & 0 \end{pmatrix}$$

for example.

The entropy of all group automorphisms (and endomorphisms) has been calculated by Yuzvinskii [11] following on from the work of Arov. For a generalized torus automorphism the result is

$$h(T) = \log s + \sum_{|\lambda_i| > 1} \log|\lambda_i|,$$

where $\{\lambda_i\}$ are the eigenvalues of U, and s is the lowest common denominator of the coefficients in the characteristic polynomial of U. s is in fact the number of points in D. Knowing the entropy of a generalized torus automorphism one can calculate the entropy of any group automorphism using Theorem A together with the addition theorem [4, 7]. Theorem A should make Yuzvinskii's result easier to prove—this is under investigation.

In Example 1 we consider the partition β of $L = [0, 1)$ into the two sets $[0, \frac{1}{2})$ and $[\frac{1}{2}, 1)$. We put $\xi = p_H^{-1}(\beta)$. ξ is a Bernoulli partition, i.e., ξ and $T^i\xi$ are independent for all i. ξ is also a generator for T. So T has a Bernoulli generator, i.e., is a Bernoulli shift. A Bernoulli generator can be obtained by partitioning L suitably whenever the eigenvalues of U all have modulus greater than one or all have modulus less than one. This condition is satisfied in Example 2—there one can take $\beta = \{[0, \frac{1}{3}), [\frac{1}{3}, \frac{2}{3}), [\frac{2}{3}, 1)\}$. A. Wilson hopes to publish some results soon on this and other properties of generalized torus automorphisms.

When the moduli of the eigenvalues of U are mixed (less than, greater than, and equal to one), then, as with tori, no such direct method works. One has to use the techniques of [9]. When there are no eigenvalues on the unit circle, one can find Markov partitions for torus automorphisms (Adler/Weiss and Sinai); the authors have failed in trying to do this for generalized torus automorphisms.

Theorem A together with this analysis of generalized torus automorphisms should provide easier proofs of the results in [3, 4], (5, 6] as well as that in [11]. We finish by making an observation which contrasts well with the sort of thing done in those papers. An automorphism of a connected finite-dimensional abelian group is a direct limit of automorphisms of generalized tori. This follows from Theorem A because such an automorphism cannot be a group Bernoulli automorphism when restricted to a closed invariant subgroup. We give an illustration of this.

Let the character group Γ of G be the rationals and let T be multiplication by 2 (on Γ). Γ is not generated by Z and T, and therefore we consider a sequence of representations $G \to G/H_i$ where the H_i are defined as follows: for each j let p_j denote the jth prime number and for each i define $\eta_i = p_2{}^i p_3^{i-1} \cdots p_i{}^2 p_{i+1}$; define H_i to be the annihilator of $(1/\eta_i)Z$. Then $G_i = \bigcap_{j=-\infty}^{\infty} T^j H_i$ is the annihilator of all rationals of the form $q\eta_i$, where q is a 2-adic rational. For each i the automorphism induced on G/G_i is isomorphic to that in Example 1. Since $\bigcap_{i=1}^{\infty} G_i = e$, T is the direct limit of these automorphisms.

Acknowledgments

The authors would like to thank S. Ginn for helpful conversations and P. Shields, W. Parry, and P. Walters for much encouragement.

References

1. D. Ornstein, Bernoulli shifts with the same entropy are isomorphic, *Advances in Math.* **4** (1970), 337–352.
2. D. Ornstein, Two Bernoulli shifts with infinite entropy are isomorphic, *Advances in Math.* **5** (1970), 339–348.
3. V. A. Rohlin, Metric properties of endomorphisms of compact commutative groups, *Izv. Akad. Nauk SSSR Ser. Mat.* **28** (1964), 867–874.
4. S. A. Yuzvinskii, Metric properties of endomorphisms of a compact group, *Izv. Akad. Nauk SSSR Ser. Mat.* **29** (1965), 1295–1328.
5. R. K. Thomas, Metric properties of transformations of G-spaces, *Trans. Amer. Math. Soc.* **160** (1971), 103–117.
6. R. K. Thomas, The addition theorem for the entropy of transformations of G-spaces, *Trans. Amer. Math. Soc.* **160** (1971), 119–130.
7. R. K. Thomas, On affine transformations of locally compact groups, *J. London Math. Soc.* **4** (1972), 599–610.
8. G. Miles and R. K. Thomas, On the polynomial uniformity of translations of the n-torus, this volume.
9. G. Miles and R. K. Thomas, Generalized torus automorphisms are Bernoullian, this volume.
10. L. S. Pontrjagin, "Topological Groups," 2nd ed., Moscow, Gostehizdal, 1954.
11. S. A. Yuzvinskii, Calculation of the entropy of a group endomorphism, *Sibirsk. Mat. Ž.* **8** (1967), 230–239.

STUDIES IN PROBABILITY AND ERGODIC THEORY
ADVANCES IN MATHEMATICS SUPPLEMENTARY STUDIES, VOL. 2

On the Polynomial Uniformity of Translations of the n-Torus

G. MILES AND R. K. THOMAS

Department of Mathematics, Birkbeck College, London, England

INTRODUCTION

For an ergodic measure preserving transformation T of a Lebesgue space M (with measure μ, $\mu(M) = 1$) it is well known that almost every orbit $(T^i x)_{i=0}^{\infty}$ is uniformly distributed in any given measureable set $V \subset M$; i.e., the number $N_k(x, V)$ of points in $\{x, Tx, \ldots, T^{k-1}x\} \cap V$ satisfies

$$\lim_{k\to\infty} {}_q \frac{N_k(x, V)}{k} = \mu(V)$$

for almost all $x \in M$.

This result is weak in the sense that it gives no indication of the speed of convergence, which can vary greatly for different types of transformations (and of course for different types of set V, though V is normally restricted to an appropriate class of sets, e.g., sets compatible with the topology of M if M is a topological space). In the present paper this speed is examined for (ergodic) translations of finite-dimensional tori and we show that for almost all translations convergence is polynomial and in particular that this result holds for algebraic translations. Translation here means translation by an element $\alpha = (\alpha_1, \alpha_2, \ldots, \alpha_n)$, $1, \alpha_1, \ldots, \alpha_n$ being rationally independent real numbers, the n-torus being represented in the usual way as the unit cube U^n with addition mod 1 (rational independence is implied by ergodicity which is assumed throughout as the nonergodic case is of no interest). Algebraic translation refers to the case when $\alpha_1, \ldots, \alpha_n$ are all algebraic numbers.

The almost all result relies on the following well-known theorem due to Khintchine [1],

THEOREM A. *Let $\Psi_1(q), \ldots, \Psi_n(q)$ be n nonnegative functions of the positive integer q and assume $\Psi(q) = \prod_i \Psi_i(q)$ is monotonically decreasing. Then the set of inequalities $|q\theta_i - p_i| < \Psi_i(q)$ has an infinity of integer solutions*

ISBN 0-12-599102-9

$q > 0$, $p_1, \ldots, p_n$ for almost all or almost no sets of numbers $\theta_1, \ldots, \theta_n$ according as $\sum_q \Psi(q)$ diverges or converges.

The result for algebraic translations on the other hand rests on the following recent theorem due to Schmidt [2].

THEOREM B. *Let 1, α_1, $\alpha_2, \ldots, \alpha_n$ be rationally independent algebraic numbers and let $\delta > 0$ be given. Then there are only a finite number of positive integers q that satisfy*

$$\left(\prod_{i=1}^{n} \|q\alpha_i\|\right) q^{1+\delta} < 1,$$

where for any real x, $\|x\|$ denotes the distance between x and the nearest integer.

Algebraic translations are important because of their association with eigenvectors of ergodic group automorphisms of compact topological groups, and in this respect the present paper is a vital step in the authors' proof that these automorphisms are measure isomorphic to Bernoulli shifts [3, 4].

We give the proof only for algebraic translations. The proof of the almost all result is essentially the same with Khintchine's theorem being used instead of Schmidt's.

1. PRELIMINARIES

U^n will denote the n-torus (n-unit cube) throughout and μ the Haar measure on U^n normalized so that $\mu(U^n) = 1$. For $a = (a_1, \ldots, a_n)$ and $b = (b_1, \ldots, b_n) \in U^n$ $(a_i, b_i \in [0, 1))$ by $a + b$ we shall mean the n-tuple $(c_1, \ldots, c_n)$, where $c_i = a_i + b_i \bmod 1$ for all i.

DEFINITION 1. Let S_r be a collection of r points in U^n and let V be a measurable subset of U^n. We say that S_r is *ϵ-uniform* with respect to V if $|\mu(V) - r_V/r| < \epsilon\mu(V)$, where r_V is the number of points in $S_r \cap V$. Note that ϵ-uniformity is additive for disjoint subsets of U^n.

DEFINITION 2. A finite collection S_r of points in U^n will be said to be ϵ-uniform with respect to a measurable partition Q of U^n if it is ϵ-uniform with respect to every element of Q.

DEFINITION 3. A closed subset V of U^n will be said to be *regular* if $\mu(V) > 0$ and the boundary of V is $n - 1$ rectifiable (i.e., there exists a

Lipschitzian function f from a bounded subset of R^{n-1} to the boundary of V) or in the case $n = 1$ whose boundary is finite. Regular sets are the "appropriate class of sets" for tori.

DEFINITION 4. Let S be an infinite sequence of points in U^n and let S_k denote the first k elements of S. We shall say that S is *polynomially uniform* (with respect to regular sets) if there is a constant K dependent only on n such that if V is a regular subset of U^n, then there exists a constant $C = C(V)$ for which, for all $0 < \epsilon < 1$, $k > C(1/\epsilon)^K$ implies that S_k is ϵ-uniform with respect to V.

$Q_n(1/q)$ will denote the partition of U^n into sets of the form:

$$Q(a_1, \ldots, a_n) = \{(x_1, \ldots, x_n) \in U^n \,|\, a_i/q \leqslant x_i < (a_i + 1)/q \text{ for all } i\},$$

where the a_i's are integers between 0 and $q - 1$ (inclusive). Note that $\mu(Q(a_1, \ldots, a_n)) = 1/q^n$.

For any real number x, $\|x\|$ will denote the distance between x and the nearest integer.

For any two points $x, y \in U^1$, $d(x, y)$ denotes the distance between x and y in U^1, i.e., $d(x, y) = \min\{|x - y|, 1 - |x - y|\}$.

For $\alpha = (\alpha_1, \ldots, \alpha_n) \in U^n$, $S(\alpha)$ will denote the orbit of the origin under the transformation of translation by α, i.e., $S(\alpha) = \{0, \alpha, 2\alpha, \ldots\}$, where $j\alpha = (\beta_1(j), \ldots, \beta_n(j))$ with $\beta_i(j) = j\alpha_i \bmod 1$. $S_k(\alpha)$ will denote the first k terms of $S(\alpha)$.

The purpose of this paper is to show that when $1, \alpha_1, \alpha_2, \ldots, \alpha_n$ are rationally independent algebraic numbers $S(\alpha)$ is polynomially uniform; clearly then all the other orbits will also be polynomially uniform giving us our "polynomial speed of convergence" (for every regular set and every orbit). We shall do this by demonstrating the existence of constants K dependent only on n and C dependent on α such that for any positive integer q, $k > Cq^K$ implies that $S_k(\alpha)$ is $1/q$ uniform with respect to $Q_n(1/q)$; the required result is then a simple corollary.

We end this section with a result we shall need.

THEOREM C. *Let $\alpha = (\alpha_1, \ldots, \alpha_n)$ be any element of U^n. Given any positive integer L there exists an integer $q \leqslant L^n$ such that $\|q\alpha_i\| < L^{-1} \leqslant q^{-1/n}$ for all i.*

Proof. (Dirichlet's box argument). We consider the set $E = \{0, \alpha, 2\alpha, \ldots, L^n\alpha\}$ which has $L^n + 1$ elements. Now the partition $Q_n(1/L)$ has only L^n elements and so two points of E, $s\alpha$ and $t\alpha$ say, must be in the same element of $Q_n(1/L)$, i.e., $\|s\alpha_i - t\alpha_i\| < L^{-1}$ for all i or, equivalently, $\|(s - t)\alpha_i\| < L^{-1}$. Putting $q = |s - t|$, we have the result.

2. Five Lemmas

The following lemmas are needed in the proof of our main result. The first two are number theoretic results based on Theorems C and B and the last three concern ϵ-uniformity.

Lemma 1. *Let $1, \alpha_1, \alpha_2, \ldots, \alpha_n$ be rationally independent algebraic numbers. There exists a positive integer M such that if L is any integer greater than M, then there exists a smallest integer q, with $L < q \leqslant L^n$, such that $\|q\alpha_i\| < L^{-1} \leqslant q^{-1/n}$ (for all i) for $n \neq 1$; $L < q \leqslant L^2$ and $\|q\alpha_1\| < L^{-2} \leqslant q^{-1}$ for $n = 1$.*

Proof. First we assume that $n > 1$. Let $1 > \delta > 0$ be given and let $m_1, m_2, \ldots, m_r$ be all the integers satisfying $(\prod_{i=1}^{n} \|m\alpha_i\|)m^{1+\delta} < 1$ (there are only a finite number by Theorem B). Let M be an integer such that $M^{-1/n} < \inf_{i,j}\{\|m_j\alpha_i\|\}$.

For any integer $L > M$ we define q to be the smallest integer for which $\|q\alpha_i\| < L^{-1}$ for all i. Theorem C tells us that $q \leqslant L^n$.

Suppose, if possible, that $q < L$. Then we have $\|q\alpha_i\| < L^{-1} < q^{-1}$ so that $(\prod_{i=1}^{n} \|q\alpha_i\|)q^{1+\delta} < q^{-n}q^{1+\delta} < 1$ implying that $q = m_j$ for some j. But in that case we have $L^{-1} > \|q\alpha_i\| = \|m_j\alpha_i\| > M^{-1/n} > L^{-1/n}$ which is a contradiction. Thus $q > L$ completing the proof for $n \neq 1$.

For $n = 1$ we modify the proof by taking q to be the smallest integer such that $\|q\alpha_1\| < L^{-2}$. Then we get $L < q \leqslant L^2$.

Note. For $n = 1$ the condition "q is the smallest integer such that $\|q\alpha_1\| < L^{-2} \leqslant q^{-1}$" implies that there exists a p relatively prime to q such that $|(p/q) - \alpha_1| < q^{-2}$.

Lemma 2. *Let $1, \alpha_1, \alpha_2, \ldots, \alpha_{m+1}$ be $m+2$ rationally independent algebraic numbers ($m \geqslant 1$). There exists a positive integer M such that if $q > M$ and $\|q\alpha_i\| < q^{-1/m}$ for $i = 1, 2, \ldots, m$, then there exists an integer q_1, with $q^{1/(2(m+1)^2)} < q_1 < q^{1/(2m)}$, and an integer p relatively prime to q_1 such that $|(p/q_1) - q\alpha_{m+1}| < 1/q_1{}^2$.*

Proof. Let $\delta = 1/\{4(m+1)(2m+1)\}$. Let M be an integer for which

(i) $r > M$ *implies that* $(\prod_{i=1}^{m+1} \|r\alpha_i\|)r^{1+\delta} \geqslant 1$ (possible because of Theorem B), and

(ii) $r > M$ implies that $r^{1/(2m)} > r^{1/(2m+1)} + 1$.

Now we suppose that q is some integer greater than M for which $\|q\alpha_i\| < q^{-1/m}$ for $i = 1, 2, \ldots, m$. Let J be the largest integer smaller than $q^{1/(2m)}$. Applying Theorem C to the set $\{q\alpha_{m+1}, 2q\alpha_{m+1}, \ldots, Jq\alpha_{m+1}\}$ we get that there is an integer q_1, which we take to be the smallest possible, such that $q_1 \leqslant J$ and $\|q_1 q\alpha_{m+1}\| < J^{-1} \leqslant q_1^{-1}$. Hence (as in the above note) there exists an integer p relatively prime to q_1 such that $|(p/q_1) - q\alpha_{m+1}| < (q_1 J)^{-1} \leqslant q_1^{-2}$.

Now suppose, if possible, that $q_1 \leqslant q^{1/(2(m+1)^2)}$. We have

$$I = \|q_1 q\alpha_{m+1}\| \left(\prod_{i=1}^{m} \|q_1 q\alpha_i\| \right) < J^{-1} q_1{}^m q^{-1} < q_1{}^m / q(q^{1/(2m)} - 1)$$

$$< q_1{}^m / q^{1+1/(2m+1)} \qquad \text{using (ii)}).$$

Applying Theorem B to I we get $I \geqslant (q_1 q)^{-(1+\delta)}$. Combining these bounds for I we get

$$(q_1 q)^{1+\delta} q_1{}^m > q^{1+1/(2m+1)}$$

implying that

$$q^{\delta + (1+\delta+m)/(2(m+1)^2)} > q^{1/(2m+1)} \qquad (\text{using } q_1 \leqslant q^{1/(2(m+1)^2)})$$

which implies that

$$\delta > \frac{m+1}{(1+2(m+1)^2)(2m+1)} > \frac{1}{3(m+1)(2m+1)}$$

which is a contradiction. Thus we have $q^{1/(2(m+1)^2)} < q_1 \leqslant J < q^{1/(2m)}$ as required.

Lemma 2 will enable induction to be used on the number of algebraic numbers. The next lemma concerns the effect of translation on ϵ-uniformity with respect to a partition.

LEMMA 3. *Let E be a collection of r points in U^n and let q be a positive integer. There exists an integer constant $J_n \geqslant 1$, dependent on n but independent of both q and E, such that if E is $(1/(Jq^2))$-uniform with respect to $Q_n(1/(Jq^2))$, J being any integer greater than J_n, then the set $x + E = \{x + e \mid e \in E\}$ is $(1/q)$-uniform with respect to $Q_n(1/q)$ for every $x \in U^n$.*

Proof. We put $J_n = 2^n 6n + 1$ and let J be some integer greater than J_n. For any partition Q, $x + Q$ denotes the partition obtained by translating every element of Q by x. We assume that E is $(1/(Jq^2))$-uniform with respect to $Q_n(1/(Jq^2))$ so that $x + E$ is $(1/Jq^2))$-uniform with respect to $x + Q_n(1/(Jq^2))$.

Let V be any element of $Q_n(1/q)$; let V' be the union of all those elements of $x + Q_n(1/(Jq^2))$ that intersect V in sets of positive measure and let V_1 be the union of all those elements of $x + Q_n(1/(Jq^2))$ that are wholly contained in V. We put $V_2 = V' \backslash V_1$.

Now $\mu(V_2) < 2^n 2n(1/q)^{n-1}(1/(Jq^2)) = M/(Jq^{n+1})$, $M = 2^n 2n$. For any set $B \subset U^n$ we let r_B be the number of points in $(x + E) \cap B$. Note that if B is a union of elements of $x + Q_n(1/(Jq^2))$, then $|\mu(B) - r_B/r| < \mu(B)/(Jq^2)$ $((1/(Jq^2))$-uniformity).

Now we have

$$\begin{aligned}|\mu(V) - r_V/r| &\leqslant |\mu(V) - r_{V_1}/r| + r_{V_2}/r \\ &\leqslant |\mu(V) - \mu(V_1)| + |\mu(V_1) - r_{V_1}/r| + |\mu(V_2) - r_{V_2}/r) + \mu(V_2) \\ &< 2\mu(V_2) + (\mu(V_1) + \mu(V_2))/(Jq^2) < 3\mu(V_2) + \mu(V_1)/(Jq^2) \\ &< q^{-n}(3M/(Jq) + 1/(Jq^2)) < q^{-n}((3M + 1)/(Jq)) < \mu(V)/q,\end{aligned}$$

i.e., $x + E$ is $(1/q)$-uniform with respect to $Q_n(1/q)$.

The next lemma will enable us to cope with ϵ-uniformity at an induction step. We consider U^{m+1} as $U^m \times U^1$ with measure $\mu = \mu_1 \times \mu_2$.

LEMMA 4. *Let J and q be positive integers. Let $D = \{x^1, x^2, \ldots, x^s\}$ be a collection of s points in U^m with $x^i = (x_1{}^i, x_2{}^i, \ldots, x_m{}^i)$ and let $E = \{y^1, y^2, \ldots, y^t\}$ be a collection of t points in U^1. Let $F = F(1) \cup F(2) \cup \cdots \cup F(s)$ be a collection of st points in U^{m+1} such that:*

(i) *$F(i) = \{z^1(i), z^2(i), \ldots, z^t(i)\}$ with $z^j(i) = (z_1{}^j(i), z_2{}^j(i), \ldots, z^j_{m+1}(i))$.*

(ii) *$d(z_k{}^j(i), x_k{}^i) < 1/(Jq^2)$ for all j and i and $k = 1, 2, \ldots, m$.*

(iii) *There exist points $y(i) \in U^1$, $i = 1, 2, \ldots, s$, such that $z^j_{m+1}(i) = y(i) + y^j$ for $j = 1, 2, \ldots, t$. Then there exists an integer constant $J' = J'(m)$ which is independent of q and F such that if D is $(1/(Jq^2))$-uniform with respect to $Q_m(1/(Jq^2))$ in U^m and E is $(1/(Jq^2))$-uniform with respect to $Q_1(1/(Jq^2))$ in U^1, then F is $(1/q)$-uniform with respect to $Q_{m+1}(1/q)$ in U^{m+1}, J being any integer greater than J'.*

Each $F(i)$ should be regarded as a "vertical" set standing over the "base" point x^i. $F(i)$ is not exactly vertical however; the base coordinates satisfy (ii). (iii) says that the vertical coordinates of the points in $F(i)$ are a translation of E.

Proof. Let V be an arbitrary element of $Q_{m+1}(1/q)$; $V = V_1 \times V_2$, where $V_1 \in Q_m(1/q)$ and $V_2 \in Q_1(1/q)$.

Let W be the union of all those elements of $Q_m(1/(Jq^2))$ that contain points distant less than $1/(Jq^2)$ from the boundary of V_1, J being some integer. We let B be the set of i's for which x^i lies in W and s_B be the number of elements in B.

Now simple calculation shows that the number of elements of $Q_m(1/Jq^2)$ contained in W is less than $4^m 2m(Jq)^{m-1}$. Let $M = \max\{4^m 2m, J_n + 1\}$, J_n as in Lemma 3. We put $J' = J'(m) = 16M$ and assume that J is some integer greater than J' so that that, in particular, $J = 16L$ for some $L > M$. We have

$$\mu_1(W) < \frac{M(Jq)^{m-1}}{(Jq^2)^m} = \frac{M}{Jq^{m+1}}. \qquad (*)$$

We let A be the set of i's for which $x^i \in V_1 \backslash W$ and s_A be the number of elements in A. Then the set $\{(z_1{}^j(i), z_2{}^j(i), \ldots, z_m{}^j(i)) | i \in A, j = 1, 2, \ldots, t\}$ is contained in V_1 (using condition (ii)).

We let t_i be the number of points in $F(i) \cap U^m \times V_2$ (=number of points in $(y(i) + E) \cap V_2$) and apply Lemma 3 (as E is $(1/(L(4q)^2))$-uniform with respect to $Q_1(1/(L(4q)^2))$) to get

$$\left|\frac{t_i}{t} - \mu_2(V_2)\right| < \frac{\mu_2(V_2)}{4q} \qquad \text{for all} \quad i. \qquad (**)$$

Using the $(1/(Jq^2))$-uniformity of D with respect to $Q_m(1/(Jq^2))$ and $(*)$ we get

$$\left|\frac{s_A}{s} - \mu_1(V_1)\right| \leqslant \left|\frac{s_A}{s} - \mu_1(V_1 \backslash W)\right| + \mu_1(W) < \mu_1(W) + \frac{\mu_1(V_1 \backslash W)}{Jq^2}$$

$$< \frac{1}{q^m}\left(\frac{1}{Jq^2} + \frac{M}{Jq}\right) < \frac{\mu_1(V_1)}{4q}. \qquad (***)$$

Now we put $r_1 = \sum_{i \in A} t_i$, $r_2 = \sum_{j \in B} t_j$ and r_V equal to the number of points in $F \cap V$. It is clear that $r_1 \leqslant r_V \leqslant r_1 + r_2$ so that we have

$$\left|\frac{r_V}{st} - \mu(V)\right| \leqslant \left|\frac{r_1}{st} - \mu(V)\right| + \frac{r_V - r_1}{st} \leqslant \left|\frac{r_1}{st} - \mu(V)\right| + \frac{r_2}{st}.$$

Now

$$\left|\frac{r_1}{st} - \mu(V)\right| = \left|\frac{1}{s}\sum_{i \in A}\left(\frac{t_i}{t} - \mu_2(V_2)\right) + \frac{s_A}{s}\mu_2(V_2) - \mu(V)\right|$$

$$< \frac{s_A}{s}\frac{\mu_2(V_2)}{4q} + \left|\frac{s_A}{s}\mu_2(V_2) - \mu(V)\right| \qquad \text{(using } (**)\text{)}$$

$$< |\mu_1(V_1)\mu_2(V_2) - \mu(V)| + \mu_1(V_1)\mu_2(V_2)\left[\frac{1}{2q} + \left(\frac{1}{4q}\right)^2\right]$$

(using $(***)$)

$$< \frac{3\mu(V)}{4q} \qquad (\text{since } \mu(V) = \mu_1(V_1)\mu_2(V_2)).$$

Also

$$\frac{r_2}{st} = \frac{1}{s}\sum_{i\in B}\frac{t_i}{t} \leqslant \frac{1}{s}\sum_{i\in B}\left|\frac{t_i}{t} - \mu_2(V_2)\right| + \frac{s_B}{s}\mu_2(V_2)$$

$$< \frac{s_B}{s}\mu_2(V_2)\left(\frac{1}{4q}+1\right) \qquad \text{(using (**))}$$

$$< 2\mu_2(V_2)\left(\left|\frac{s_B}{s} - \mu_1(W)\right| + \mu_1(W)\right)$$

$$< 2\mu_2(V_2)\mu_1(W)\left(\frac{1}{Jq^2}+1\right)$$

(using $(1/(Jq^2))$-uniformity of D w.r.t. W)

$$< 4\mu_2(V_2)\frac{M}{Jq^{m+1}} \qquad \text{(using (*))}$$

$$= \frac{4M}{Ja^{m+2}} < \frac{\mu(V)}{4q}.$$

So finally

$$\left|\frac{r_V}{st} - \mu(V)\right| < \frac{\mu(V)}{q},$$

i.e., F is $(1/q)$-uniform with respect to V. As V is arbitary, F is $(1/q)$-uniform with respect to $Q_{m+1}(1/q)$ as required.

The final lemma will enable us to pass from an orbit of fixed length to any sufficiently long orbit.

LEMMA 5. *Let β be any fixed element of U^n. Suppose that there exist integer constants P, A, and B, dependent only on n and β, such that for any integer p greater than P there exists an integer $k' \leqslant Ap^B$ for which $S_{k'}(\beta)$ (notation as in Section 1) is $(1/p)$-uniform with respect to $Q_n(1/p)$. Then there exist constants C and K such that for all integers q, $k \geqslant Cq^K$ (integer) implies that $S_k(\beta)$ is $(1/q)$-uniform with respect to $Q_n(1/q)$.*

Proof. (i) For $q > P$ we have $4Jq^2 > P$, J being some integer greater than J_n as in Lemma 3, and so, by supposition, we have $S_{k'}(\beta)$ $(1/(4Jq^2))$-uniform with respect to $Q_n(1/(4Jq^2))$ for some $k' \leqslant A'q^{B'}$, where $A' = A/(4J)^B$ and $B' = 2B$. So for every $\gamma \in U^n$ the set $\gamma + S_{k'}(\beta)$ is $(1/(2q))$-uniform with respect to $Q_n(1/(2q))$ (by Lemma 3). Now for any integer $k \geqslant 2qq^nk'$ we can consider $S_k(\beta)$ as a union of translates of $S_{k'}(\beta)$ with less than k' points left over and deduce that $S_k(\beta)$ is $(1/q)$-uniform with respect to $Q_n(1/q)$.

(ii) Now, using (i), let D be such that $k \geqslant D$ implies that $S_k(\beta)$ is $(1/(JP^2))$-uniform with respect to $Q_n(1/(JP^2))$. Then Lemma 3 implies that $S_k(\beta)$ is $(1/q)$-uniform with respect to $Q_n(1/q)$ for all $q \leqslant P$ and all $k \geqslant D$.

Now $2qq^n k' \leqslant 2A'q^{n+1+B'}$. So if we put $C = \max\{2A', D\}$ and $K = n + 1 + B'$ we have our result.

3. Main Result

THEOREM. *Let $\alpha = (\alpha_1, \alpha_2, \ldots, \alpha_n)$ be an element of U^n with $1, \alpha_1, \alpha_2, \ldots, \alpha_n$ rationally independent algebraic numbers and q and k be positive integers. There exist constants C and K independent of q such that $k \geqslant Cq^K$ implies that $S_k(\alpha)$ (and the first k terms of every other orbit) is $(1/q)$-uniform with respect to $Q_n(1/q)$.*

Proof. We start with the case $n = 1$. Let p be an integer such that $2p^2 > M$ with M as in Lemma 1 so that (see note following Lemma 1) there exist relatively prime integers a and b such that $|(a/b) - \alpha_1| < 1/b^2$ with $2p^2 < b \leqslant 4p^4$. We consider the set $S_b(\alpha_1)$ (the first b terms of the orbit of the origin) and the partition $Q_1(1/p)$. Let V be any element of $Q_1(1/p)$ and r_V be the number of points in $S_b(\alpha_1) \cap V$. Let R_b be the set $\{(i/b) | i = 0, 1, \ldots, b-1\} \subset U^n$ (R_b is the rational approximation to $S_b(\alpha_1)$) and r_V' be the number of points in $R_b \cap V$. Simple consideration shows that $|r_V'/b - \mu(V)| < 1/b$ and $|r_V' - r_V| < 1$ and so

$$|r_V/b - \mu(V)| < 2/b < 1/p^2 = \mu(V)/p, \tag{†}$$

i.e., $S_b(\alpha_1)$ is $(1/p)$-uniform with respect to $Q_1(1/p)$. We apply Lemma 5 now (with $P = (M/2)^{1/2}$, $A = 4$ and $k' = b$) to get the result of the theorem.

Now we proceed by induction: we assume that the theorem has been proved for $\alpha' = (\alpha_1, \alpha_2, \ldots, \alpha_m) \in U^m$ with associated constants C' and K' and we prove the theorem for $\alpha = (\alpha_1, \alpha_2, \ldots, \alpha_{m+1}) \in U^{m+1}$.

The result for α' will still hold if we increase C' and K' and so we assume that $C' > 2^{2(m+1)^2}$ and $K' > 2(m+1)^2$.

For a positive integer p we put $s_1 = Jp^2$, where J is some integer greater than $J'(m)$ as in Lemma 4. For p large enough (greater than some P_1) we can apply Lemma 1 with $L = C's_1^{K'}$ to obtain an integer s, with $C's_1^{K'} < s \leqslant (C's_1^{K'})^{m+1}$, such that $\|s\alpha_i\| < s^{-1/m}$ for $i = 1, 2, \ldots, m$.

If p, and hence s, is sufficiently large ($p > P_2$) we can apply Lemma 2 to obtain relatively prime integers a and t such that $s^{1/(2(m+1)^2)} < t \leqslant s^{1/(2m)}$ and

$|a/t - s\alpha_{m+1}| < t^{-2}$. Now $t > s^{1/(2(m+1)^2)} > (C's_1^{K'})^{1/(2(m+1)^2)} > 2s_1{}^2$ so that we can use the same argument as for (†) above to show that the set

$$E = \{0, s\alpha_{m+1}, 2s\alpha_{m+1}, \ldots, (t-1)s\alpha_{m+1}\}$$

is $(1/s_1 = 1/(Jp^2))$-uniform with respect to $Q_1(1/(Jp^2))$.

The set

$$D = \{0, \alpha', 2\alpha', \ldots, (s-1)\alpha'\} \qquad (= S_s(\alpha') \subset U^m)$$

is $(1/s_1 = 1/(Jp^2))$-uniform with respect to $Q_m(1/(Jp^2))$ because $s > C's_1^{K'}$. We take

$$F = S_{st}(\alpha) = F_0 \cup F_1 \cup \cdots \cup F_{s-1},$$

where $F_i = \{i\alpha, (i+s)\alpha, (i+2s)\alpha, \ldots, (i+(t-1)s)\alpha\}$. Now we have

$$\begin{aligned} d(i\alpha_k, (i+js)\alpha_k) &= \|js\alpha_k\| < ts^{-1/m} \leqslant s^{-1/(2m)} \\ &< (C's_1^{K'})^{-1/(2m)} < s_1^{-1} = 1/(Jp^2). \end{aligned}$$

So all thc requirements of Lemma 4 are satisfied and consequently $S_{st}(\alpha)$ is $(1/p)$-uniform with respect to $Q_{m+1}(1/p)$.

Now $st < s^2 \leqslant (C's_1^{K'})^{2(m+1)} = (C'(Jp^2)^{K'})^{2(m+1)}$, i.e., $st = k < Ap^B$, where $A = (C'J^{K'})^{2(m+1)}$ and $B = 4K'(m+1)$. Application of Lemma 5 gives the theorem for α thus completing the induction step. As the case $n = 1$ has been proved, the proof of the theorem is complete for $S_k(\alpha)$. Application of Lemma 3 gives the result for every $S_k(\alpha) + x$.

Finally we have the result stated in the introduction:

COROLLARY. *Let α be as in the theorem. Then $S(\alpha)$ (and every other orbit) is polynomially uniform with respect to any regular set $V \subset U^n$.*

Proof. Let W_q denote the union of all the elements of $Q_n(1/q)$ (q a positive integer) that intersect the boundary of V. There exists a constant $m(V)$, not depending on q, such that $\mu(W_q) < m(V)/q$ for all q.

For q fixed we take k to be some integer greater than Cq^K (as in the theorem) so that $S_k(\alpha)$ is $(1/q)$-uniform with respect to $Q_n(1/q)$. We put $V_q = V \backslash W_q$ and let r_V, r_{V_q}, and r_{W_q} be the number of elements of $S_k(\alpha)$ that lie in V, V_q, and W_q, respectively. $S_k(\alpha)$ is $(1/q)$-uniform with respect to both V_q and W_q and so we have

$$\begin{aligned} |r_V/k - \mu(V)| &< r_{W_q}/k + |r_{V_q}/k - \mu(V_q)| + \mu(W_q) \\ &< 2\mu(W_q) + \mu(W_q)/q + \mu(V_q)/q \\ &< 3\mu(W_q) + \mu(V_q)/q < (3m(V) + \mu(V))/q. \end{aligned}$$

Let c be some integer greater than $(3m(V) + \mu(V))/\mu(V)$. For a given $0 < \epsilon < 1$ we take p to be the smallest integer for which $1/p < \epsilon$. Then for $q = cp$ we have $S_k(\alpha)\,(1/p)$-uniform with respect to V. Thus $k' > (C(2c)^K)(1/\epsilon)^K$ implies that $S_{k'}(\alpha)$ is $(1/\epsilon)$-uniform with respect to V, i.e., $S(\alpha)$ is polynomially uniform.

Acknowledgments

The authors would like to thank J. St. C. L. Sinnadurai for many helpful conversations and for bringing Schmidt's work to their attention.

References

1. A. Khintchine, Zur metrischen Theorie der diophantischen Approximationen, *Math. Z.* **24** (1926), 706–714.
2. W. M. Schmidt, Simultaneous approximation to algebraic numbers by rationals, *Acta Math.* **125** (1970), 189–201.
3. G. Miles and R. K. Thomas, The breakdown of automorphisms of compact topological groups, this volume.
4. G. Miles and R. K. Thomas, Generalized torus automorphisms are Bernoullian, this volume.

STUDIES IN PROBABILITY AND ERGODIC THEORY
ADVANCES IN MATHEMATICS SUPPLEMENTARY STUDIES, VOL. 2

Generalized Torus Automorphisms Are Bernoullian

G. MILES AND R. K. THOMAS

Department of Mathematics, Birkbeck College, London, England

INTRODUCTION

Generalized torus was introduced in [1]. It was shown in that paper that ergodic group automorphisms of compact topological groups are Bernoullian (measure isomorphic to Bernoulli shifts) if generalized torus automorphisms are Bernoullian—we prove now that this is so. The main part of the proof is a combination of the result of [2] with geometrical techniques (also used by Ornstein and Weiss in [2]). So [1, 3] and the present paper together prove that ergodic group automorphisms of compact groups are Bernoullian.

PRELIMINARIES

We recall the definition of generalized torus automorphism:

Let G be a compact separable topological group with Haar measure μ normalized so that $\mu(G) = 1$ and let T be a group automorphism of G (note that T preserves μ). We say that T is a generalized torus automorphism (or G is a generalized torus) if there exists a closed normal subgroup H of G such that:

(i) G/H is an n-torus U^n (with Haar measure ν; $\nu(U^n) = 1$);

(ii) $\bigcap_{-\infty}^{\infty} T^i H = e$, the identity of G;

(iii) $p_H(\bigcap_{i=1}^{m} T^i H) = p_H(TH) = D$, a finite subgroup of U^n, and $p_H(\bigcap_{i=1}^{m} T^{-i}H) = p_H(T^{-1}H) = C$, a finite subgroup of U^n, for all integers $m > 0$; p_H denotes the natural projection of G onto $G/H = U^n$;

(iv) $C \cap D$ is the identity of U^n.

In the above situation we let S_H denote the closed subgroup of the two-way infinite product of copies of U^n consisting of all sequences $(x_i)_{-\infty}^{\infty}$ with $x_i = p_H(T^i x)$ for some $x \in G$. We give S_H the induced topology and measure.

ISBN 0-12-599102-9

T acting on G is then isomorphic to the left shift on S_H which we shall also refer to as T; we shall also identify G and S_H and write $x = (x_i)_{-\infty}^{\infty}$.

In [1] it was proved that these sequences satisfy a Markov property: there exists a homomorphism $\tau: U^n \to U^n/D$ such that S_H consists of all sequences $(x_i)_{-\infty}^{\infty}$ for which $x_{i+1} \in \tau(x_i)$; the kernel of τ is C.

We regard U^n in the usual way as R^n factored by the integer lattice—that is we consider U^n as a unit cube with addition mod 1. The projection $U^n \to U^n/D$ can then be represented by some integer matrix P. The map τ is a torus endomorphism and so is represented by an integer matrix V. The rational matrix $M = P^{-1}V$ will be used a great deal—we call it "the matrix associated with T". We describe its significance in some detail now.

We regard U^n as the unit cube sitting in R^n, i.e., we identify a point $x \in U^n$ with the corresponding point in the unit cube which is regarded as a subset of R^n. So M acts on a point $x \in U^n$ to give a point $Mx \in R^n$. To get back to U^n we reduce the coordinates mod 1—we use square brackets to denote this. Thus $x \to [Mx]$ is a well defined map on U^n. Now the condition $x_{i+1} \in \tau(x_i)$ becomes $x_{i+1} = [Mx_i] + d$, where d is some element of D. D is in fact the set $\{[Mz] \mid z \in Z^n\}$ (Z^n being the integer lattice in R^n) and C the set $\{[M^{-1}z] \mid z \in Z^n\}$. In this way T is completely determined by M. Generalized torus automorphisms include torus automorphisms (when M is an integer matrix and $C = D = 0$); in general here M is a rational matrix—hence the term "generalized torus automorphism." Our result thus includes Katznelson's result [4] that torus automorphisms are Bernoullian—our method of proof is very different however.

The ergodicity condition for group automorphisms here becomes: T is ergodic if and only if M has no eigenvalues that are roots of unity. This follows from the fact that M coincides with the matrix representing the induced automorphism on the character group of G. We shall always assume that T is ergodic.

We consider the action of M further and introduce some notation. We put $D_1 = D$ and for $i = 1, 2, \ldots$ we put $D_{i+1} = \tau(D_i)$. The D_i's are an increasing sequence of finite subgroups of U^n. We define C_i similarly by putting $C_1 = C$ and $C_{i+1} = \tau^{-1}(C_i)$. The elements of S_H satisfy the condition: $x_{i+k} = [M^k x_i] + y$, with $y \in D_k$. So for $x = (x_i)_{-\infty}^{\infty}$ we define $d_j(x, i) \in D_j$ by

$$x_{i+j} = [M^j x_i] + d_j(x, i), \qquad j \text{ any positive integer.}$$

Similarly we define $c_j(x, i) \in C_j$ by

$$x_{i-j} = [M^{-j} x_i] + c_j(x, i), \qquad j \text{ any positive integer.}$$

PARTITIONS. Let $(X, \mathscr{B}, \mu)$ be a measure space with $\mu(X) = 1$ and T be an invertible measure-preserving transformation of X. Given any two partitions $\xi = \{A_1, A_2, \ldots, A_a\}$ and $\eta = \{B_1, B_2, \ldots, B_b\}$ of X, we denote by $\xi \vee \eta$ the partition of X into sets of the form $A_i \cap B_j$ with lexicographic ordering. If $(\xi_i)_{-\infty}^{\infty}$ is an infinite sequence of partitions, then $\bigvee_{-\infty}^{\infty} \xi_i$ denotes the smallest partition that is a refinement of every ξ_i. $\mathscr{A}(\xi)$ denotes the σ-algebra generated by ξ. A partition ξ is said to be a *generator* if $\mathscr{A}(\bigvee_{-\infty}^{\infty} T^i\xi) = \mathscr{B}$, where $T^i\xi = \{T^iA_1, T^iA_2, \ldots, T^iA_a\}$. A partition ξ is said to be *independent* for T if $\mu(\bigcap_{j=-m}^{m} T^jA_{i_j}) = \prod_{j=-m}^{m} \mu(T^jA_{i_j})$ for all choices of i_j and all m. T is said to be *Bernoullian* (or isomorphic to a Bernoulli shift) if there exists an independent generator for T.

Given a point $x \in X$ and a partition $\xi = \{A_1, A_2, \ldots, A_a\}$, the *$(T, \xi)$-name* of x is the sequence $(a_i)_{-\infty}^{\infty}$ such that $x \in T^{-i}A_{a_i}$ for all i.

Given two measure spaces X_1 and X_2 with measures μ_1 and μ_2 with $\mu_1(X_1) = \mu_2(X_2) = 1$, a map $F: X_1 \to X_2$ is said to be *ϵ-measure-preserving* if there is a set $E_1 \subset X_1$ such that $\mu_1(E_1) < \epsilon$ and $|\mu_2(FA)/\mu_1(A) - 1| < \epsilon$ for every measurable set $A \subset X_1 \backslash E_1$.

If a property P holds for all the elements of a partition ξ with the possible exception of a set of elements whose union has measure less than ϵ, then we say P holds for *ϵ-almost* every element of ξ. If ξ is the partition of X into points, we say P for ϵ-almost all $x \in X$.

We shall use the following definition for very weak Bernoulli (V.W.B.) which is not quite the same as the one originally given by Ornstein but is sufficient for his proofs—it is a little simpler to handle. The authors thank P. Shields for pointing it out to them.

DEFINITION. A partition ξ is said to be *very weak Bernoulli* (with respect to T) if given an $\epsilon > 0$, there exists an integer $N = N(\epsilon)$ such that for all $m \geq 1$ and ϵ-almost every element A of $\bigvee_{i=0}^{m} T^i\xi$ there exist ϵ-measure-preserving maps $s_A: A \to X$ (it is assumed here that A has its conditional measure μ_A with $\mu_A(A) = 1$) such that for ϵ-almost all $x \in A$ the N terms $a_1, a_2, \ldots, a_N$ in the (T, ξ)-name of x agree with the corresponding terms in the (T, ξ)-name of s_Ax for all but at most ϵN of the indices $1, 2, \ldots, N$.

We state the two theorems of Ornstein that we shall need. They are taken from [5, 6].

THEOREM I. *If ξ is V.W.B. for T, then T restricted to $\mathscr{A}(\bigvee_{i=-\infty}^{\infty} T^i\xi)$ is Bernoullian.*

THEOREM II. *If $\mathscr{A}_1 \subset \mathscr{A}_2 \subset \cdots$ is an increasing sequence of T invariant σ-algebras such that $\bigvee_{i=1}^{\infty} \mathscr{A}_i = \mathscr{B}$ and T restricted to each $\mathscr{A}_i$ is Bernoullian, then T is Bernoullian (on X).*

We return now to G. Throughout the rest of this paper ξ will denote a partition of G of the form $\xi = p_H^{-1}(\eta)$, where $\eta = \{Y_1, Y_2, \ldots, Y_a\}$ is a partition of U^n into regular sets (i.e., simply connected sets with $(n-1)$-rectifiable boundaries) which are small enough so that:

(i) for all $y \in U^n$, $d, d' \in D$, $c, c' \in C$, $d \neq d'$, $c \neq c'$, $y + d$, and $y + d'$ are in different elements of η and $y + c$ and $y + c'$ are in different elements of η;
(ii) for every element A of $\xi \vee T\xi$ and every element B of $\xi \vee T^{-1}\xi$, $p_H(A)$ and $p_H(B)$ are single connected components of U^n;
(iii) η induces a nontrivial partition of every subtorus of U^n.

It is easy to check that the conditions on η imply that ξ is a generator—remember that T is assumed to be ergodic. We note that the (T, ξ)-name of a point $x = (x_i)_{-\infty}^{\infty}$ can be obtained from η: the name is $(a_i)_{-\infty}^{\infty}$, where $x_i \in Y_{a_i}$.

We let c be the order of the subgroup C. For an element A of $\bigvee_{i=0}^{N} T^i\xi$ we have $\mu(A) = c^{-N}\nu(p_H(A))$ (μ and ν are the measures in G and U^n, respectively, as above).

Let $v_1, v_2, \ldots,$ be some basis for R^n. Let p be an integer and $z = (z_1, z_2, \ldots, z_n) \in z^n$. We put $R(z, p) = \{y = \sum e_i v_i \in R^n \,|\, z_i/p \leqslant e_i < (z_i + 1)/p$ for all $i\}$ and $R'(p) = \{y \in U^n \,|\, y \in R(z, p)$ implies $R(z, p) \not\subset U^n\}$—here U^n is regarded as a subset of R^n as before. Now we denote by $R(p)$ the partition of U^n consisting of the set $R'(p)$ and all the sets $R(z, p)$ that are contained in U^n.

We shall also use the partition $\hat{R}(p)$ of R^n obtained from $R(p)$: an element $\hat{R}_i$ of $\hat{R}(p)$ is of the form $\hat{R}_i = \bigcup_{z \in Z^n} (z + R_i)$, where R_i is an element of $R(p)$.

We shall need a modified version of the result in [3]. First some notation. As in [3] $Q_n(1/q)$ (q an integer) denotes the partition of U^n into cubes of edge $1/q$. Let $\alpha = (\alpha_1, \alpha_2, \ldots, \alpha_n)$ be an element of U^n such that $1, \alpha_1, \alpha_2, \ldots, \alpha_n$ are rationally independent algebraic numbers. We take a basis $v_1, v_2, \ldots, v_n$ for R^n with $v_1 = \alpha$ (U^n is being regarded as a subgroup of R^n now). Let K_1 and K_2 be integer constants such that for all integers $p > 0$.

(i) the set $I(K_1 p) =$ union of all elements of $Q_n(1/(K_1 p))$ that are adjacent to the faces of U^n satisfies $\nu(I(K_1 p)) < 1/(2p)$;
(ii) $R'(K_2 p) \subset I(K_1 p)$.

As in [3], $S_k(\alpha)$ denotes the first k terms of the orbit of the origin under translation in U^n by α, i.e., the set $\{0, \alpha, 2\alpha, \ldots, (k-1)\alpha\}$ (reduced mod 1). We shall call such a set a partial orbit. $y + S_k(\alpha)$ is the partial orbit of y (of length k). A finite set W is said to be ϵ-*uniform* with respect to a set $A \subset U^n$ if $|r_A/w - \sigma(A)| < \epsilon . \nu(A)$, where r_A and w are the numbers of points in $w \cap A$ and W, respectively.

THEOREM III. *Using the above notation, there exist positive integer constants A and B such that for every $y \in U^n$ and every integer $k \geqslant Ap^B$, the set*

$y + S_k(\alpha)$ is $1/p$-uniform with respect to every element of $R(K_2p)$ except the element $R'(K_2p)$; the number of points in $(y + S_k(\alpha)) \cap R'(K_2p)$ is less than k/p.

Proof. The main result of [3] is that there exist constants C and K independent of q such that $k \geqslant Cq^K$ implies that $S_k(\alpha) + y$ is $(1/q)$-uniform with respect to $Q_n\,(1/q)$.

We use the same technique as was used in [3] a number of times. We put $q = Jp^2$, where J is some integer multiple of K_1. Now we consider an arbitrary element $R(z, K_2p)$; $R(z, K_2p) = R_1 \cup R_2$, where R_1 is the union of all those elements of $Q_n(1/q)$ that are wholly contained in $R(z, K_2p)$ and $R_2 = R(z, K_2p)\backslash R_1$. Now as the shape of $R(z, K_2p)$ is the same for all p there exists a J such that R_2 is sufficiently small in comparison to R_1 for $S_k(\alpha) + y$ to be $(1/p)$-uniform with respect to $R(z, K_2p)$ for any p, i.e., $k \geqslant C(Jp^2)^K = Ap^B$ implies $S_k(\alpha) + y$ $(1/p)$-uniform with respect to every $R(z, K_2p)$ (for more details of this type of argument see some of the proofs in [3]).

This leaves the set $R'(K_2p) \subset I(K_1p)$. Now $S_k(\alpha) + y$ is $(1/q)$-uniform with respect to $I(K_1p)$ and so, putting $r =$ number of points in $(S_k(\alpha) + y) \cap I(K_1p)$, we have $|r/k - v(I(K_1p))| < v(I(K_1p))/q$ implying that $r < k/p$. The required result follows.

Note. We conclude this section by noting that if S is a union of partial orbits, each of which satisfies the requirements of Theorem III, then S itself is $(1/p)$-uniform with respect to every element of $R(K_2p)$ except $R'(K_2p)$ and the number of points in $S \cap R'(K_2p)$ is less than s/p, where s is the total number of points in S.

Section I

As before we let M be the matrix associated with T; M acts on R^n. $R^n = V_1 \oplus V_2$, where V_1 and V_2 are M-invariant subspaces corresponding to eigenvalues greater than 1 in modulus and less than or equal to 1 in modulus, respectively. In this section we shall prove our result under the assumption that $V_1 \neq 0$ and $V_2 \neq 0$.

There exist a $\rho > 1$ and an s_0 such that $\|M^s v\| \geqslant \rho^s\|v\|$ and $\|M^{-s}v\| \leqslant \rho^{-s}\|v\|$ for all $v \in V_1$ and all $s > s_0$. Also there exists an $h > 0$ such that $\|M^r w\| \leqslant hr^n\|w\|$ for all $w \in V_2$ and all $r \geqslant 1$.

Let $v_1, v_2, \ldots, v_n$ be a basis for R^n with $v_1, v_2, \ldots, v_j$ a basis for V_1 and $v_{j+1}, v_{j+2}, \ldots, v_n$ a basis for V_2. Throughout this section we shall assume that $v_1 = \alpha = (\alpha_1, \alpha_2, \ldots, \alpha_n)$, where $1, \alpha_1, \alpha_2, \ldots, \alpha_n$ are rationally independent algebraic numbers.

NOTATION. Given a point $y = y_1 + y_2$ with $y_1 \in V_1$ and $y_2 \in V_2$, we denote by $E_r(y)$ the set of points $\{x \in R^n | x = x_1 + y_2, x_1 \in V_1, d(x_1, y_1) < r\}$ ("the expanding r-neighborhood of y"); d here is the normal euclidean metric.

Let $\xi = p_H^{-1}(\eta)$ be a partition of $S_H = G$ as described in the preliminaries. We shall show that ξ is V.W.B. For ϵ-almost every element A of $\bigvee_{i=0}^{m} T^i\xi$ we shall construct an ϵ-measure-preserving map s_A of A onto S_H. To do this we shall regard $p_H(A)$ as a subset of R^n (regarding U^n as the unit cube $\subset R^n$). Then we apply M^K for a suitably large integer K and use our geometrical technique and number theory (Theorem III) to obtain an ϵ-measure-preserving map φ from $M^K p_H(A)$ to U^n—this map has the property that $M^i[M^K x]$ and $M^i\varphi(M^K x)$ are close together for ϵ-almost all $x \in A$ and for $i = 1, 2, \ldots, Q$, here Q is large enough for V.W.B. We then make use of the $d_j(x, i)$'s and $c_j(x, i)$'s in the preliminaries to obtain $s_A : A \to S_H$ with the property that $p_H(T^{K+i}y)$ and $p_H(T^{K+i}s_A(y))$ are close (in U^n) for ϵ-almost all $y \in A$ and $i = 1, 2, \ldots, Q$. The closeness of these points implies good (but not total) agreement among the corresponding terms in the (T, ξ)-names of y and $s_A(y)$. We get V.W.B. by showing that Q is large enough (compared with K). In fact, if M does not have repeated eigenvalues of modulus 1 with an off-diagonal 1 in its Jordan canonical form, Q can be taken to be infinity and an easier argument used—in particular the number theory need not be used; it can be replaced by the mixing property. When M does have repeated modulus one eigenvalues with an off-diagonal 1 in the Jordan form, however, Q is finite—the points slowly wander apart. The number theory shows that the speed of drift is slow compared with the speed of mixing of T so that we can obtain a sufficiently large Q.

LEMMA 1. *For ϵ'-almost every element A of $\bigvee_{i=0}^{m} T^i\xi$ there is a set $B \subset p_H(A)$ with $v(B)/v(p_H(A)) > 1 - \epsilon'$ and a constant $r = r(\epsilon)$ independent of m such that $E_r(y) \subset p_H(A)$ for all $y \in B$.*

Proof. Let m be arbitrary and fixed. We define $Y_r \subset G$ by: $x \in Y_r$ if $E_r(x_0) \subset p_H(A_m(x))$, where $x_0 = p_H(x)$ as before and $A_m(x)$ is the element of $\bigvee_{i=0}^{m} T^i\xi$ that contains x.

Let L_i be the set of points $y \in U^n$ such that the distance from y to the nearest boudary point of η is less than ρ^{-i} (ρ as in the second paragraph of this section). There exists a constant Q such that $v(L_i) \leqslant Q/\rho^i$ for all i. We find a k such that $\sum_{i=k}^{\infty} v(L_i) < \delta/2$; δ will be determined later.

We choose u such that $v(\{y \in U^n|$ distance from y to nearest boundary point of $\eta < u\}) < \delta/2(s_0 + 1)$ (s_0 as in the second paragraph of this section) and u' such that $d(y, y') < u'$ implies $d(M^{-s}y, M^{-s}y') < u$ for all $0 \leqslant s \leqslant s_0$ and all $y, y' \in R^n$.

Now we put $r = \min\{\rho^{-k}, u'\}$. Then $x = (x_i)_{-\infty}^{\infty} \notin Y_r$ only if either

(i) for some $0 \leqslant s \leqslant s_0$ the distance of x_{-s} from the boundary of some element of η is less than u or

(ii) for some $s > s_0$ the distance from x_{-s} to the boundary of some element of η is less than $\rho^{-s}r$ in which case $x_{-s} \in L_{k+s}$.

It follows that $\mu(G \backslash Y_r) < [(s_0 + 1)\delta/2(s_0 + 1)] + (\delta/2) = \delta$.

Let the elements of $\bigvee_{i=0}^{m} T^i\xi$ be $A_1, A_2, \ldots, A_t$. Now $\sum_{i=1}^{t} \mu(A_i \backslash Y_r) = \mu(G \backslash Y_r) < \delta$. So if we put $\delta = \epsilon'^2$, there is a set of indices I such that $\sum_{i \in I} \mu(A_i) < \epsilon'$ and $\mu(A_i \backslash Y_r)/\mu(A_i) > \epsilon'$ for $i \in I$. Thus $\mu(A_i \cap Y_r)/\mu(A_i) > 1 - \epsilon'$ for $i \notin I$.

Now, since $\mu(A_i) = c^{-m}\mu(p_H(A_I))$ (c being the order of the subgroup C as in the preliminaries) and $\mu(A_i \cap Y_r) = c^{-m}\nu(p_H(A_i \cap Y_r))$, putting $B_i = p_H(A_i \cap Y_r)$ we have $E_r(y) \subset p_H(A_i)$ for all $y \in B_i$ with $\nu(B_i)/\nu(p_H(A_i)) > 1 - \epsilon'$ for $i \notin I$ (i.e., for ϵ'-almost all A_i) completing the proof.

For the next lemma we need to introduce some notation and concepts.

Definition. A set $L \subset R^n = V_1 \oplus V_2$ will be called an *expanding sheet* if it has the form $L = L_0 + w$, where L_0 is a regular subset of V_1 and w is an element of V_2.

In order to construct the mapping φ mentioned above we shall require the expanding sheets we consider to intersect the elements of $\hat{R}(p)$ in a special way (for a particular p). We shall say that an expanding sheet $L = L_0 + w$ *cleanly* intersects a regular subset Q of R^n if either (i) $L \cap Q = \varnothing$ or (ii) $L \cap Q = (V_1 + w) \cap Q$. This means that L either misses Q or passes completely through Q. We shall say that a set P cleanly intersects Q if P is a union of expanding sheets each of which cleanly intersects Q. An element $\hat{R}$ of $\hat{R}(p)$ is of the form $\hat{R} = \bigcup_{z \in Z} (z + R)$, where Z is the integer lattice in R^n and R is an element of $R(p)$. We shall say that P cleanly intersects $\hat{R}$ if P cleanly intersects every component $z + R$.

Now if $\hat{R} \neq \hat{R}'(p)$ and P cleanly intersects $\hat{R}$ in a set of positive measure, then we can define a map $\varphi_{\hat{R}}$ from $P \cap \hat{R}$ onto R with the properties we require. R is of the form $R_1 + R_2$, where R_1 and R_2 are regular subsets of V_1 and V_2, respectively. Now if $P \cap (z + R)$ has positive measure, we first subtract z (i.e., reduce mod 1) to obtain $(P - z) \cap R$ which (because P intersects $z + R$ cleanly) is of the form $R_1 + P_z$, where P_z is a subset of V_2 of positive measure. If P met only a single component $z + R$ of $\hat{R}$ we would obtain $\varphi_{\hat{R}}$ now by mapping R_1 to R_1 identically and expanding P_z to R_2 in any manner that increases measure uniformly (i.e., increases the measure of every subset of P_z by the same factor). When P meets more than one component of $\hat{R}$ we have the problem that the P_z's obtained may overlap. We overcome this, if necessary, by first shrinking all the P_z's and then moving them around so that they no longer overlap (they must of course remain within R_2 and the shrinking must decrease measure uniformly); then the union of these sets is expanded to R_2 (uniformly increasing the measure). R_1 is mapped identically onto R_1 as before.

For $P = M^k p_H(A)$ (as above) we obtain the map φ required by combining all the $\varphi_{\hat{R}}$'s for a suitable $\hat{R}(p)$. We must show that φ is ϵ-measure-preserving. First we extend the definition of ϵ-uniformity to continuous subsets of R^n:

DEFINITION. A set $E \subset R^n$ with an induced measure λ_E will be said to be *ϵ-uniform* with respect to a set $F \subset R^n$, $0 < \lambda(F) < \infty$, (λ = measure in R^n) if

$$|\lambda_E(E \cap F)/\lambda_E(E) - \lambda(F)| < \epsilon\lambda(F).$$

We shall also find it convenient to talk of ϵ-uniformity with respect to an element $\hat{R}$ of $\hat{R}(p)$. As it stands this has no meaning because $\lambda(\hat{R}) = \infty$. We use $\lambda(R)$ instead, i.e., E is ϵ-uniform with respect to an element $\hat{R} = \bigcup_{z \in Z} (z + R)$ of $\hat{R}(p)$ if

$$|\lambda_E(E \cap \hat{R})/\lambda_E(E) - \lambda(R)| < \epsilon\lambda(R).$$

Clearly as with simple ϵ-uniformity, this property extends to unions of disjoint sets, i.e., E_1 and E_2 ϵ-uniform with respect to F and $E_1 \cap E_2 = \varnothing$ imply that $E_1 \cup E_2$ is ϵ-uniform with respect to F. This extends to an arbitrary number of sets: if E has a measurable decomposition $E = \bigcup_{x \in X} E_x$ with $E_x \cap E_y = \varnothing$ for $x \neq y$ and every E_x is ϵ-uniform with respect to F, then E is ϵ-uniform with respect to F.

NOTATION. (i) $\delta(p)$ will denote the diameter of an element of $R(p)$ other than $R'(p)$. Note that $\delta(p) = f/p$ for all p, where f is some constant.

(ii) $l_1(x)$ will denote the line in R^n through x in the direction of the vector v_1.

(iii) $G(p)$ will denote the lattice of points in V_1:

$$G(p) = \left\{ y \in V_1 \,\middle|\, y = \sum_{i=1}^{j} (z_i/p)v_i, \quad z_i \in Z \text{ (integers)} \right\}.$$

(iv) For $u \in R^n$, p-box (u) will denote the set:

$$p\text{-box }(u) = \left\{ u + y \,\middle|\, y \in V_1, \, y = \sum_{i=1}^{j} a_i v_i, \, 0 \leqslant a_i < 1/p \text{ for all } i \right\}.$$

LEMMA 2. *Let $\epsilon' > 0$ be given and p be a positive integer. Let L be an expanding sheet in R^n with induced measure λ_L. If L contains a subset L_1*

such that the expanding neighborhood $E_a(y)$ is contained in L for all $y \in L_1$ and $\lambda_L(L_1)/\lambda_L(L) > 1 - \epsilon'$ with $a = a' + 3\delta(K_2 p)$ and $a' > |\alpha| A p^B$, then there exists a subset L_2 of L such that $\lambda_L(L_2)/\lambda_L(L) > 1 - \epsilon' - (1/p)$, L_2 cleanly intersects $\hat{R}(K_2 p)$ and is $1/p$-uniform with respect to every element of $\hat{R}(K_2 p)$ except the element $\hat{R}'(K_2 p)$.

A, B, α, and K_2 are all as in Theorem III—remember that we are taking $v_1 = \alpha$.

Proof. $L = L_0 + w$ for some $L_0 \subset V_1$ and $w \in V_2$. We put

$$L_3 = \{y \mid d(x, y) \leqslant \delta(K_2 p) \text{ for some } x \in L_1\},$$
$$L_4 = \{y \mid y \in l_1(x) \text{ for some } x \in L_3 \text{ with } d(x, y) \leqslant a' + \delta(K_2 p)\},$$
$$G_1 = (G(K_2 p) + w) \cap L_4 \quad \text{and} \quad G_2 = G_1 \backslash R'(K_2 p).$$

We let s_1 and s_2 be the numbers of points in G_1 and G_2, respectively.

Now G_1 is a union of partial orbits under translation by α and each partial orbit contains more than Ap^B points. So on applying Theorem III (and the note following it) we get G_4 is $1/p$-uniform with respect to every element of $\hat{R}(K_2 p)$ except $R'(K_2 p)$ ($1/p$-uniformity with respect to elements of $\hat{R}(K_2 p)$ is equivalent to $1/p$-uniformity with respect to the elements of $R(K_2 p)$—we are just not reducing mod 1 here) and the number of points in $G_1 \cap R'(K_2 p)$ is less than s_1/p, i.e., $s_2 > s_1(1 - (1/p))$.)

We put $L_2 = (\bigcup_{z+R \in I} (z + R)) \cap (V_1 + w)$, where $z \in Z^n$, $R \in R(K_2 p)$, and $z + R \in I$ if $(z + R) \cap G_2 \neq \varnothing$. It is clear that $L_2 \cap \hat{R}'(K_2 p) = \varnothing$ and that L_2 cleanly intersects every other element of $\hat{R}(K_2 p)$. It is also clear that $L_2 \subset L$ and the $1/p$-uniformity of G_2 implies that L_2 is $1/p$-uniform with respect to every element of $\hat{R}(K_2 p)$ except $\hat{R}'(K_2 p)$.

It remains to estimate the measure of L_2. For every $z + R \in I$, $(z + R) \cap L_2$ is a $K_2 p$-box containing a single element of G_2. It follows that $\lambda_L(L_2) = \lambda_L(\bigcup_{g \in G_2} K_2 p\text{-box}(g))$. Putting $L_5 = \bigcup_{g \in G_1} K_2 p\text{-box}(g)$ we have

$$\lambda_L(L_2) = \frac{s_2}{s_1} \lambda_L(L_5) > \left(1 - \frac{1}{p}\right) \lambda_L(L_5).$$

It is clear that L_5 covers L_1. Therefore

$$\lambda_L(L_2) > \left(1 - \frac{1}{p}\right)(1 - \epsilon)\lambda_L(L) > \left(1 - \epsilon' - \frac{1}{p}\right) \lambda_L(L)$$

completing the proof.

Now we tie Lemmas 1 and 2 together.

LEMMA 3. *Let $1 > \epsilon > 0$ be given and m be an arbitrary positive integer. There exists a constant J depending only on ϵ such that if p is any integer greater than $3/\epsilon$ and k is an integer greater than $J \log p$, then for ϵ-almost every $A \in \bigvee_{i=0}^{m} T^i\xi$, the map $\varphi: M^k p_H(A) \to U^n$ described above is ϵ-measure-preserving.*

Proof. The set $p_H(A)$ has a measurable decomposition: $p_H(A) = \bigcup_{x \in X} F_x$, where each F_x is an expanding sheet.

We apply Lemma 1 with $\epsilon' = \epsilon^2/9$ to get a set $B \subset p_H(A)$ and an r. We put $F_x' = F_x \cap B$ for all $x \in X$. We regard B and $p_H(A)$ as subsets of R^n so that we have

$$\lambda(p_H(A) \backslash B)/\lambda(p_H(A)) < \epsilon^2/9.$$

Hence there exists a set $Y \subset X$ such that

$$\lambda_{F_x}(F_x')/\lambda_{F_x}(F_x) > 1 - (\epsilon/3) \qquad \text{for all} \quad x \in Y$$

and $B' = \bigcup_{x \in Y} F_x$ is measurable with $\lambda(B')/\lambda(p_H(A)) > 1 - (\epsilon/3)$.

For $k > s_0$ we get that $E_a(y) \subset M^k F_x$ for all $y \in M^k F_x'$, where $a = \rho^k r$ (s_0 and ρ as in the second paragraph of this section). To apply Lemma 2 we must have $a > Ap^B + 3\delta(K_2 p)$. To achieve this we put

$$J = \max\left\{\frac{\log(A + f) - \log r + B}{\log \rho}, s_0\right\}$$

(f as in Notation (i) preceeding Lemma 2). Then $k > J \log p$ implies that $k > s_0$ (since $p > 3$ and $\log 3 > 1$) and

$$k \log \rho > \log\left(\frac{A + f}{r}\right)\log p + B \log p > \log\left(\frac{A + f}{r}\right) + B \log p.$$

Hence $a = \rho^k r > (A + f)p^B > Ap^B + 3\delta(K_2 p)$ as required.

Now we apply Lemma 2 (with $\epsilon' = \epsilon/3$, $L = M^k F_x$ and $L_1 = M^k F_x'$) to get a set $L_2 = M^k F_x''$ for each $x \in Y$ with all the properties stated in Lemma 2. We put $B'' = \bigcup_{x \in Y} F_x''$. It is clear that $M^k B''$ cleanly intersects $\hat{R}(K_2 p)$, $M^k B'' \cap R'(K_2 p) = \varnothing$ and

$$\frac{\lambda(B'')}{\lambda(p_H(A))} = \frac{\lambda(B'')}{\lambda(B')}\frac{\lambda(B')}{\lambda(p_H(A))} > \left(1 - \frac{\epsilon}{3} - \frac{1}{p}\right)\left(1 - \frac{\epsilon}{3}\right)$$

$$> \left(1 - \frac{2\epsilon}{3}\right)\left(1 - \frac{\epsilon}{3}\right) > 1 - \epsilon$$

so that $\lambda(M^k B'')/\lambda(M^k p_H(A)) > 1 - \epsilon$. This together with the fact that $M^k B''$ is ϵ-uniform (in fact $1/p$ uniform but $1/p < \epsilon$) with respect to every element of

$\hat{R}(K_2 p)$ except $\hat{R}'(K_2 p)$ implies that the map φ constructed as above is ϵ-measure-preserving.

Now the map φ defined above satisfies:

$$P_{V_1}([y]) = P_{V_1}(\varphi(y))$$

and

$$d([y], \varphi(y)) < \delta(p) = f/p \qquad \text{for all} \quad y \in M^k B'',$$

where $[\cdot]$ denotes reduction mod 1, P_{V_1} is the natural projection onto V_1 with respect to the decomposition $R^n = V_1 + V_2$ and d is the euclidean metric.

We define $k(\epsilon, \delta)$ to be the smallest integer k such that a map φ constructed as above is both ϵ-measure-preserving and satisfies $d([y], \varphi(y)) < \delta$ (for all y for which $\varphi(y)$ is defined).

Let ϵ, δ be given $\delta < \epsilon/3$. Let J be as in Lemma 3, and let p be an integer such that $(\delta/f)^2 \leqslant 1/p < \delta/f$. It is clear that if k is the smallest integer $> J \log p$, then we have $k(\epsilon, \delta) \leqslant k \leqslant J \log p^2 \leqslant 4J \log(f/\delta)$. In particular, if $\delta = e^{-s} f$ (which implies that $s > \log(3f/\epsilon)$), then we have $k(\delta, e^{-s} f) \leqslant 4Js$. We now use this estimate to prove Lemma 4 from which we obtain the required result almost immediately.

LEMMA 4. *Let ϵ' be given. Then there exist constants K and K′ such that for ϵ'-almost every atom $A \in \bigvee_{i=0}^{m} T^i \xi$ there is an ϵ'-measure-preserving map φ_A from $M^K p_H(A)$ to U^n and*

(i) $2K/K' \leqslant \epsilon'$,

(ii) $K' \mu\{y \in U^n | d(y, \partial\eta) < 1/(K')^2\} < \epsilon'/2$ *($\partial\eta$ denotes the union of the boundaries of the elements of η),*

(iii) *for $(\epsilon'/2)$-almost every $y \in M^K p_H(A)$, $d(M^J[y], M^J \varphi_A(y)) < (1/K')^2$, for $1 \leqslant J \leqslant K'$.*

Proof. Let J be the constant of Lemma 3 for $\epsilon = \epsilon'/2$.

(1) Let $s' = 8Js(2/\epsilon')$. Then $8Js/s' = \epsilon'/2$.

(2) $I(s) = s'\mu\{y \in U^n | d(y, \partial\eta) < 1/s'^2\} = \text{constant}/s' \to 0$ as $s \to \infty$. Therefore for s greater than some s_2, $I(s) < \epsilon'/2$.

(3) If $d(y, y') < e^{-S} f$ and $P_{V_1}(y) = P_{V_1}(y')$, then for $1 \leqslant j \leqslant s'$, $d(M^j y, M^j y') < h s'^n e^{-s} f$ (h as in second paragraph of this section). But $s'^2 h s'^n e^{-s} f = (\text{constant})\, s^{n+2}/e^s \to 0$ as $s \to \infty$. Thus for s greater than some s_3, $h s'^n e^{-s} f < 1/s'^2$.

We put $s = \max\{s_1, s_2, s_3\}$, where s_1 is the smallest integer greater than $\log(3f/\epsilon)$, $K' = s'$ and $K = k(\epsilon'/2, e^{-s} f) = k(\epsilon, e^{-s} f) \leqslant 4Js$.

By Lemma 3 and the definition of $k(\epsilon, \delta)$ it follows that for ϵ-almost every atom A of $\bigvee_{i=0}^{m} T^i$ there is a map $\varphi_A : M^K p_H(A) \to U^n$ which is ϵ-measure-preserving and such that $d([y], \varphi_A(y)) < e^{-s} f$ and $P_{V_1}([y]) = P_{V_1}(\varphi_A(y))$

(for all y for which φ_A is defined). Properties (i), (ii), and (iii) now follow immediately from (1), (2), and (3), respectively, and the proof is complete.

THEOREM 1. *Under the assumptions made in this section (i.e., V_1 and V_2 both nontrivial and $1, \alpha_1, \ldots, \alpha_n$ rationally independent algebraic numbers) the partition ξ is* V.W.B.

Proof. Let ϵ' be given. Then for ϵ'-almost every element A of $\bigvee_{i=0}^{m} T^i\xi$ we can construct a map $s_A : A \to S_H$ by using the map φ_A of Lemma 4. Specifically s_A takes $x = (x_i)_{-\infty}^{\infty}$ to $x' = (x_i')_{-\infty}^{\infty}$, where

$$
\begin{aligned}
x_K' &= \varphi_A(M^K x_0) + d_K(x, 0) & & \\
x'_{K+j} &= [M^j x_K'] + d_j(x, K) & &\text{for} \quad 1 \leqslant j \leqslant K' - K \\
x'_{K'+j} &= [M^j x'_{K'}] + d_j(x, 0) & &\text{for} \quad 1 \leqslant j \leqslant K \\
\left.\begin{aligned} x'_{K+K'+j} &= [M^j x'_{K+K'}] + d_j(x, K') \\ x'_{K-j} &= [M^{-j} x_K'] + c_j(x, -m) \end{aligned}\right\} & j \geqslant 1
\end{aligned}
$$

(see preliminaries for the definition of $d_j(x, i)$ and $c_j(x, i)$; K and K' are as in Lemma 4).

As φ_A is measurable and ϵ'-measure-preserving, it follows that s_A also has these properties.

Let $(a_i)_{-\infty}^{\infty}$ and $(a_i')_{-\infty}^{\infty}$ be the (T, ξ)-names of x and $x' = s_A(x)$, respectively. If $K < i \leqslant K'$ and $a_i \neq a_i'$, then either $d(x_i, x_i') \leqslant 1/K'^2$ or else one of the two points x_i, x_i' is closer than $1/K'^2$ to $\partial\eta$.

Hence from (ii) and (iii) of Lemma 4 it follows that for ϵ'-almost all $x \in A$, $a_i = a_i'$ for $K < i \leqslant K'$. Thus by (i) of Lemma 4, for these "good" x, $a_i = a_i'$ for all but at most $\epsilon' K'$ of the indices i, $0 \leqslant i \leqslant K'$. Thus ξ is V.W.B. as required.

COROLLARY. *Under all the assumptions of this section, T is Bernoullian.*

Proof. As well as being V.W.B., ξ is a generator (see above). Thus T is Bernoullian by Theorem I of the preliminaries.

SECTION II

THEOREM 2. *Using the same notation as in Section I, if either V_1 or V_2 is equal to zero, then T is Bernoullian.*

Proof. We need only prove the theorem for $V_1 = 0$—if $V_2 = 0$ we consider T^{-1} instead of T. So we assume $V_1 = 0$.

For ρ and s_0 as in the second paragraph of Section I we have $d(M^s y, M^s y') < \rho^{-s} d(y, y')$ for $s > s_0$ and all $y, y' \in R^n$. Let ϵ be given and let ψ_A be an arbitrary measure-preserving map from $p_H(A)$ onto U^n, $A \in \bigvee_{i=0}^{m} T^i \xi$.

It is not hard to see that we have the following analog of Lemma 4:

LEMMA 4′. *Given ϵ there exist integers K and K' and a measure-preserving map $\varphi_A: M^K p_H(A) \to U^n$, $A \in \bigvee_{i=0}^{m} T^i \xi$, $\varphi_A = M^K \circ \psi_A \circ M^{-K}$ such that:*

(i) $2K/K' < \epsilon$;

(ii) $K' \mu\{ y \in U^n \mid d(y, \partial\eta) < 1/K'^2 \} < \epsilon/2$;

(iii) *for $\epsilon/2$-almost every $y \in M^K p_H(A)$, $d(M^j[y], M^j \varphi_A(y)) < 1/K'^2$ for $1 \leqslant j \leqslant K$.*

This lemma is a consequence of the fact that separate points are rapidly brought together under the action of M. The proof is easy and is ommitted.

The remainder of the proof of this theorem is just a repetition of the proof of Theorem 1 with Lemma 4′ replacing Lemma 4.

SECTION III

In this section we shall deal with the case where all the eigenvalues of the matrix M associated with the transformation T have modulus one. Here if we write $R^n = V_1 + V_2$ as before, then we have $V_1 = 0$, i.e., $V_2 = R^n$ and so there is a constant h such that $\|M^s v\| < hs^n \|v\|$ for all $v \in R^n$.

Given ϵ we shall show that there is a constant $K(\epsilon)$ such that for all $A \in \bigvee_{i=0}^{m} T^i$ there is a measure-preserving map $\varphi_A: M^{K(\epsilon)} p_H(A) \times D_{K(\epsilon)} \to U^N$ ($D_{K(\epsilon)}$ the group as described in the preliminaries) such that the distance between $\varphi_A(y, d)$ and $[y] + d$ is less than ϵ. Using this map together with an estimate of $K(\epsilon)$ we shall be able to deduce that ξ is V.W.B. in a similar fashion to that of Section I.

The crucial lemma is Lemma 1 in which we deduce certain properties of the uniformity of the sets C_s. We remark again that the ergodicity condition implies that no eigenvalue of M is a root of unity.

LEMMA 1. *Given an $n \times n$ matrix M all of whose eigenvalues lie on the unit circle and none of which are roots of unity, and an $\epsilon > 0$, there exists a $K' = K'(\epsilon)$ such that given any linearly independent vectors $v_1, v_2, \ldots, v_n$, the group $\Delta(K')$ generated by $M^j v_i$, $i = 1, 2, \ldots, n$, $j = 1, 2, \ldots, K'$ contains n linearly independent vectors $u_1, u_2, \ldots, u_n$ with $\|u_i\| < \epsilon \|v_i\|$.*

Proof. Let L be such that the eigenvalues $\lambda_1, \lambda_2, \ldots, \lambda_n$ of M^L satisfy $\|\lambda_i - 1\| < 1$ for all i. For each v_j we have $(M^L - I)v_j = M^L v_j - v_j \in \Delta(L)$

and so for any integer k, $(M^L - I)^k v_j \in \Delta(kL)$. But all the eigenvalues of $M^L - I$ have modulus less than some $\rho < 1$. Hence there exists a k_0 (see second paragraph of Section I) such that for all $k > k_0$, $\|(M^L - I)^k v_j\| < \rho^k v_j$. We put $K' = kL$ and $u_i = (M^L - I)^k v_k$ for all i, k chosen greater than k_0, and such that $\rho^k < \epsilon$. The independence of the u_i's follows from the independence of the v_i's and the invertibility of $(M^L - I)^k$.

From now on v_i will denote the unit vector $(0, 0, \ldots, 1, \ldots, 0) \in R^n$ the 1 being in the ith position, $i = 1, 2, \ldots, n$. Thus in particular $\Delta(K) = D_K$.

We define $K(\epsilon)$ to be the smallest integer such that $D_{K(\epsilon)}$ contains n linearly independent vectors u_i with $\|u_i\| < \epsilon/n$. The next lemma gives an estimate for the size of $K(\epsilon)$.

LEMMA 2. *There exists an s_1 such that for all $s > s_1$ we have $K(e^{-s}) \leqslant bs$ where b is a constant independent of s.*

Proof. Let ρ and k_0 be as in the proof of Lemma 1 (for v_i as defined above). Let s_1 be the smallest integer $> \log(n)$ such that we can find an integer $k' > k_0$ which satisfies $\rho^{k'} < e^{-s_1}/n \leqslant \rho^{k'/2}$.

Let $s > s_1$ and let k_1 be an integer that satisfies $\rho^{k_1} < e^{-s}/n \leqslant \rho^{k_1/2}$. Now from the proof of Lemma 1 it follows that $K(e^{-s}) < kL$ for all integers $k > k_0$ for which $\rho^k < e^{-s}/n$. But $e^s \geqslant (\rho^{-k/2})/n$. Thus taking logs we have $s \geqslant k_1 \log(\rho^{-1})/2 - \log(n)$. Hence $k_1 < 4s/\log(\rho^{-1})$ and so we get $K(e^{-s}) < k_1 L < 4sL/\log(\rho^{-1})$. Putting $b = 4L/\log(\rho^{-1})$ gives the required result.

Let $w_1, w_2, \ldots, w_n$ be chosen from $D_{K(\epsilon)}$ so that

(i) $\|w_1\| \leqslant \|w\|$ for all $w \in D_{K(\epsilon)}$;

(ii) for $n \geqslant i > 1$ w_i is linearly independent of $w_1, w_2, \ldots w_{i-1}$ and $\|w_i\| \leqslant \|w\|$ for all $w \in D_{K(\epsilon)}$ which are also independent of $w_1, w_2, \ldots, w_{i-1}$.

Since $D_{K(\epsilon)}$ contains n linearly independent vectors u_i with $\|u_i\| < \epsilon/n$ it follows that $\|w_i\| < \epsilon/n$, $i = 1, 2, \ldots, n$.

We define $R(w) = R(w_1, w_2, \ldots, w_n)$ to be the partition of U^n (considered as the n-dimensional torus) whose elements are sets of the form:

$$R(z, w) = \{y = \textstyle\sum a_i w_i \in U^n \mid z_i \leqq a_i < (z_i + 1), \qquad z = (z_1, z_2, \ldots z_n) \in Z\}.$$

Observe that for each element W of $R(w)$ $W \cap D_{K(\epsilon)}$ contains one and only one point. Also for any element W of $R(w)$ we have $\operatorname{diam}(W) < n\epsilon/n = \epsilon$.

In the next two lemmas we make use of the partition $R(w)$ in showing how to construct the map φ_A mentioned at the beginning of this section.

LEMMA 3. *Let ϵ be given. Let $B' \subset U^n$ be small enough so that for all $d, d' \in D_{K(\epsilon)}$, $d \neq d'$, $(d + B') \cap (d' + B')$ is empty. Then there is a measure* *preserving map* $g_B: B = \bigcup_{d \in D_{K(\epsilon)}} (d + B') \to U^n$ *such that for all* $y \in B$, $d(y, g_B(y)) < \epsilon$.

Proof. Note that $\mu(B \cap W)$ is the same for all elements W of $R(w)$. Let g_B be an arbitrary measure preserving map from B to U^n such that $g_B(B \cap W) = W$ for all $W \in R(w)$. Since $\operatorname{diam}(W) < \epsilon$ the required result follows.

LEMMA 4. *Let* $A \in \bigvee_{i=0}^{m} T^i \xi$. *Then there exists a measure preserving map* $\varphi_A: M^{K(\epsilon)} p_H(A) \times D_{K(\epsilon)} \to U^n$ *such that for all points* $(y, d) \in M^{k(\epsilon)} p_H(A) \times D_{K(\epsilon)}$ *we have* $d(\varphi_A(y, d), [y] + d) < \epsilon$.

Proof. $M^{K(\epsilon)} p_H(A)$ may be expressed as a finite union of sets B_i: $M^{K(\epsilon)} p_H(A) = \bigcup_{i \in I} B_i$, where for each $i \in I$ $[B_i]$ satisfies the conditions of Lemma 3 and if $y, y' \in B_i$, $y \neq y'$ then $[y] \neq [y']$. Thus with each B_i there is an associated map $g_{B_i}: \bigcup_{d \in D_{K(\epsilon)}} (d + [B_i]) \to U^n$ which keeps points and their images within ϵ of each other.

Partition each $W \in R(w)$ into I sets: $W_1, W_2, \ldots, W_{|I|}$ with $\mu(W_i)/\mu(W) = \mu(B_i)/\mu(M^{K(\epsilon)} p_H(A))$, ($|I|$ denotes the number of elements of I). Let $\psi_1, \psi_2, \ldots, \psi_{|I|}$ be arbitrary measure preserving maps $\psi_i: U^n \to \bigcup_{W \in R(w)} W_i$; i.e., ψ_i uniformly contracts W onto W_i for each $W \in R(w)$.

We now define φ_A as follows: if $y \in B_i$ then $\varphi_A(y, d) = \psi_i \circ g_{B_i}([y] + d)$. Since $[y] + d$ and $\varphi_A(y, d)$ are both in the same component of $R(w)$ the distance between them is $< \epsilon$. This completes the proof.

We now have the following analog of Lemma 4 of Section I:

LEMMA 4″. *Let* ϵ *be given. Then there exist constants* K, K' *such that for all* $A \in \bigvee_{i=0}^{m} T^i \xi$ *there is a measure-preserving map* $\varphi_A: M^K p_H(A) \times D_K \to U^n$ *of the type described above (i.e.,* $K = K(\delta)$ *for some* δ*), and*

(i) $K/K' < \epsilon$,
(ii) $K' \mu\{y \in U^n \mid d(y, \partial\eta) < 1/K'^2\} < \epsilon/2$,
(iii) *for all* $(y, d) \in M^K p_H(A) \times D_K$, $d(M^j([y] + d), M_A{}^j(y, d)) < 1/K'^2$ *for* $K \leqslant j \leqslant K'$.

Proof. Using the estimate for $K(e^{-s})$ given in Lemma 2 the proof is similar to that of Lemma 4 of Section I and is therefore ommitted.

THEOREM 3. *If the matrix associated with* T *has all its eigenvalues of unit modulus and none roots of unity then* T *is* V.W.B. *and hence* T *is Bernoullian.*

Proof. Let ϵ be given. We define a measure preserving map $s_A : A \to S_H$ by $s_A(x) = x'$, $x = (x_i)_{-\infty}^{\infty}$, $x' = (x_i')_{-\infty}^{\infty}$, where

$$\begin{aligned} x_K' &= \varphi_A([M^K(x_0)], d_K(x, 0)) \\ \left.\begin{aligned} x'_{K+j} &= [M^j(x_K')] + d_j(x, K) \\ x'_{K-j} &= [M^{-j}(x_K')] + c_j(x, K) \end{aligned}\right\} & \; j \geq 1 \end{aligned}$$

(φ_A, K, and K' as in Lemma 4″).

Having defined the map s_A the proof is the same as the proof of Theorem 1 (Section I) with Lemma 4″ replacing Lemma 4, and so is ommitted.

Section IV

In this section we tie together the results of the three previous sections to obtain the desired result that generalized torus automorphisms are Bernoullian. We begin with two preliminary lemmas and then give the main theorem.

Lemma 1. *Let M, the matrix associated with the generalized torus automorphism T, be of the form:*

$$M = \begin{pmatrix} P & I & & & & \\ & P & I & & 0 & \\ & & P & I & & \\ & & & \ddots & & \\ & 0 & & & P & I \\ & & & & & P \end{pmatrix}$$

where I is the identity matrix and P is the companion matrix of a monic polynomial which is irreducible over the rationals. If all the eigenvalues of P have modulus one or if P has at least one eigenvalue of modulus greater than one, then T is Bernoullian.

Proof. If all the eigenvalues of P have modulus > 1 or if all the eigenvalues of P have modulus $= 1$ then the lemma follows immediately from the results of Sections II and III, respectively.

Let P have at least one eigenvalue of modulus > 1 and one of modulus ≤ 1. Let

$$P = \begin{bmatrix} 0 & 1 & 0 & 0 & \cdots & 0 \\ 0 & 0 & 1 & 0 & \cdots & 0 \\ \vdots & & & & & \vdots \\ 0 & 0 & 0 & 0 & \cdots & 1 \\ -a_0 & -a_1 & -a_2 & & \cdots & -a_{m-1} \end{bmatrix}$$

so that the characteristic polynomial is $a_0 y + a_1 y + \cdots + y^m$. Two possibilities arise:

(i) If λ is a real eigenvalue of modulus > 1 then $v = (1, \lambda, \lambda^2, \ldots, \lambda^{m-1})$ is an eigenvector and $1, \lambda, \lambda^2, \ldots, \lambda^{m-1}$ are rationally independent since otherwise we contradict the irreducibility of the characteristic polynomial.

Let there be k blocks P in M. Then choose algebraic numbers $\alpha_1, \alpha_2, \ldots, \alpha_k$ such that $1, \alpha_1, \alpha_1\lambda, \ldots, \alpha_1\lambda^{m-1}, \alpha_2, \alpha_2\lambda, \ldots, \alpha_2\lambda^{m-1}, \ldots, \alpha_k, \alpha_k\lambda, \ldots, \alpha_k\lambda^{m-1}$ are rationally independent. Then $v_1 = (\alpha_1, \alpha_1\lambda, \ldots, \alpha_1\lambda^{m-1}, \alpha_2, \alpha_2\lambda, \ldots, \alpha_2\lambda^{m-1}, \ldots, \alpha_k, \alpha_k\lambda, \ldots, \alpha_k\lambda^{m-1})$ satisfies the conditions of Section I and so in this case T is Bernoullian.

(ii) If λ and $\bar{\lambda}$ are complex eigenvalues of modulus > 1 then $v' = (2, \lambda + \bar{\lambda}, \lambda^2 + \bar{\lambda}^2, \ldots, \lambda^{m-1} + \bar{\lambda}^{m-1})$ and $v'' = (0, i(\lambda - \bar{\lambda}), i(\lambda^2 - \bar{\lambda}^2), \ldots, i(\lambda^{m-1} + \bar{\lambda}^{m-1}))$ generate an expanding eigenplane. Choose α (algebraic) so that no nonzero rational combination of the numbers $2, (\lambda + \bar{\lambda}), \ldots, (\lambda^{m-1} + \bar{\lambda}^{m-1})$ is a rational multiple of a rational combination of the numbers $i\alpha(\lambda - \bar{\lambda}), \ldots, i\alpha(\lambda^{m-1} - \bar{\lambda}^{m-1})$. Then $2, \{\lambda + \bar{\lambda} + i\alpha(\lambda - \bar{\lambda})\}, \ldots, \{\lambda^{m-1} + \bar{\lambda}^{m-1} + i\alpha(\lambda^{m-1} - \bar{\lambda}^{m-1})\}$ must be rationally independent since otherwise we contradict the irreducibility of the characteristic polynomial. Hence as in (i) we can construct a v_1 which satisfies the conditions of Section I and so in this case also T is Bernoullian.

LEMMA 2. *Let T be a generalized torus automorphism and M the associated matrix. Let M be $n \times n$. Let $\hat{T}_1 : Q^n \to Q^n$ be defined by $(y_1, y_2, \ldots, y_n) \xrightarrow{T_1} (y_1, y_2, \ldots, y_n)M$ and write G for the character group of Q^n. Then the automorphism $T_1 : G \to G$ dual to $\hat{T}_1$ is Bernoullian.*

Proof. Let Γ be the character group of the generalized torus. Then the dual to T is represented by M acting on the right and Γ will be the smallest subgroup of Q^n containing Z^n and invariant under M (see [1] for details).

Let $Z^n \subset A_1 \subset A_2 \subset A_3 \subset \cdots$ be a sequence of subgroups of Q^n such that:

(i) $A_i = (1/a_i)Z^n$ for some integer a_i, and

(ii) if B_i denotes the smallest subgroup of Q^n containing A_i and invariant under M then for any $y \in Q^n$ there exists an $i(y)$ such that $y \in B_{i(y)}$.

Clearly $B_1 \subset B_2 \subset B_3 \subset \cdots$.

If H_i is the annihilator of A_i then $G_i = \bigcap_{-\infty}^{\infty} T^j H_i$ is the annihilator of B_i. But for each i the automorphism induced by T_1 on G/G_i is isomorphic to T and so is Bernoullian. Therefore, since $\bigcap_1^{\infty} G_i = e$, T_1 is the direct limit of these automorphisms and so by Theorem II of the preliminaries is Bernoullian.

THEOREM. *If T is a generalized torus automorphism, then T is Bernoullian.*

Proof. Let M be the matrix associated with T and let G, T_1, and $\hat{T}_1$ be defined as in Lemma 2 above. We now change basis to get M into a more tractable form. Specifically M is similar to a block diagonal matrix

$$M' = \begin{bmatrix} M_1 & & & \\ & M_2 & & 0 \\ & & \ddots & \\ & 0 & & M_k \end{bmatrix}$$

where each M_i is of the form given in Lemma 1 except that there is no restriction on the modulus of the eigenvalues (i.e., M' is the generalized Jordan form of M, see e.g., [7], for details).

We may assume without loss of generality that $M_1, M_2, \ldots, M_j$ are the only blocks which have eigenvalues of modulus less than one but no eigenvalues of modulus greater than one (if any such exist). Let

$$M(1) = \begin{bmatrix} M_1 & & 0 \\ & \ddots & \\ 0 & & M_j \end{bmatrix},$$

and let the generalized Jordan form of $M(1)^{-1}$ be

$$M'(1)^{-1} = \begin{bmatrix} M_1' & & & \\ & M_2' & & 0 \\ & & \ddots & \\ & 0 & & M'_{j'} \end{bmatrix}$$

Let

$$M(2) = \begin{bmatrix} M_{j+1} & & 0 \\ & \ddots & \\ 0 & & M_k \end{bmatrix}.$$

Note that each of the blocks $M_1', M_2', \ldots, M'_{j'}, M_{j+1}, M_{j+2}, \ldots, M_k$ satisfies the conditions of Lemma 1 of this section.

Let $M'(1)^{-1}$ be $m \times m$. Then $M'(1)^{-1}$ determines an automorphism $\hat{T}(1)$ of Q^m ($M'(1)^{-1}$ acting on the right). Now applying Lemmas 1 and 2 we see that $T(1)$ the dual of $\hat{T}(1)$ is isomorphic to a product of Bernoullian automorphisms and so is itself Bernoullian. Similarly $M(2)$ determines a Bernoullian automorphism $T(2)$. But T_1 is isomorphic to the product of $T(1)^{-1}$ and $T(2)$ and hence is Bernoullian. Finally since T is a factor transformation of T_1 it follows that, by a theorem of Ornstein's [8], T is Bernoullian.

References

1. G. Miles and R. K. Thomas, The breakdown of automorphisms of compact topological groups, this volume.
2. D. Ornstein and B. Weiss, Geodesic flows are Bernoullian, *Israel J. Math.* **14** (1973), 184–198.
3. G. Miles and R. K. Thomas, On the polynomial uniformity of translations of the n-torus, this volume.
4. Y. Katznelson, Ergodic automorphisms of T^n are Bernoulli shifts, *Israel J. Math.* **10** (1971), 186–195.
5. D. Ornstein, Imbedding Bernoulli shifts in flows, *in* "Contributions to Ergodic Theory and Probability," Lecture Notes in Mathematics, pp. 178–218, Springer–Verlag, Berlin and New York, 1970.
6. D. Ornstein, Two Bernoulli shifts with infinite entropy are isomorphic, *Advances in Math.* **5** (1970), 339–348.
7. A. I. Mal'cev, "Foundations of Linear Algebra," Freeman, San Francisco, 1963.
8. D. Ornstein, Factors of Bernoulli shifts are Bernoulli shifts, *Advances in Math.* **5** (1970), 349–364.

STUDIES IN PROBABILITY AND ERGODIC THEORY
ADVANCES IN MATHEMATICS SUPPLEMENTARY STUDIES, VOL. 2

The Isomorphism Theorem for Generalized Bernoulli Schemes

J. C. KIEFFER

Department of Mathematics, University of Missouri—Rolla, Rolla, Missouri

It is shown that actions of an arbitrary infinite abelian group on a probability space are isomorphic if they have the same entropy, provided the actions are generalized Bernoulli schemes. This result generalizes the isomorphism theorem of Ornstein. In obtaining the result, generalizations of the Shannon–McMillan theorem and Sinai's theorem are used.

Let G be an infinite abelian group and let $(\Omega, \mathscr{F}, m)$ be a probability space. We say that T is an action of G on Ω if T is a homomorphism of G into the automorphism group of Ω. For $g \in G$, let T^g be the image of g under T; T^g is an invertible measure-preserving transformation of Ω and $T^{g_1+g_2} = T^{g_1}T^{g_2}$, $g_1, g_2 \in G$. Following Kirillov [4], we define an action T of G on $(\Omega, \mathscr{F}, m)$ to be a generalized Bernoulli scheme if there exists a nontrivial partition P of Ω such that $\{T^iP : i \in G\}$ are independent and generate $\mathscr{F}$. (In this paper, all partitions are finite.)

Let T_1 be an action of G on $(\Omega_1, \mathscr{F}_1, m_1)$. Let T_2 be an action of G on $(\Omega_2, \mathscr{F}_2, m_2)$. For $i = 1, 2$, let $(\bar{\mathscr{F}}_i, \bar{m}_i)$ be the measure albegra corresponding to $(\Omega_i, \mathscr{F}_i, m_i)$. For each $g \in G$, T_i^g induces an automorphism $\bar{T}_i^g$ of $(\bar{\mathscr{F}}_i, \bar{m}_i)$ in the natural way, $i = 1, 2$. We say that T_1 and T_2 are isomorphic if there exists an isomorphism $\Phi : \bar{\mathscr{F}}_1 \to \bar{\mathscr{F}}_2$ such that $\bar{T}_2^g\Phi = \Phi\bar{T}_1^g$, $g \in G$.

In [3] the entropy of an action T of G on Ω is defined. It is the purpose of this paper to prove that actions of an arbitrary infinite abelian group are isomorphic if they have the same entropy, provided the actions are generalized Bernoulli schemes (Theorem 2, below). Ornstein has shown that Theorem 2 is true for the special case where G is the group of integers Z. Our method of proof generalizes upon the method used by Ornstein [5]. A partial proof of Theorem 2 for the case where G is the direct product of finitely many copies of Z is given in [2]. It is stated without proof in [8] that Theorem 2 holds where G is the group of dyadic rationals modulo one.

ISBN 0-12-599102-9

DEFINITIONS. For the rest of the paper, G is a fixed arbitrary infinite abelian group. If E, F are subsets of G let $E + F = \{e + f : e \in E, f \in F\}$. Similarly we define $E - F$. If S is a set, let $|S|$ denote the cardinality of S. If A, B are sets let $A \backslash B = \{x \in A : x \notin B\}$. An action T of G on $(\Omega, \mathscr{F}, m)$ is said to be aperiodic if for any $A \in \mathscr{F}$ of positive measure, and any $g \neq 0$ in G, there exists a measurable subset B of A, of positive measure, such that $T^g B \cap B = \varnothing$. We define a finite nonempty subset K of G to be an R-set if for any aperiodic action T of G on any probability space Ω, and for any $\epsilon > 0$, there exists a measurable subset F of Ω such that $\{T^i F : i \in K\}$ are pair-wise disjoint and $m(\bigcup_{i \in K} T^i F) > 1 - \epsilon$. (We named the R-set after Rokhlin [6], who showed that if T is an aperiodic action of Z then any finite set of consecutive integers is an R-set.)

The proof of the isomorphism theorem for generalized Bernoulli schemes depends on showing the existence of a net $\{K_\alpha\}$ of finite nonempty subsets of G such that:

(1.1) $\lim_\alpha |K_\alpha|^{-1} |K_\alpha + E| = 1$, for every finite nonempty subset E of G;
(1.2) $0 \in K_\alpha$ for each α, and $\bigcup_\alpha K_\alpha = G$;
(1.3) G is partitioned by translates of K_α, for each α;
(1.4) $\lim_\alpha |K_\alpha| = \infty$;
(1.5) each K_α is an R-set.

We will now show that such a net exists. We employ a technique used by Conze [1] to show that certain subsets of Z^2 are R-sets. Let T be an arbitrary aperiodic action of G on $(\Omega, \mathscr{F}, m)$.If E is a finite nonempty subset of G, let $\mathscr{D}(E)$ be the set of all $A \in \mathscr{F}$ such that $\{T^g A : g \in E\}$ are pairwise disjoint. Let $\phi_E = \sup_{A \in \mathscr{D}(E)} m(\bigcup_{g \in E} T^g A)$. Note that E is an R-set if and only if $\phi_E = 1$.

The following lemma, whose proof we omit, is an easy consequence of the aperiodicity of T.

LEMMA 1. *If $A \in \mathscr{F}$ and $m(A) > 0$, there exists $B \in \mathscr{F}$ such that $B \subset A$, $m(B) > 0$, and $B \in \mathscr{D}(E)$.*

LEMMA 2. *If $B \in \mathscr{F}$, there exists a measurable subset A of B such that $A \in \mathscr{D}(E)$ and $m(\bigcup_{g \in E} T^g A) \geqslant |E - E|^{-1} |E| m(B)$.*

Proof. Choose $A \in \mathscr{D}(E)$, $A \subset B$, which is maximal with respect to inclusion in $(\bar{\mathscr{F}}, \bar{m})$. Let $C = \bigcup_{g \in E - E} T^g A$. Suppose $m(B \backslash C) > 0$. Choose D such that $D \subset B \backslash C$, $D \in \mathscr{D}(E)$, $m(D) > 0$. It is easily checked that $D \cup A \in \mathscr{D}(E)$. Since $m(D \cup A) > m(A)$ this contradicts the maximality of A. Thus $m(B \backslash C) = 0$, and so $m(B) \leqslant m(C) \leqslant |E - E| m(A) = |E - E|\, |E|^{-1} m(\bigcup_{g \in E} T^g A)$.

LEMMA 3. *Let E, F be finite nonempty subsets of G such that E is partitioned by translates of F. Then*

(a) $\phi_E \leqslant \phi_F$;
(b) $\phi_F \geqslant |F - F|^{-1}|F| + (1 - |F - F|^{-1}|F|)\phi_E - |F - F|^{-1}|F|(|E|^{-1}|E + F - F| - 1)$.

Proof. Without loss of generality assume $0 \in F$. Let $\{F + g : g \in C\}$ partition E. If $A \in \mathscr{D}(E)$, then $\bigcup_{g \in C} T^g A$ is in $\mathscr{D}(F)$. From this follows (a). Now to prove (b). Let $A \in \mathscr{D}(E)$. Let $A' = \bigcup_{g \in E} T^g A$. Let S be the complement of $\bigcup_{g \in F - F + E} T^g A$. Choose B such that $B \subset S$, $B \in \mathscr{D}(F)$, and $m(\bigcup_{g \in F} T^g B) \geqslant |F - F|^{-1}|F| m(S)$. Let $D = B \cup [\bigcup_{g \in C} T^g A]$. It can be shown that $D \in \mathscr{D}(F)$. Thus

$$\phi_F \geqslant m\left[\bigcup_{g \in F} T^g D\right] = m\left[\bigcup_{g \in F} T^g B\right] + m(A')$$
$$\geqslant |F - F|^{-1}|F| m(S) + m(A'). \tag{1}$$

Now

$$m([\Omega \backslash A'] \backslash S) = m\left(\left[\bigcup_{g \in F - F + E} T^g A\right] \Big\backslash \left[\bigcup_{g \in E} T^g A\right]\right)$$
$$\leqslant m\left[\bigcup_{g \in (F - F + E) \backslash E} T^g A\right] \leqslant (|F - F + E| - |E|) m(A)$$
$$\leqslant |E|^{-1}|F - F + E| - 1.$$

Thus, $m(S) \geqslant 1 - m(A') - (|E|^{-1}|F - F + E| - 1)$. Putting in this lower bound for $m(S)$ in inequality (1) and simplifying, we obtain

$$\phi_F \geqslant |F - F|^{-1}|F| + (1 - |F - F|^{-1}|F|) m(A')$$
$$- |F - F|^{-1}|F|(|E|^{-1}|F - F + E| - 1).$$

Taking the supremum over $A \in \mathscr{D}(E)$, we obtain (b).

LEMMA 4. *There exists a net $\{K_\alpha\}$ of finite nonempty subsets of G satisfying* (1.1)–(1.5).

Proof. For each pair (S, ϵ) where S is a finite subset of G and $\epsilon > 0$, we construct an R-set $K = K_{(S,\epsilon)}$ containing $S \cup \{0\}$ such that G is partitioned by translates of K and $|K|^{-1}|K + S| < 1 + \epsilon$. Let G' be the subgroup of G generated by S. If G' is finite, then G' is an R-set by Lemma 2, so we may take $K = G'$. If G' is infinite, then it is isomorphic to a group $C \times Z^k$, where C is a finite group and k is a positive integer. Let $\Psi : G' \to C \times Z^k$ be the isomorphism. For $n = 1, 2, \ldots$, let $E_n = C \times \{(x_1, x_2, \ldots, x_k) \in Z^k : -2^n \leqslant x_i < 2^n,\ i = 1, 2, \ldots, k\}$. Let $E_n' = \Psi^{-1}(E_n)$. It can be seen that for every finite subset B of G' we have $\lim_{n\to\infty} |E_n'|^{-1}|E_n' + B| = 1$. If each E_n' is an R-set we may take K to be some E_n' for n sufficiently large. Now if $n > m$, E_n' is partitioned by translates of E_m'. Thus by Lemma 3, $\{\phi_{E_n'}\}$ is a nonincreasing sequence as $n \to \infty$. Let $\lim_{n\to\infty} \phi_{E_n'} = \beta$. If $\beta = 1$, then each E_n' is an R-set. If $n > m$ we have by Lemma 3

$$\begin{aligned}\phi_{E_m'} \geqslant{} & |E_m' - E_m'|^{-1}|E_m'| + (1 - |E_m' - E_m'|^{-1}|E_m'|)\phi_{E_n'} \\ & - |E_m' - E_m'|^{-1}|E_m'|(|E_n'|^{-1}|E_n' + E_m' - E_m'| - 1).\end{aligned}$$

Letting $n \to \infty$, we have

$$\phi_{E_m'} \geqslant |E_m' - E_m'|^{-1}|E_m'| + (1 - |E_m' - E_m'|^{-1}|E_m'|)\beta.$$

Now $|E_m' - E_m'|^{-1}|E_m'| = (2^{m+2} - 1)^{-k} 2^{mk+k}$.

Letting $m \to \infty$, we obtain $\beta \geqslant 2^{-k} + (1 - 2^{-k})\beta$, from which it follows that $\beta = 1$.

Direct the pairs (S, ϵ) so that $(S, \epsilon) > (S', \epsilon')$ if $S \supset S'$ and $\epsilon < \epsilon'$. Then the net $\{K_{(S,\epsilon)}\}$ satisfies (1.1)–(1.5).

DEFINITION. If f is a real-valued function with domain G, and α is a real number, we say that $\lim_{g\to\infty} f(g) = \alpha$ if for every $\epsilon > 0$ there exists a finite subset E of G such that $|f(g) - \alpha| < \epsilon$ for $g \notin E$. We say that an action T of G on $(\Omega, \mathscr{F}, m)$ is mixing if $\lim_{g\to\infty} m(A \cap T^g B) = m(A)m(B)$, $A, B \in \mathscr{F}$.

LEMMA 5. *Let the action T of G on Ω be a generalized Bernoulli scheme. Then*

(a) *Ω is nonatomic;*
(b) *T is mixing;*
(c) *T is aperiodic.*

Proof. The proof of (a) is routine and is omitted. To prove (b), let P be a nontrivial partition such that $\{T^i P : i \in G\}$ are independent and generate $\mathscr{F}$. If E is any finite nonempty subset of G, and A, B are unions of atoms of $\bigvee_{i\in E} T^i P$, then it is easily seen that

$$\lim_{g\to\infty} m(A \cap T^g B) = m(A)m(B). \tag{1}$$

It follows that (1) holds for any $A, B \in \mathscr{F}$. Now to prove (c). Let $P = \{P^1, P^2, \ldots, P^n\}$. For each $g \in G$, let $X(g)$ be the random variable with range $\{1, 2, \ldots, n\}$ defined as follows: for each $\omega \in \Omega$, $X(g)(\omega) = j$ if and only if $T^g\omega \in P^j$. Then $X(g)T^{g'} = X(g + g')$, $g, g' \in G$, and $\{X(g) : g \in G\}$ are independent, nontrivial, identically distributed random variables. Fix $g \in G$, $g \neq 0$. Let $A \in \mathscr{F}$ have positive measure. Let G' be the subgroup of G generated by g. If G' is infinite, it follows that $\bigcap_{k=-\infty}^{\infty} \{X(kg) = X(kg+g)\}$ has zero measure. Thus, some set B of form $B = A \cap \{X(kg) = i, X(kg+g) = j\}$ has positive measure for some k, i, j, where $i \neq j$. We have $B \cap T^{-g}B = \phi$. If G' is finite of order a, let S be a countably infinite set of coset representatives of G' in G. It is easily seen that $\bigcap_{s \in S} \bigcap_{k=0}^{a-2} \{X(s+kg) = X(s+kg+g)\}$ has zero measure. Thus, some set B of form $B = A \cap \{X(s+kg) = i, X(s+kg+g) = j\}$ has positive measure for some s, k, i, j, where $i \neq j$. We have $B \cap T^{-g}B = \phi$.

DEFINITIONS. We adopt the following definitions, many of which come from [5]. Let $P = \{P^j : j \in D\}$ be a partition of $(\Omega, \mathscr{F}, m)$. (The index set D is finite, since we only consider finite partitions in this paper.) If $A \in \mathscr{F}$, then $P \cap A$ is defined to be the partition of A such that $P \cap A = \{P^j \cap A : j \in D\}$. If $P = \{P^j : j \in D\}$ is a partition of $A \in \mathscr{F}$, where $m(A) > 0$, define dist $P = \{m(P^j)/m(A) : j \in D\}$. If $Q = \{Q^j : j \in D\}$ is a partition of $B \in \mathscr{F}$, defined over the same index set D, then define $|\text{dist } P - \text{dist } Q| = \sum_j |(m(P^j)/m(A)) - (m(Q^j)/m(B))|$ and if $A = B$ define $|P - Q| = m(A)^{-1} \sum_j m(P^j \triangle Q^j)$. Let $I = \{\alpha_j : j \in D\}$, where each α_j is nonnegative and $\sum_j \alpha_j = 1$. We call I an abstract partition. If $P = \{P^j : j \in D\}$ is a partition of $A \in \mathscr{F}$, define $|\text{dist } P - \text{dist } I| = \sum_j |(m(P^j)/m(A)) - \alpha_j|$. If P, Q are partitions of $A \in \mathscr{F}$, then $P \supset Q$ means P refines Q. If $\epsilon > 0$, then $P \overset{\epsilon}{\supset} Q$ means that there exists a partition Q' such that $P \supset Q'$ and $|Q' - Q| < \epsilon$. If $\mathscr{F}'$ is some sub-sigmafield of $\mathscr{F}$ and Q is a partition of Ω, then $\mathscr{F}' \overset{\epsilon}{\supset} Q$ means there exists a partition Q' whose sets are in $\mathscr{F}'$ such that $|Q' - Q| < \epsilon$.

Let $\{P_i : i \in E\}$ be a finite collection of partitions of some set. Suppose for each i, $P_i = \{P_i^j : j \in D_i\}$. Let D be the cartesian product $D = \prod_{i \in E} D_i$. Then $\bigvee_{i \in E} P_i$ is the partition $\{\bigcap_{i \in E} P_i^{k(i)} : k \in D\}$. If E is a finite subset of G, T is an action of G on Ω, and P is a partition of Ω, define $PE = \bigvee_{i \in E} T^{-i}P$. Define PG to be the sigmafield generated by $\{T^iP : i \in G\}$. If $\{P^j : j \in D\}$ and $\{Q^j : j \in D\}$ are partitions of Ω and E is a finite subset of G, there is a natural correspondence between the sets of PE and the sets of QE, since both partitions have the same index set D^E. If dist P = dist Q, $\{T^iP : i \in G\}$ are independent, and $\{T^iQ : i \in G\}$ are independent, the natural correspondence between PE and QE for finite E induces a correspondence between PG and QG which preserves measure. (More precisely, the correspondence is an isomorphism of the measure algebras $(\overline{PG}, \overline{m})$ and $(\overline{QG}, \overline{m})$.)

If P, Q are partitions of $A \in \mathscr{F}$, we say that P is ϵ-independent of Q if there is a collection $\mathscr{C}$ of atoms of Q such that the measure of the union of the atoms in $\mathscr{C}$ is at least $(1 - \epsilon)m(A)$, and if $B \in \mathscr{C}$ then $|\text{dist}(P \cap B) - \text{dist } P| < \epsilon$. If $\{P_i : i \in E\}$ are partitions of $A \in \mathscr{F}$, where E is a finite set, we say these partitions are ϵ-independent if there exists an ordering $i_1 < i_2 < \cdots < i_t$ of the elements of E (where $|E| = t$), such that for $2 \leqslant j \leqslant t$, P_{i_j} is ϵ-independent of $\bigvee_{k=1}^{j-1} P_{i_k}$.

If E is a finite subset of G and $P = \{P^j : j \in D\}$ is a partition of Ω, then the P-E-name of a point $\omega \in \Omega$ is defined to be the function $\phi : E \to D$ such that $\phi(g) = j$ if and only if $T^g\omega \in P^j$. If $A \in \mathscr{F}$ is a set all of whose points have the same P-E-name, we refer to this common name as the P-E-name of A.

If I is an abstract partition, let $H(I)$ denote the entropy of I. If P is a partition of Ω, let $H(P)$ denote the entropy of P. If Q is also a partition of Ω, let $H(P|Q)$ denote the conditional entropy of P given Q.

THEOREM 1 (Generalized Shannon–McMillan theorem). *Let T be an action of G on Ω. Let P be a partition of Ω.*

(a) *There exists a number $H(P, T)$ such that for every net $\{K_\alpha\}$ of finite nonempty subsets of G satisfying* (1.1), *we have* $\lim_\alpha |K_\alpha|^{-1} H(PK_\alpha) = H(P, T)$.

(b) *If $\{K_\alpha\}$ satisfies* (1.1) *and* (1.3), *then* $H(P, T) = \inf_\alpha |K_\alpha|^{-1} H(PK_\alpha)$.

(c) *If T is mixing and $\{K_\alpha\}$ satisfies* (1.1), *then for any β, $\epsilon > 0$ there exists an index α' such that for $\alpha > \alpha'$ there is a collection of atoms of PK_α of combined measure at least $1 - \epsilon$, satisfying the property that each atom in the collection has measure between $2^{-|K_\alpha|(H(P,T)\pm\beta)}$.*

Proof. For (a) and (c), consult [3]. To prove (b), let $\{K_\alpha\}$ satisfy (1.1) and (1.3). Let K be any K_α. Then translates $\{K + g : g \in E\}$ of K partition G, for some E. For each α, let $E_\alpha = \{g \in E : (K + g) \cap K_\alpha \neq \phi\}$. Then $KE_\alpha \subset K - K + K_\alpha$, and so $|E_\alpha| \leqslant |K|^{-1}|K - K + K_\alpha|$. Now $H(PK_\alpha) \leqslant H(PKE_\alpha) \leqslant |E_\alpha| H(PK) \leqslant |K|^{-1}|K - K + K_\alpha| H(PK)$. Dividing by $|K_\alpha|$ and taking the limit we get $H(P, T) \leqslant |K|^{-1} H(PK)$. Now (b) follows.

DEFINITION. If T is an action of G on Ω, define the entropy $H(T)$ of T to be the supremum of $H(P, T)$ over all partitions P of Ω.

Isomorphic actions of G have the same entropy, and we wish to show that the converse is true if the actions are generalized Bernoulli schemes. This will follow from the following theorem, which will be proved at the end of the paper.

THEOREM 2 (Generalization of the Ornstein isomorphism theorem). *Let the action T of G on $(\Omega, \mathscr{F}, m)$ be a generalized Bernoulli scheme. Then if I is an abstract partition such that $H(I) = H(T)$, then there exists a partition Q of Ω such that* dist Q = dist I *and $\{T^i Q : i \in G\}$ are independent and generate $\mathscr{F}$.*

(For the proof of the following lemma, see [7, p. 24].)

LEMMA 6. *Let n be a positive integer. Given* $\epsilon > 0$, *there exists* $\delta = \delta(\epsilon, n) > 0$ *such that if* $|P - Q| < \delta$, *where* P, Q *are partitions of* Ω *each consisting of n sets, then there exists a partition S such that* $H(S) < \epsilon$ *and* $P \vee S \supset Q$.

LEMMA 7. *Let T be an action of G on* $(\Omega, \mathscr{F}, m)$.

(a) $H(P, T) \leqslant H(P)$; $H(P \vee Q, T) \leqslant H(P, T) + H(Q, T)$; $H(P, T) \geqslant H(Q, T)$ *if* $P \supset Q$; $H(PE, T) = H(P, T)$ *if E is a finite nonempty subset of G.*

(b) (Generalization of Kolmogorov–Sinai theorem). *If* $\{T^i P : i \in G\}$ *generate* $\mathscr{F}$, *then* $H(P, T) = H(T)$.

(c) *If T is a generalized Bernoulli shift and P is such that* $\{T^i P : i \in G\}$ *are independent and generate* $\mathscr{F}$, *then* $H(T) = H(P)$.

Proof. First we prove the last part of (a). (The other parts are easy.) Let E be a finite nonempty subset of G, and let $\{K_\alpha\}$ satisfy (1.1). Then the net $\{K_\alpha + E\}$ also satisfies (1.1). Thus, $H(PE, T) = \lim_\alpha |K_\alpha|^{-1} H(P[K_\alpha + E], T) = \lim_\alpha |K_\alpha + E|^{-1} H(P[K_\alpha + E], T) = H(P, T)$. To prove (b), let Q be any partition of Ω. Suppose Q consists of n sets. Fix $\epsilon > 0$. Choose $\delta = \delta(\epsilon, n)$ according to Lemma 6. Suppose $\{T^i P : i \in G\}$ generate $\mathscr{F}$. Then there exists a finite nonempty subset E of G such that $PE \overset{\delta}{\supset} Q$. Pick Q' such that $PE \supset Q'$ and $|Q' - Q| < \delta$. There exists by Lemma 6 a partition S such that $H(S) < \epsilon$ and $S \vee Q' \supset Q$. Hence $H(P, T) = H(PE, T) \geqslant H(Q', T) \geqslant H(S \vee Q', T) - H(S, T) > H(Q, T) - \epsilon$. Now (b) follows, and (c) follows from (b).

LEMMA 8. *Let T be an action of G on* Ω. *Let n be a positive integer. Given* $\epsilon > 0$, *there exists* $h(\epsilon) > 0$ (*depending on n*) *such that if* P, Q *are partitions of* Ω *satisfying* $|P - Q| < h(\epsilon)$, *where* P, Q *each consist of n sets, then* $|H(P, T) - H(Q, T)| < \epsilon$.

Proof. Let $h(\epsilon) = \delta(\epsilon, n)$, where $\delta(\epsilon, n)$ is given by Lemma 6. Then if $|P - Q| < h(\epsilon)$, where P, Q consist of n sets, there exists S with $H(S) < \epsilon$ such that $S \vee P \supset Q$. Thus $H(P, T) \geqslant H(S \vee P, T) - H(S, T) > H(Q, T) - \epsilon$. By symmetry, we also have $H(Q, T) > H(P, T) - \epsilon$. The lemma follows.

The following lemma may be found in [7, p. 22].

LEMMA 9. *For any* $\epsilon > 0$, *there exists* $\delta > 0$ *such that if* P, Q *are any partitions of* Ω *satisfying* $H(P) - H(P|Q) < \delta$, *then P is* ϵ-*independent of Q.*

LEMMA 10. *Let T be an action of G on* Ω. *Let* $\{K_\alpha\}$ *satisfy* (1.1)–(1.5). *Let I be an abstract partition with n elements such that* $H(I) = H(T)$. *Then* *for any $\epsilon > 0$, there exists $g(\epsilon) > 0$ (depending on n) such that if P is a partition of Ω with n sets satisfying $|\text{dist } P - \text{dist } I| < g(\epsilon)$ and $H(T) - H(P, T) < g(\epsilon)$, then for each index α there exists $K_\alpha' \subset K_\alpha$ such that $\{T^{-i}P : i \in K_\alpha'\}$ are ϵ-independent and $|K_\alpha'|/|K_\alpha| > 1 - \epsilon$.*

Proof. Fix $\epsilon > 0$. From Lemma 9, there exists $\delta < \epsilon$ such that $H(P) - H(P|Q) < \delta$ implies that P is ϵ-independent of Q. There exists δ' such that $|\text{dist } P - \text{dist } I| < \delta'$ implies that $|H(P) - H(I)| < \delta^2/2$. Let $g(\epsilon) = \min(\delta', \delta^2/2)$. Suppose $|\text{dist } P - \text{dist } I| < g(\epsilon)$ and $H(T) - H(P, T) < g(\epsilon)$. Then $H(P) - H(P, T) < \delta^2$. Fix K_α. We have by Theorem 1(b) that $H(P) - |K_\alpha|^{-1}H(PK_\alpha) < \delta^2$. Let $K_\alpha = \{g_1, g_2, \ldots, g_t\}$. Now $H(P) - |K_\alpha|^{-1}H(PK_\alpha) = |K_\alpha|^{-1}\sum_{i=1}^{t} H(Pg_i) - H(Pg_i|P[g_1, g_2, \ldots, g_{i-1}])$, where we interpret the $i = 1$ term of the sum to be zero. There exists $K_\alpha' \subset K_\alpha$ such that $|K_\alpha'|/|K_\alpha| > 1 - \delta > 1 - \epsilon$ and $H(Pg_i) - H(Pg_i|P[g_1, \ldots, g_{i-1}]) < \delta$ if $g_i \in K_\alpha'$. Let the elements of K_α' be $g_{i_1}, g_{i_2}, \ldots, g_{i_s}$, where $i_1 < i_2 < \cdots < i_s$. Then $H(Pg_{i_j}) - H(Pg_{i_j}|P[g_{i_1}, g_{i_2}, \ldots, g_{i_{j-1}}]) < \delta, j = 2, \ldots, s$. It follows that $\{T^{-i}P : i \in K_\alpha'\}$ are ϵ-independent.

LEMMA 11. *Assume $(\Omega, \mathscr{F}, m)$ is nonatomic. Let K', K be finite sets with $K' \subset K$. Let $\{P_i : i \in K\}$ be a collection of partitions of $F \in \mathscr{F}$, where $m(F) > 0$. Let I be an abstract partition such that $|\text{dist } P_i - \text{dist } I| < \epsilon$, $i \in K$. Suppose $\{P_i : i \in K'\}$ are ϵ-independent. Then there exists a collection $\{\bar{P}_i : i \in K\}$ of independent partitions of F such that dist $\bar{P}_i =$ dist I, $i \in K$, and $|\bar{P}_i - P_i| < 3\epsilon$, $i \in K'$.*

Proof. By Lemma 3 [5] wc may find a collection $\{\bar{P}_i : i \in K'\}$ of independent partitions of F such that dist $\bar{P}_i =$ dist I and $|\bar{P}_i - P_i| < 3\epsilon$, $i \in K'$. Since Ω is nonatomic, one can construct partitions $\{\bar{P}_i : i \in K \backslash K'\}$ of F such that dist $\bar{P}_i =$ dist I, $i \in K \backslash K'$, and $\{\bar{P}_i : i \in K\}$ are independent.

LEMMA 12. *Let T be a mixing, aperiodic action of G on a nonatomic space $(\Omega, \mathscr{F}, m)$. Let $I = \{\alpha_1, \alpha_2, \ldots, \alpha_n\}$ be an abstract partition such that $H(I) = H(T)$. Suppose $P = \{P^1, P^2, \ldots, P^n\}$ is a partition of Ω such that $|\text{dist } P - \text{dist } I| < \min[\epsilon^2/10, g(\epsilon^2/10)]$ and $0 < H(T) - H(P, T) < g(\epsilon^2/10)$. If $\epsilon < \frac{1}{2}$ or P is a trivial partition, then given $\epsilon' > 0$ there exists a partition $\tilde{P}$ of Ω such that (1) $|\text{dist } \tilde{P} - \text{dist } I| < \epsilon'$, (2) $H(T) - H(\tilde{P}, T) < \epsilon'$, and (3) $|\tilde{P} - P| < 6\epsilon$.*

Proof. We have $H(T) - H(P, T) = \sigma > 0$. Let $\{K_\alpha\}$ be a fixed net of finite subsets of G satisfying (1.1)–(1.5). In this proof, if $\mathscr{P}$ is a collection of atoms of some partition, let $\hat{\mathscr{P}}$ denote the union of the atoms in $\mathscr{P}$.

(1) Pick $Q \supset P$ such that $0 < H(T) - H(Q, T) < \epsilon'/10$. Let $\beta = 1/100[H(T) - H(Q, T)]$.

(2) Pick $\epsilon'' > 0$ so that $\epsilon'' < \min(\beta, \epsilon, \frac{1}{2})$ and if $|Q' - Q| < \epsilon''$, then $H(Q', T) > H(T) - 2\epsilon'/10$.

(3) Pick K_α so large that the following hold, where in the rest of the proof we set $K = K_\alpha$ and $k = |K|$ to ease the notation:

(a) If $\{\bar{P}_i : i \in K\}$ are independent partitions of $F \in \mathscr{F}$ with dist $\bar{P}_i =$ dist I for each i, then there exists a collection $\bar{\mathscr{P}}_F$ of atoms of $\bigvee_{i \in K} \bar{P}_i$ such that $m(\hat{\bar{\mathscr{P}}}_F) > m(F)(1 - \epsilon''/10)$ and $B \in \bar{\mathscr{P}}_F$ implies that $m(B)$ is between $m(F)2^{-k(H(I) \pm \beta)}$ and $k^{-1}|\{i \in K : B \in \bar{P}_i^{\,j}\}|$ is with ϵ'/n of α_j, $j = 1, 2, \ldots, n$.

(b) There exists a collection $\mathscr{P}$ of atoms of PK such that $m(\hat{\mathscr{P}}) > 1 - \epsilon''/10$, and $B \in \mathscr{P}$ implies $m(B)$ is between $2^{-k(H(P,T) \pm \beta)}$.

(c) There exists a collection $\mathscr{Q}$ of atoms of QK such that $m(\hat{\mathscr{Q}}) < 1 - \epsilon''/100$ and $B \in \mathscr{Q}$ implies $m(B)$ is between $2^{-k(H(Q,T) \pm \beta)}$.

(d) $k\beta > 1$, $k \geqslant 2$, and $-k^{-1} \log k^{-1} - (1 - k^{-1}) \log(1 - k^{-1}) < \epsilon'/10$.

(4) Using Lemma 10, pick $K' \subset K$ so that $|K'|/k > 1 - \epsilon^2/10$ and $\{T^{-i}P : i \in K'\}$ are $\epsilon^2/10$-independent.

(5) Now pick $F \in \mathscr{F}$ so that the following hold:

(a) $\{T^i F : i \in K\}$ are pairwise disjoint.

(b) $m(\bigcup_{i \in K} T^i F) > 1 - \epsilon''/100$.

(c) $m(\hat{\mathscr{P}} \cap F) > (1 - 2\epsilon''/100)m(F)$, and if A is an atom in $\mathscr{P}$, then $m(A \cap F)$ is between $m(F)2^{-k(H(P,T) \pm 2\beta)}$.

(d) $m(\hat{\mathscr{Q}} \cap F) > (1 - 2\epsilon''/100)m(F)$, and if A is an atom in $\mathscr{Q}$, then $m(A \cap F)$ is between $m(F)2^{-k(H(Q,T) \pm 2\beta)}$.

(e) $|\text{dist } P_i - \text{dist } I| < \epsilon^2/9$, $i \in K$, where P_i is the partition of F such that $P_i = T^{-i}P \cap F$.

(f) $\{P_i : i \in K'\}$ are $\epsilon^2/9$-independent partitions of F.

(To accomplish this, since K is an R-set, there exists $F' \in \mathscr{F}$ satisfying (a) and (b) in place of F. Then using the fact that T is mixing, choose $F = T^g F'$, for appropriate $g \in G$, so that (a)–(f) hold.)

(6) By Lemma 11, we may find independent partitions $\{\bar{P}_i : i \in K\}$ of F such that dist $\bar{P}_i =$ dist I, $i \in K$, and $|\bar{P}_i - P_i| < \epsilon^2/3$, $i \in K'$. Let $\bar{P}$ be the partition of $\bigcup_{i \in K} T^i F$ such that $\bar{P}^j = \bigcup_{i \in K} T^i \bar{P}_i^{\,j}$, $j = 1, 2, \ldots, n$. Note that $\bar{P}K \cap F = \bigvee_{i \in K} \bar{P}_i$.

(7) Let $\mathscr{P}_F$ be the collection of atoms of $PK \cap F$ of measure between $m(F)2^{-k(H(P,T) \pm 2\beta)}$. We have $m(\hat{\mathscr{P}}_F) > (1 - 2\epsilon''/100)m(F)$, by 5(c). Let $\bar{\mathscr{P}}_F$ be the collection of atoms of $\bar{P}K \cap F$ given by 3(a). Let $\mathscr{S}$ be the collection of those atoms A of $\mathscr{P}_F$ such that more than half of A is covered by atoms B in $\bar{\mathscr{P}}_F$ such that the $\bar{P}$-K-name of B differs from the P-K-name of A in less than ϵk places.

(a) Let C be the set of points in F whose P-K-name differs from its $\bar{P}$-K-name in less than ϵk places. Now $m(\Omega \backslash F) \leqslant m(F)(2k\epsilon)^{-1} \sum_{i \in K} |P_i - \bar{P}_i| \leqslant m(F)(2k\epsilon)^{-1}[|K'|(\epsilon^2/3) + 2(k - |K'|)] \leqslant m(F)(\epsilon/6 + \epsilon/10) < m(F)(\epsilon/3)$, by (4), (6).

(b) We have $\mathscr{S} = \{A \in \mathscr{P}_F : m(C \cap \hat{\bar{\mathscr{P}}}_F | A) > \frac{1}{2}\}$. A simple calculation shows that $m(\hat{\mathscr{S}}) \geqslant m(\hat{\mathscr{P}}_F) - 2m(F \backslash (C \cap \hat{\bar{\mathscr{P}}}_F)) > (1 - \epsilon)m(F)$.

(8) If A' is a union of r atoms of $\mathscr{S}$, then more than half of A' is covered by atoms B of $\bar{\mathscr{P}}_F$ such that the $\bar{P}$-K-name of each such atom B differs from the P-K-name of some atom in A' in less than ϵk places. The number of such atoms B is at least $(r/2)(2^{-k(H(P,T)+2\beta)}/2^{-k(H(I)-\beta)}) \geqslant r2^{k(\sigma - 4\beta)}$. Thus, from a marriage lemma (Lemma 2 [5]) we conclude that we may assign each atom of $\mathscr{S}$ $[2^{k(\sigma-4\beta)}]$ atoms of $\bar{\mathscr{P}}_F$ (where $[\cdot]$ denotes the greatest integer function), such that

(a) If $A \in \mathscr{S}$, each atom B assigned to A has a $\bar{P}$-K-name which differs from the P-K-name of A in less than ϵk places.

(b) No atom B is assigned to two different A's.

(9) It is now possible to assign each atom in $\mathscr{P}_F$ $[2^{k(\sigma-10\beta)}]$ atoms in $\bar{\mathscr{P}}_F$ such that

(a) No atom B of $\bar{\mathscr{P}}_F$ is assigned to more than one atom A of $\mathscr{P}_F$.

(b) If $A \in \mathscr{S}$, then each atom B assigned to A has a $\bar{P}$-K-name differing from the P-K-name of A in less than ϵk places.

To see this, suppose $\epsilon < \frac{1}{2}$. For each $A \in \mathscr{S}$, take $[2^{k(\sigma-10\beta)}]$ of the $[2^{k(\sigma-4\beta)}]$ atoms assigned to A in (8) and keep them assigned to A. We have at least $2^{k(\sigma-5\beta)}$ left over atoms for each $A \in \mathscr{S}$. Since the number of A in $\mathscr{S}$ is at least $(1-\epsilon)2^{k(H(P,T)-2\beta)} > (\frac{1}{2})2^{k(H(P,T)-2\beta)} > 2^{k(H(P,T)-3\beta)}$, this gives us at least $2^{k(H(I)-8\beta)}$ left over atoms. There are at most $2^{k(H(P,T)+2\beta)}$ atoms in $\mathscr{P}_F$. Thus it is possible to assign each A in $\mathscr{P}_F \backslash \mathscr{S}$ $[2^{k(\sigma-10\beta)}]$ left over atoms.

On the other hand, if P is trivial, then $\mathscr{P}_F$ consists of one atom alone. If $\mathscr{S} = \mathscr{P}_F$, we are done. If $\mathscr{S} = \phi$, the only other possibility, the assignment can be made because the number of atoms in $\bar{\mathscr{P}}_F$ is at least $(1 - \epsilon''/100)2^{k(H(I)-\beta)} > 2^{k(H(I)-2\beta)} > [2^{k(\sigma-10\beta)}]$.

(10) Let $\mathscr{Q}_F$ be the collection of atoms of $QK \cap F$ that have measure between $m(F)2^{-k(H(Q,T)\pm 2\beta)}$ and lie in an atom of $\mathscr{P}_F$.

(a) $m(\hat{\mathscr{Q}}_F) \geqslant m(\hat{\mathscr{P}}_F \cap \hat{\mathscr{Q}} \cap F) > m(F)(1 - 4\epsilon''/100)$, by 5(c), (d).

(b) Each atom of $\mathscr{P}_F$ contains no more than $2^{k(H(Q,T)-H(P,T)+4\beta)} = 2^{k(\sigma-96\beta)}$ atoms of $\mathscr{Q}_F$.

(11) We have thus assigned in (9) more B's in $\bar{\mathscr{P}}_F$ to each A in $\mathscr{P}_F$ than there are atoms of $\mathscr{Q}_F$ in A. Thus there is a one to one map τ of $\mathscr{Q}_F$ into $\bar{\mathscr{P}}_F$

such that if $C \in \mathcal{Q}_F$ lies in an atom of $\mathcal{S}$, then $\tau(C)$ has a $\bar{P}$-K-name which disagrees with the P-K-name of C in less than ϵk places.

(12) Let $\Omega' = \bigcup_{i \in K} T^i \hat{\mathcal{Q}}_F$. Define partitions $\{\tilde{P}_i : i \in K\}$ of $\hat{\mathcal{Q}}_F$ such that $\tilde{P}_i^j = \bigcup\{C \in \mathcal{Q}_F : \tau(C) \subset \bar{P}_i^j\}$, $j = 1, 2, \ldots, n$. Since τ is one to one, $\bigvee_{i \in K} \tilde{P}_i = \mathcal{Q}_F$. Define a partition $\tilde{P}$ of Ω so that $\tilde{P}^j \cap \Omega' = \bigcup_{i \in K} T^i \tilde{P}_i^j$, $j = 1, 2, \ldots, n$, and $\text{dist}[\tilde{P} \cap (\Omega \backslash \Omega')] = \text{dist } I$. Observe that $\tilde{P}K \cap \hat{\mathcal{Q}}_F = \bigvee_{i \in K} \tilde{P}_i = \mathcal{Q}_F$, and $QK \cap \hat{\mathcal{Q}}_F = \mathcal{Q}_F$. Let $\bar{F}$ be the partition of Ω consisting of F and $\Omega \backslash F$. Now if $i \in K$, we have

(a) $\tilde{P}[K - K] \cap T^i \hat{\mathcal{Q}}_F \supset \tilde{P}[K - i] \cap T^i \hat{\mathcal{Q}}_F = T^i[\tilde{P}K \cap \hat{\mathcal{Q}}_F] = T^i \mathcal{Q}_F = T^i[QK \cap \hat{\mathcal{Q}}_F] \supset Q \cap T^i \hat{\mathcal{Q}}_F$.

Consider the collection $\mathcal{M}$ of subsets of Ω' which are unions of atoms of $(\tilde{P} \vee \bar{F})[K - K] \cap \Omega'$. The sets $\{T^i \hat{\mathcal{Q}}_F : i \in K\}$ are in $\mathcal{M}$. Thus, from (a), if $Q^1, Q^2, \ldots, Q^t$ are the elements of Q, the sets $\{Q^j \cap T^i \hat{\mathcal{Q}}_F : i \in K, j = 1, 2, \ldots, t\}$ are in $\mathcal{M}$. Thus $\{Q^j \cap \Omega' : j = 1, \ldots, t\}$ are in $\mathcal{M}$, and it follows that

(b) $(\tilde{P} \vee \bar{F})[K - K] \cap \Omega' \supset Q \cap \Omega'$.

Now from 10(a) and 5(b), we have $m(\Omega \backslash \Omega') > 1 - 5\epsilon''/100$. Thus (b) implies $(\tilde{P} \vee \bar{F})[K - K] \overset{\epsilon'/10}{\supset} Q$, and so $H(\tilde{P}, T) \geqslant H(\tilde{P} \vee \bar{F}, T) - H(\bar{F}, T) \geqslant -\epsilon'/10 + H((\tilde{P} \vee \bar{F})[K - K], T)) > H(T) - 3\epsilon'/10$, by (2), 3(d). Thus, condition (2) of the lemma holds.

(13) We have $m(\tilde{P}^j) = \sum_{C \in \mathcal{Q}_F} m(C)|\{i \in K : \tau(C) \subset \bar{P}_i^j\}| + \alpha_j m(\Omega \backslash \Omega')$. Thus $|\text{dist } \tilde{P} - \text{dist } I| = \sum_{j=1}^n |\sum_{C \in \mathcal{Q}_F} m(C)(\alpha_j k - |\{i \in K : \tau(C) \subset \bar{P}_i^j\}|)| < m(\mathcal{Q}_F) k \epsilon' \leqslant \epsilon'$, by 3(a). Condition (1) holds.

(14) Let $\mathcal{Q}_F'$ be the atoms of $\mathcal{Q}_F$ which are contained in an atom of $\mathcal{S}$. Let $\Omega'' = \bigcup_{i \in K} T^i \hat{\mathcal{Q}}_F'$. Now $P^j \cap \Omega'' = \bigcup_{C \in \mathcal{Q}_F'} [\bigcup_{i \in K,\, T^iC \subset P^j} T^i C]$, and $\tilde{P}^j \cap \Omega'' = \bigcup_{C \in \mathcal{Q}_F'} [\bigcup_{i \in K,\, T^iC \subset P^j} T^i C]$. It is not hard to see that $\sum_j m[(P^j \cap \Omega'') \triangle (\tilde{P}^j \cap \Omega'')] \leqslant 2 \sum_{C \in \mathcal{Q}_F'} m(C)|\{i \in K$: the P-K-name of C and the$\tilde{P}$-K-name of C do not agree at $i\}| < 2 \sum_C m(C) k\epsilon \leqslant 2\epsilon$. (The $\tilde{P}$-K-name of C and the $\bar{P}$-K-name of $\tau(C)$ are identical by (12), and then (11) is used.) Hence, $|\tilde{P} - P| < 2\epsilon + 2m(\Omega \backslash \Omega'')$. We have $m(\hat{\mathcal{Q}}_F') = m(\hat{\mathcal{Q}}_F \cap \mathcal{S}) > m(F)(1 - \epsilon - 4\epsilon''/100)$, by 10(a) and 7(b). Thus $m(\Omega'') > m(\bigcup_{i \in F} T^i F)(1 - \epsilon - 4\epsilon''/100) > 1 - \epsilon - 5\epsilon''/100 > 1 - 2\epsilon$. Condition (3) of the lemma follows.

LEMMA 13. *Let T be an action of G on a nonatomic space Ω. Assume $H(T) > 0$. Let I be an abstract partition such that $H(I) = H(T)$. Then if $H(T) - H(P, T) < \epsilon$ and $|\text{dist } I - \text{dist } P| < \min(\epsilon, h(\epsilon))$, there exists a partition P' such that $|P' - P| < \epsilon, 0 < H(T) - H(P', T) < \epsilon, |\text{dist } I - \text{dist } P'| < \epsilon$.*

Proof. Let P satisfy the above hypotheses. If $H(P, T) < H(T)$ we are through. Otherwise, if $H(P, T) = H(T)$, since Ω is nonatomic we may find P' such that $\text{dist } P' = \text{dist } I$ and $|P - P'| = |\text{dist } I - \text{dist } P|$. Then,

$|\text{dist } P' - \text{dist } I| < \epsilon$, $|P - P'| < \epsilon$, and $H(T) - H(P', T) < \epsilon$, by Lemma 8. If $H(P', T) < H(T)$, we are through. Otherwise, find $\delta > 0$ so small that $|P' - P''| < \delta$ implies $H(T) - H(P'', T) < \epsilon$, $|\text{dist } P'' - \text{dist } I| < \epsilon$, and $|P'' - P| < \epsilon$. There exists P'' such that $|P' - P''| < \delta$ and $H(P'') < H(P')$. (This follows because $H(\cdot)$ has a local minimum only at trivial partitions and P' is not trivial because $H(P') = H(T) > 0$.) Since $H(P'', T) \leqslant H(P'') < H(T)$, we are through.

LEMMA 14. *Let T be a mixing, aperiodic action of G on a nonatomic space Ω. Assume $H(T) > 0$. Let I be an abstract partition such that $H(I) = H(T)$. For any $\epsilon > 0$, $\eta > 0$, there exists $\delta > 0$ such that if $|\text{dist } P - \text{dist } I| < \delta$ and $H(T) - H(P, T) < \delta$ then there exists a partition $\tilde{P}$ satisfying $|\text{dist } \tilde{P} - \text{dist } I| < \eta$, $H(T) - H(\tilde{P}, T) < \eta$, $|\tilde{P} - P| < \epsilon$.*

Proof. We may assume $\epsilon < \frac{7}{2}$. Let $\delta = \min[\epsilon^2/490, g(\epsilon^2/490), h(\min(\epsilon^2/490, g(\epsilon^2/490)))]$. Assume $|\text{dist } P - \text{dist } I| < \delta$ and $H(T) - H(P, T) < \delta$. By Lemma 13, pick P' so that $|\text{dist } P' - \text{dist } I| < \min(\epsilon^2/490, g(\epsilon^2/490))$, $0 < H(T) - H(P', T) < g(\epsilon^2/490)$, and $|P' - P| < \epsilon^2/490 < \epsilon/7$. Then by Lemma 12 (with ϵ in Lemma 12 replaced by $\epsilon/7$), we may find $\tilde{P}$ so that $|\text{dist } \tilde{P} - \text{dist } I| < \eta$, $H(T) - H(\tilde{P}, T) < \eta$, and $|\tilde{P} - P'| < 6\epsilon/7$. Then $|\tilde{P} - P| < \epsilon$.

LEMMA 15. *T is a mixing, aperiodic action of G on a nonatomic space Ω. Assume $H(T) > 0$. I is an abstract partition such that $H(I) = H(T)$. For any $\epsilon > 0$ there exists $\delta > 0$ such that $|\text{dist } P - \text{dist } I| < \delta$ and $H(T) - H(P, T) < \delta$ imply that there exists a partition P' satisfying* dist P' = dist I, $\{T^i P' : i \in G\}$ *are independent, and $|P' - P| < \epsilon$.*

Proof. Pick a positive sequence $\{\epsilon_i\}_1^\infty$ such that $\sum_1^\infty \epsilon_i < \epsilon$. Pick a positive sequence $\{\delta_i\}_1^\infty$, converging to zero, such that for each i, $|\text{dist } P - \text{dist } I| < \delta_i$ and $H(T) - H(P, T) < \delta_i$ imply that there exists P' such that $|\text{dist } P' - \text{dist } I| < \delta_{i+1}$, $H(T) - H(P', T) < \delta_{i+1}$, and $|P' - P| < \epsilon_i$. Suppose P_1 satisfies $|\text{dist } P_1 - \text{dist } I| < \delta_1$ and $H(T) - H(P_1, T) < \delta_1$. We show that there exists P such that dist P = dist I, $\{T^i P : i \in G\}$ are independent, and $|P - P_1| < \epsilon$. We may find $P_2, P_3, \ldots$ such that $|\text{dist } P_i - \text{dist } I| < \delta_i$, $H(T) - H(P_i, T) < \delta_i$ and $|P_i - P_{i-1}| < \epsilon_{i-1}$, $i = 2, 3, \ldots$. Since $\sum_2^\infty |P_i - P_{i-1}| < \infty$, there exists P such that $|P_i - P| \to 0$. It follows easily that $H(P, T) = H(T)$ and dist P = dist I, and hence $\{T^i P : i \in G\}$ are independent by Lemma 10. Also $|P_1 - P| \leqslant \sum_2^\infty |P_i - P_{i-1}| < \epsilon$.

DEFINITION. We say that the abstract partition $I_1 = \{p_i : i \in S_1\}$ refines the abstract partition $I_2 = \{q_j : j \in S_2\}$ if S_1 may be partitioned into sets $\{S_1{}^j : j \in S_2\}$ such that $\sum_{i \in S_1{}^j} P_i = q_j$, $j \in S_2$.

LEMMA 16. *Given an abstract partition* $I_2 = \{q_j : j \in S_2\}$ *and a number* α *such that* $\alpha \geqslant H(I_2)$, *there exists an abstract partition* I_1 *which refines* I_2 *such that* $H(I_1) = \alpha$.

Proof. Choose n so large that $\log n + H(I_2) \geqslant \alpha$. Let $\{S_1{}^j : j \in S_2\}$ be a system of pairwise disjoint sets, each set consisting of n elements. Let $S_1 = \bigcup_{j \in S_2} S_1{}^j$. Consider the collection $\mathscr{S}$ of all abstract partitions $I_1 = \{p_i : i \in S_1\}$ such that $\sum_{i \in S_1{}^j} p_i = q_j, j \in S_2$. Consider $I_1{}' = \{p_i : i \in S_1\}$ in $\mathscr{S}$ such that $p_i = q_j/n$ on $S_1{}^j$, $j \in S_2$. Then $H(I_1{}') = \log n + H(I_2) \geqslant \alpha$. Consider $I_2'' = \{p_i : i \in S_1\}$ in $\mathscr{S}$ such that for each $j \in S_2$, $p_i = q_j$ for exactly one $i \in S_1{}^j$. Then $H(I_1'') = H(I_2) \leqslant \alpha$. Since $\mathscr{S}$ is a convex set and the entropy function H is continuous, there exists $I_1 \in \mathscr{S}$ such that $H(I_1) = \alpha$.

THEOREM 3 (Generalization of Sinai's theorem). *Let* T *be a mixing aperiodic action of* G *on a nonatomic space* Ω. *Let* I *be an abstract partition such that* $H(I) \leqslant H(T)$. *Then there exists a partition* Q *of* Ω *such that* dist $Q =$ dist I *and* $\{T^i Q : i \in G\}$ *are independent.*

Proof. If $H(T) = 0$, we can take Q to be an appropriate trivial partition. If $H(T) > 0$ and $H(I) = H(T)$, then from Lemma 15, there exists $\delta > 0$ such that $|\text{dist}\,\tilde{P} - \text{dist}\,I| < \delta$, $H(T) - H(\tilde{P}, T) < \delta$ imply that there exists a partition Q satisfying dist $Q =$ dist I, $\{T^i Q : i \in G\}$ are independent, and $|Q - \tilde{P}| < 2$. In Lemma 12, choose P to be a trivial partition and ϵ so large that $|\text{dist}\ P - \text{dist}\ I| < \min(\epsilon^2/10, g(\epsilon^2/10))$ and $0 < H(T) - H(P, T) < g(\epsilon^2/10)$. Then there exists $\tilde{P}$ such that $|\text{dist}\ \tilde{P} - \text{dist}\ I| < \delta$ and $H(T) - H(\tilde{P}, T) < \delta$. Q therefore exists. If $H(T) > 0$ and $H(I) < H(T)$, using Lemma 16, find an abstract partition I' which refines I so that $H(I') = H(T)$. Then, as we have just shown, there exists Q' such that dist $Q' =$ dist I', and $\{T^i Q' : i \in G\}$ are independent. It follows that there is a partition Q such that $Q' \supset Q$ and that dist $Q =$ dist I. Automatically $\{T^i Q : i \in G\}$ are independent.

LEMMA 17. *Let* T *be a mixing aperiodic action of* G *on the nonatomic space* $(\Omega, \mathscr{F}, m)$. *Let* $\bar{P}, Q, P$ *be partitions of* Ω *such that*:

(1) *Q is nontrivial and $\{T^i Q : i \in G\}$ are independent.*
(2) *$PE \overset{\epsilon}{\supset} Q$ and $PE' \overset{\epsilon'}{\supset} Q$, where E, E' are finite nonempty subsets of G.*
(3) $\bar{P} \subset QG$.
(4) $|P - \bar{P}| < \beta$.

Then there exists a partition $\tilde{P}$ *such that*

(a) $\tilde{P} \subset QG$.
(b) dist $\tilde{P} =$ dist P.

(c) $\tilde{P}E' \overset{2\epsilon'}{\supset} Q$.

(d) $\tilde{P}E \overset{2\epsilon}{\supset} Q$, *and furthermore if* L *is a partition such that* $PE \supset L$ *and* $|L - Q| < \epsilon$ *(whose existence is given by* (2)*), then* $\tilde{P}E \supset \tilde{L}$ *and* $|\tilde{L} - Q| < 2\epsilon$ *where* $\tilde{L}$ *is the partition corresponding to* L *under the natural correspondence between* $P\tilde{E}$ *and* PE.

(e) $|\tilde{P} - \bar{P}| < 2\beta$.

Proof. (1) Let Ω_Q be the space (Ω, QG, m) and let T_Q be the action of G on Ω_Q induced by the action T on Ω. Because of (1), Ω_Q is a nonatomic space.

(2) Let $\{K_\alpha\}$ satisfy (1.1)–(1.5). For each index α, let $K_\alpha' = \bigcap_{e \in E} (K_\alpha - e)$, $K_\alpha'' = \bigcap_{e \in E'} (K_\alpha - e)$. It follows from (1.1) that $\lim_\alpha |K_\alpha'|/|K_\alpha| = 1$ and $\lim_\alpha |K_\alpha''|/|K_\alpha| = 1$. Pick $K = K_\alpha$ so large that if $K' = K_\alpha'$ and $K'' = K_\alpha''$, then $|K'|/|K| > 1 - \epsilon/10$, $|K''|/|K| > 1 - \epsilon'/10$.

(3) Since T_Q is a generalized Bernoulli scheme, T_Q is aperiodic by Lemma 5, so since K is an R-set, we may find $F \in QG$ such that $\{T^iF : i \in K\}$ are pairwise disjoint and $m(\bigcup_{i \in K} T^iF) > 1 - \epsilon''$, where $\epsilon'' > 0$ is chosen so that $\epsilon'' < \beta/10, \epsilon/10, \epsilon'/10, \frac{1}{2}$.

(4) Let $\{R_i : i \in K\}$ be the collection of QG-measurable partitions of F such that $R_i = T^{-i}(\bar{P} \vee Q) \cap F$, $i \in K$. Let $\{P_i : i \in K\}$ be the partitions of F such that $P_i = T^{-i}P \cap F$, $i \in K$. Since Ω_Q is nonatomic it is possible to pick a collection $\{\tilde{P}_i : i \in K\}$ of QG-measurable partitions of F such that $\operatorname{dist}[\bigvee_{i \in K} (R_i \vee P_i)] = \operatorname{dist}[\bigvee_{i \in K} (R_i \vee \tilde{P}_i)]$. Suppose $P = \{P^j : j = 1, 2, \ldots, n\}$. Set $\Omega' = \bigcup_{i \in K} T^iF$. Define $\tilde{P} = \{\tilde{P}^j : j = 1, 2, \ldots, n\}$ to be a partition of Ω_Q such that $\tilde{P}^j \cap \Omega' = \bigcup_{i \in K} T^i\tilde{P}_i^j$, and $\operatorname{dist}[\tilde{P} \cap (\Omega \backslash \Omega')] = \operatorname{dist}[P \cap (\Omega \backslash \Omega')]$.

(5) It is easily seen that $\operatorname{dist}[(\bar{P}K \vee QK \vee PK) \cap F] = \operatorname{dist}[\bigvee_{i \in K} (R_i \vee P_i)] = \operatorname{dist}[\bigvee_{i \in K} (R_i \vee \tilde{P}_i)] = \operatorname{dist}[(\bar{P}K \vee QK \vee \tilde{P}K) \cap F]$. In particular, for $i \in K$, $\operatorname{dist}[Pi \cap F] = \operatorname{dist}[\tilde{P}i \cap F]$, and thus $\operatorname{dist}[P \cap T^iF] = \operatorname{dist}[\tilde{P} \cap T^iF]$. Therefore, $\operatorname{dist}[P \cap \Omega'] = \operatorname{dist}[\tilde{P} \cap \Omega']$. Condition (b) of the lemma thus holds. Condition (a) follows because $\tilde{P}$ was constructed to be QG-measurable.

(6) We show now that (d) holds. Choose a partition L such that $PE \supset L$ and $|L - Q| < \epsilon$. Set $\Omega'' = \bigcup_{i \in K'} T^iF$. Now if $i \in K'$, then $i + E \subset K$. Thus, from (5) it follows that $\operatorname{dist}[P(i + E) \vee Qi] \cap F = \operatorname{dist}[\tilde{P}(i + E) \vee Qi] \cap F$, $i \in K'$. Hence $\operatorname{dist}(PE \vee Q) \cap T^iF = \operatorname{dist}(\tilde{P}E \vee Q) \cap T^iF$ and then $\operatorname{dist}(L \vee Q) \cap T^iF = \operatorname{dist}(\tilde{L} \vee Q) \cap T^iF$, $i \in K'$. Therefore $\operatorname{dist}(L \vee Q) \cap \Omega'' = \operatorname{dist}(\tilde{L} \vee Q) \cap \Omega''$, and so $|(L \cap \Omega'') - (Q \cap \Omega'')| = |(\tilde{L} \cap \Omega'') - (Q \cap \Omega'')|$. This implies $|\tilde{L} - Q| \leqslant m(\Omega'')|(\tilde{L} \cap \Omega'') - (Q \cap \Omega'')| + 2m(\Omega \backslash \Omega'') \leqslant |L - Q| + 2m(\Omega \backslash \Omega'')$. Now $m(\Omega'') = |K'|m(F) = (|K'|/|K|)m(\Omega') > 1 - 2\epsilon/10$, by (2), (3). We conclude $|\tilde{L} - Q| < \epsilon + 4\epsilon/10 < 2\epsilon$ and so (d) follows.

(7) From (5), for each $i \in K$, $\operatorname{dist}(\bar{P}i \vee \tilde{P}i) \cap F = \operatorname{dist}(\bar{P}i \vee Pi) \cap F$, and so $\operatorname{dist}(\bar{P} \vee \tilde{P}) \cap T^iF = \operatorname{dist}(\bar{P} \vee P) \cap T^iF$. This gives $\operatorname{dist}(\bar{P} \vee \tilde{P}) \cap \Omega' =$

$\text{dist}(\bar{P} \vee P) \cap \Omega'$, and thus $|(\tilde{P} \cap \Omega') - (\bar{P} \cap \Omega')| = |(P \cap \Omega') - (\bar{P} \cap \Omega')|$. We conclude $|\tilde{P} - \bar{P}| \leqslant m(\Omega')|(P \cap \Omega') - (\bar{P} \cap \Omega')| + 2m(\Omega \backslash \Omega') < |P - \bar{P}| + 2\beta/10 < 2\beta$, and (e) holds.

LEMMA 18. *Let T be an action of G on $(\Omega, \mathscr{F}, m)$. Let P be a nontrivial partition of Ω such that $\{T^i P : i \in G\}$ are independent and generate $\mathscr{F}$. Let Q be a nontrivial partition of Ω such that $\{T^i Q : i \in G\}$ are independent and $H(Q) = H(P)$. Given ϵ, $\epsilon' > 0$ there exists a partition Q' of Ω such that* (a) $\text{dist } Q' = \text{dist } Q$, (b) *$\{T^i Q' : i \in G\}$ are independent,* (c) $|Q - Q'| < \epsilon$, *and* (d) $Q'G \overset{\epsilon'}{\supset} P$.

Proof. We will start the proof with a sub-lemma.

SUB-LEMMA. *There is a partition P_1 of Ω such that* (a) $P_1 \subset QG$, (b) *$\{T^i P_1 : i \in G\}$ are independent,* (c) $\text{dist } P_1 = \text{dist } P$, *and* (d) *if Q_1 is the partition corresponding to Q under the natural correspondence between PG and $P_1 G$, then $|Q - Q_1| < 4\epsilon/10$.*

Proof of the Sub-Lemma. As in the proof of Lemma 17, if R is a partition of Ω, let Ω_R be the space (Ω, RG, m) and let T_R be the action of G on Ω_R induced by the action T on Ω.

(1) Pick a finite subset E of G so that $PE \overset{\epsilon/10}{\supset} Q$. Let $\alpha = \epsilon/30|E|$.

(2) Now $H(T_Q) = H(Q)$, and $H(Q) = H(P)$ by assumption. Thus by Lemma 15, there exists $\delta > 0$ such that if $\tilde{P}$ is a partition of Ω_Q satisfying $|\text{dist } \tilde{P} - \text{dist } P| < \delta$ and $H(T_Q) - H(\tilde{P}, T_Q) < \delta$, then there exists a partition P_1 of Ω_Q such that $\text{dist } P_1 = \text{dist } P$, $\{T^i P_1 : i \in G\}$ are independent, and $|P_1 - \tilde{P}| < \alpha$.

(3) By Lemma 8, pick $\epsilon'' > 0$ so that if R is a partition of Ω_Q satisfying $|R - Q| < 2\epsilon''$, then $H(R, T_Q) > H(T_Q) - \delta$. Pick a finite subset E' of G such that $PE' \overset{\epsilon''}{\supset} Q$.

(4) Applying Lemma 17, we may find a partition $\tilde{P}$ such that $\tilde{P} \subset QG$, $\text{dist } \tilde{P} = \text{dist } P$, $\tilde{P}E \overset{2\epsilon/10}{\supset} Q$, and $\tilde{P}E' \overset{2\epsilon''}{\supset} Q$. Furthermore let L be a partition such that $PE \supset L$ and $|L - Q| < \epsilon/10$. Then if $\tilde{L}$ is the partition such that $\tilde{P}E \supset \tilde{L}$ and $\tilde{L}$ corresponds to L under the correspondence between PE and $\tilde{P}E$, we have $|\tilde{L} - Q| < 2\epsilon/10$.

(5) Now, $H(\tilde{P}, T_Q) = H(\tilde{P}E', T_Q) > H(T_Q) - \delta$, since $\tilde{P}E' \overset{2\epsilon''}{\supset} Q$. Also $|\text{dist } \tilde{P} - \text{dist } P| < \delta$. Thus by (2), there exists a partition P_1 of Ω_Q such that $\text{dist } P_1 = \text{dist } P$, $\{T^i P_1 : i \in G\}$ are independent, and $|P_1 - \tilde{P}| < \alpha = \epsilon/30|E|$. Let $\tilde{L}$ be the partition given in (4). Let L_1 be the partition such that $P_1 E \supset L_1$ and L_1 corresponds to $\tilde{L}$ under the correspondence between $\tilde{P}E$ and $P_1 E$.

Then $|\tilde{L} - L_1| \leqslant |E|\,|\tilde{P} - P_1| < \epsilon/30$, which implies $|L_1 - Q| < \epsilon/30 + 2\epsilon/10 < 3\epsilon/10$. Now L and L_1 correspond under the correspondence between PG and P_1G. Thus, since the correspondence preserves measure, $|L - Q| < \epsilon/10$ implies $|L_1 - Q_1| < \epsilon/10$. In conclusion, $|Q_1 - Q| \leqslant |L_1 - Q| + |L_1 - Q_1| < 4\epsilon/10$.

Proof of Lemma 18 *from the Sub-Lemma.* Pick a finite subset E of G such that $QE \overset{\epsilon'/10}{\supset} P_1$. Let $\alpha = \min(\epsilon/10, \epsilon'/30|E|)$. From Lemma 15, since $H(Q) = H(T_{P_1})$, there exists δ such that if $\tilde{Q}$ is a partition of Ω_{P_1} with $|\text{dist}\,\tilde{Q} - \text{dist}\,Q| < \delta$ and $H(T_{P_1}) - H(\tilde{Q}, T_{P_1}) < \delta$, then there exists a partition Q'' of Ω_{P_1} such that dist $Q'' =$ dist Q, $\{T^iQ'':i \in G\}$ are independent, and $|Q'' - \tilde{Q}| < \alpha$. Pick ϵ'' so that if R is a partition of Ω_{P_1} such that $|R - P_1| < 2\epsilon''$, then $H(R, T_{P_1}) > H(T_{P_1}) - \delta$. Pick E'' so that $QE'' \overset{\epsilon''}{\supset} P_1$.

If we apply Lemma 17 to T_Q acting on Ω_Q (with $\bar{P}, Q, P$ replaced by Q_1, P_1, Q) we may find a partition $\tilde{Q}$ of Ω_Q such that $\tilde{Q} \subset P_1G$, dist $\tilde{Q} =$ dist Q, $\tilde{Q}E \overset{2\epsilon'/10}{\supset} P_1$, $\tilde{Q}E'' \overset{2\epsilon''}{\supset} P_1$, and $|\tilde{Q} - Q_1| < 8\epsilon/10$. Thus $H(\tilde{Q}, T_{P_1}) = H(\tilde{Q}E'', T_{P_1}) > H(T_{P_1}) - \delta$. Also, $|\text{dist}\,\tilde{Q} - \text{dist}\,Q| < \delta$. Therefore there exists a partition Q'' of Ω_{P_1} such that dist $Q'' =$ dist Q, $\{T^iQ'':i \in G\}$ are independent, and $|Q'' - \tilde{Q}| < \alpha$. Now $\tilde{Q}E \overset{2\epsilon'/10}{\supset} P_1$ and $|Q'' - \tilde{Q}| < \alpha$ imply that $Q''E \overset{\epsilon'}{\supset} P_1$. Since $|\tilde{Q} - Q_1| < 8\epsilon/10$ and $|Q'' - \tilde{Q}| < \alpha \leqslant \epsilon/10$, we have $|Q_1 - Q''| < \epsilon$. Now Q'' corresponds to some partition Q' of Ω under the correspondence between PG and P_1G. Since Q_1 also corresponds to Q, we have $|Q' - Q| = |Q'' - Q_1| < \epsilon$. We have $Q'E \overset{\epsilon'}{\supset} P$ since $Q''E \overset{\epsilon'}{\supset} P_1$. Also, $\{T^iQ':i \in G\}$ are independent and dist $Q' =$ dist Q since the same statement holds for Q''.

Proof of Theorem 2. The action T of G on $(\Omega, \mathscr{F}, m)$ is a generalized Bernoulli shift. We have an abstract partition I such that $H(I) = H(T)$. We wish to find a partition Q such that dist $Q =$ dist I, and $\{T^iQ:i \in G\}$ are independent and generate $\mathscr{F}$. Let P be a nontrivial partition such that $\{T^iP:i \in G\}$ are independent and generate $\mathscr{F}$. By Theorem 3, there exists a partition Q_1 such that dist $Q_1 =$ dist I and $\{T^iQ_1:i \in G\}$ are independent. By Lemma 18, there exists Q_2 such that dist $Q_2 =$ dist I, $\{T^iQ_2\}$ are independent, and $Q_2G \overset{2^{-2}}{\supset} P$. Choose a finite subset E_2 of G such that $Q_2E_2 \overset{2^{-2}+2^{-3}}{\supset} P$. Applying Lemma 18 again, find Q_3 such that dist $Q_3 =$ dist I, $\{T^iQ_3\}$ are independent, $Q_3G \overset{2^{-3}}{\supset} P$, and $|Q_2 - Q_3| < \epsilon_2$, where $\epsilon_2 < \frac{1}{2}$ is so small that $Q_3E_2 \overset{2^{-2}+2^{-3}+2^{-4}}{\supset} P$. Choose a finite subset E_3 of G such that $Q_3E_3 \overset{2^{-3}+2^{-4}}{\supset} P$. Suppose we have partitions $Q_2, Q_3, \ldots, Q_n$ and finite subsets of G $E_2, E_3, \ldots, E_n$ such that $|Q_{j+1} - Q_j| < 2^{-j}$, $j = 2, \ldots, n-1$, and for $j = 2, \ldots, n$, $Q_nE_j \overset{2^{-j}+\cdots+2^{-n-1}}{\supset} P$, dist $Q_j =$ dist I, and $\{T^iQ_j\}$

are independent. Pick Q_{n+1} so that dist Q_{n+1} = dist I, $\{T^i Q_{n+1}\}$ are independent, $Q_{n+1}G \overset{2^{-n-1}}{\supset} P$, and $|Q_{n+1} - Q_n| < \epsilon_n$, where $\epsilon_n < 2^{-n}$ is so small that $Q_{n+1}E_j \overset{2^{-j}+\cdots+2^{-n-2}}{\supset} P$, $j = 2, \ldots, n$. Choose E_{n+1} so that $Q_{n+1}E_{n+1} \overset{2^{-n-1}+2^{-n-2}}{\supset} P$. Consider the sequence $\{Q_n\}$ constructed in this way. Since $\sum_n |Q_{n+1} - Q_n| < \infty$, there exists a partition Q such that $\lim_n |Q_n - Q| = 0$. We have dist Q = dist I and $\{T^i Q : i \in G\}$ are independent, since this statement holds for each Q_n. For fixed j, $Q_{n+1}E_j \overset{2^{-j}+\cdots+2^{-n-2}}{\supset} P$ for n sufficiently large. This implies $QE_j \overset{2^{-j+2}}{\supset} P$ for each j. Hence $P \subset QG$ and so $\{T^i Q\}$ generate $\mathscr{F}$.

References

1. J. P. Conze, Entropie d'un groupe abelien de transformations, *Z. Wahrscheinlichkeitstheorie Verw. Geb.* **25** (1972), 11–30.
2. Y. Katznelson and B. Weiss, Commuting measure-preserving transformations, *Israel J. Math.* **12** (1972), 161–173.
3. J. C. Kieffer, A generalized Shannon-McMillan theorem for the action of an amenable group on a probability space, *Ann. Probability* **3** (1975), 1031–1037.
4. A. A. Kirillov, Dynamical systems, factors, and representations of groups, *Russian Math. Surveys* **22** (1967), 63–75.
5. D. S. Ornstein, Bernoulli shifts with the same entropy are isomorphic, *Advances in Math.* **4** (1970), 337–352.
6. V. A. Rokhlin, A general measure-preserving transformation is not mixing, *Dokl. Akad. Nauk SSSR* **60** (1948), 349–351.
7. M. Smorodinsky, Ergodic theory, entropy, "Lecture Notes in Mathematics," Vol. 214, Springer-Verlag, Berlin and New York, 1971.
8. A. M. Stepin, On entropy invariants of decreasing sequences of measurable partitions, *Functional Anal. Appl.* **3** (1971), 237–246.

AMS (MOS) 1970 subject classification: primary, 28A65.

STUDIES IN PROBABILITY AND ERGODIC THEORY
ADVANCES IN MATHEMATICS SUPPLEMENTARY STUDIES, VOL. 2

Measurable Transformations on Homogeneous Spaces

J. R. CHOKSI†

Department of Mathematics, McGill University, Montreal, Quebec, Canada

AND

R. R. SIMHA

School of Mathematics, Tata Institute of Fundamental Research, Bombay, India

For an arbitrary σ-finite Baire measure μ on an arbitrary locally compact, σ-compact homogeneous space M, under the action of a locally compact, σ-compact group G, it is shown that every automorphism of the measure algebra of μ can be induced by an invertible, completion Baire measurable point transformation of M. If M is of the form G/L, and μ is taken to be the essential Baire measure, then the result holds without any assumption of σ-compactness.

1. INTRODUCTION

If $(X, \mathscr{S}, \mu)$ is a measure space and ϕ is an automorphism of its measure algebra, it is well known that ϕ is not in general induced by an invertible, measurable point transformation of $(X, \mathscr{S}, \mu)$. However, it was shown by von Neumann [14] that ϕ is so induced if μ is a finite (or σ-finite) Borel measure on a Polish space. This was generalized by Dorothy Maharam [9] to the direct product of normalized measures on a possibly uncountable product of Polish spaces, and in [2, 3] to an arbitrary finite (or σ-finite) measure μ on the completion of the product σ-algebra of such a product. If the factor spaces are all compact, and μ is finite, then μ becomes a Radon measure restricted to the Baire σ-algebra of the product space. However,

† The first author wishes to acknowledge support by a grant from the National Research Council of Canada, and the hospitality of the Tata Institute of Fundamental Research, Bombay, during the preparation of this work.

ISBN 0-12-599102-9

if one considers the Borel extension $\bar{\mu}$ of μ (the measure algebras of μ and $\bar{\mu}$ are necessarily isomorphic), the result is false for $\bar{\mu}$. Thus a measure algebra automorphism for a regular Borel measure $\bar{\mu}$ on the product of compact metric spaces is not in general induced by a completion Borel measurable point transformation, whereas it is induced by a completion Baire measurable point transformation of the Baire restriction μ of $\bar{\mu}$. One can ask the same questions for Radon measures on arbitrary compact spaces, here of course both Baire and Borel results are false. (See the Introductions in [2, 3]; see also Panzone and Segovia [10, Sect. 5, Example (c)] and [4, Section 2]). However it was shown in [4], that the Baire result is true for an *arbitrary*, finite Radon measure on a compact group. Roughly speaking, it appears that for most spaces the truth of the Borel result is entirely a property of the individual measure on the given space, for very few spaces does it hold for all Borel measures on the space; the Baire result seems however to depend on topological properties of homogeneity of the space, for such spaces it holds for *all* Baire measures. In this paper we justify this vague assertion, we show in fact that for an arbitrary, Radon measure μ on an arbitrary locally compact, σ-compact homogeneous space M, acted upon by a locally compact, σ-compact group G, every automorphism of the measure algebra of μ is induced by an invertible, completion Baire measurable, point transformation of M, (the σ-compactness of G implies that of M). Some generalization is possible to the non-σ-compact case, e.g., in the case when M is itself the quotient of a locally compact group, the result still holds (if μ is taken to be the essential measure). Our method is to note that in the σ-compact case M is necessarily of the form G/L, L a closed subgroup and then to generalize and adapt the argument in [4] in two ways: first to get rid of the unnecessary assumption of compactness (this is fairly easy) and second to introduce and prove some odd group theoretic lemmas to deal with problems caused by the lack of distributivity in the lattice of subgroups of the form LK, K compact normal in G. To avoid duplication, we assume the reader to be familiar with [4] which we hereafter denote by CG, and to which we refer freely.

2. Homogeneous Spaces and Projective Limits

Let M be a locally compact, homogeneous space, acted upon by a locally compact, σ-compact group G, it follows easily that M is σ-compact. Further, it is known (see [11, Chapter III, Section 24, Theorem 20, p. 148]) that M is then of the form G/L, where L is a closed subgroup of G; in the sequel we shall therefore assume that M *is* G/L. Let $\mathfrak{H}$ denote as in CG, the set of compact normal subgroups H of G such that G/H is metrisable (necessarily

σ-compact and so with a countable basis); let $\mathfrak{K}$ denote, again as in CG, the set of *all* compact normal subgroups of G. Both $\mathfrak{H}$, $\mathfrak{K}$ are σ-directed under the relation $H_1 < H_2$ if $H_2 \subset H_1$, further both are lattices in this order relation with sup, $H_1 \cap H_2$, inf, $H_1 H_2$. In CG we frequently used the result that a continuous bijection of compact spaces is a homeomorphism. In its place we here use the following: *A continuous, proper bijection of locally compact spaces is a homeomorphism.* (A map is *proper* if the inverse image of every compact set is compact). We also need the following:

The natural projection map $\pi_K: G/L \to G/LK$ is proper for each $K \in \mathfrak{K}$.

Proof. G/LK can be identified in a natural way with $G/L/K$, i.e., the quotient of G/L under the natural action of K on G/L. Further, since K is compact, the saturation of any compact set in G/L by K is compact. Now the result follows from [1, Chapter I, Section 10, No. 4, Proposition 9] (equivalence of (a) and (b)).

A theorem of Kakutani and Kodaira [7] (see also [6, Theorem 6, p. 287]) states that every neighborhood of the identity of G contains an element of $\mathfrak{H}$ (σ-compactness of G is essential here). It follows that the obvious continuous map from G to $G_0 = \text{proj lim}\{G/H : H \in \mathfrak{H}\}$ is injective. However, the map $G \to G/H$ is proper, by the property mentioned above. It follows easily that $G \to G_0$ is surjective. It also follows that $G_0 \to G/H$ is proper for each $H \in \mathfrak{H}$, consequently G_0 is locally compact, and also $G \to G_0$ is proper. It follows that the map $G \to G_0$ is a homeomorphism, i.e., that $G = \text{proj lim}\{G/H : H \in \mathfrak{H}\}$. By the same argument, for $K \in \mathfrak{K}$,

$$\begin{aligned} G/K &= \text{proj lim}\{G/H : H \in \mathfrak{H}, H \supset K\} \\ &= \text{proj lim}\{G/K/H/K : H \in \mathfrak{H}, H \supset K\}. \end{aligned}$$

In CG we used the Peter–Weyl theorem instead of the theorem of Kakutani–Kodaira. This gave us approximation by Lie groups, for which we had no real use, and which in fact complicated the issue.

We next consider the homogeneous space G/L. Note that $L = \bigcap\{LH : H \in \mathfrak{H}\}$. In fact, more generally if $\{K_i : i \in I\}$ is any decreasing directed family of groups in $\mathfrak{K}$, then $L \bigcap_I K_i = \bigcap_I LK_i$. For clearly $L \bigcap_I K_i \subset \bigcap_I LK_i$. Now if $g \in \bigcap_I LK_i$, then $g = l_i k_i$, where $l_i \in L$, $k_i \in K_i$ for each $i \in I$. There exists $k \in \bigcap_I K_i$, adherent to $\{k_i\}$, so since $l_i = g k_i^{-1} \in L$ and gk^{-1} is adherent to $\{g k_i^{-1}\}$, we have $gk^{-1} \in L$, since L is closed. (We shall need this remark later on).

Still using our original partial ordering on $\mathfrak{H}$, namely $H_1 < H_2$ if $H_2 \subset H_1$ (and not just if $LH_2 \subset LH_1$) we obtain a projective system $\{G/LH : H \in \mathfrak{H}\}$. If H, $K \in \mathfrak{K}$, $H \supset K$, $\pi_{H,K}$ denotes the projection $G/LK \to G/LH$, π_H the projection $G/L \to G/LH$. In some cases, the projection map may be the identity. Next, we have already noted that the map $G/L \to G/LK$, $K \in \mathfrak{K}$ is proper. We note also that the fibres (inverse images of points) are homeomorphic to LK/L, which is homeomorphic to $K/L \cap K$ which is of course compact. Similarly if $K_1 \supset K_2$, $K_1, K_2 \in \mathfrak{K}$, the map $G/LK_2 \to G/LK_1$ is continuous and proper. Hence as in the case of a group,

$$G/L = \text{proj lim}\{G/LH : H \in \mathfrak{H}\},$$

and for any $K \in \mathfrak{K}$,

$$G/LK = \text{proj lim}\{G/LH : H \in \mathfrak{H}, H \supset K\}.$$

Note that G/LH is metrisable, with a denumerable base, if $H \in \mathfrak{H}$. Also, with the above ordering $\mathfrak{H}$ and $\mathfrak{K}$ still form σ-directed sets and LK_1K_2 still gives an inf for LK_1 and LK_2. Problems arise however because in general $L(K_1 \cap K_2) \subset LK_1 \cap LK_2$ and there is no equality. [In fact $LK_1 \cap LK_2$ need not be a group in our directed system (i.e., one of the form LN, $N \in \mathfrak{K}$). A counter example was shown to us by Amit Roy in the finite group $S_3 \times S_3$]. The difficulties in extending the results of CG to homogeneous spaces spring almost entirely from this lack of equality. We overcome this by proving two key results showing that for our purposes, there are "enough" groups where equality holds.

Our plan is as follows. We give nine lemmas each of which is the precise generalization of the lemma with the same number in CG. In the proof of each, only those steps which differ from or are additional to the proof in CG are given, otherwise the reader is referred to CG. (Tedious repetition is thus avoided). In addition our two key group theoretic results are stated and proved as Propositions A and B. Our two main theorems are also generalizations of the corresponding theorems in CG. In the last section we generalize these to non-σ-compact groups.

LEMMA 1. *If $K \in \mathfrak{K}$ and $LK \neq L$, then there exists $H \in \mathfrak{H}$ such that $LK \not\subset LH$.*

Proof. Trivial.

LEMMA 2. *If $H, K \in \mathfrak{K}$, with $LH \cap LK = L$, then*

(i) *the natural map θ from $G/L \to G/LH \times G/LK$ is injective and is a homeomorphism onto the image;*

(ii) *if LHx, LKy are cosets of LH and LK, then $LHx \cap LKy$ contains at most one coset of L;*

(iii) *$LHx \cap LKy \neq \varnothing$ if and only if*

$$\pi_{HK,H}(LHx) = \pi_{HK,K}(LKy);$$

(iv) *if T_H, T_K are bijections of G/LH and G/LK, respectively, and $\tilde{T}$ is the bijection of $G/LH \times G/LK$ given by $\tilde{T} = (T_H, T_K)$ and if further*

$$\pi_{HK,H} \circ T_H \circ \pi_H = \pi_{HK,K} \circ T_K \circ \pi_K.$$
$$\pi_{HK,H} \circ T_H^{-1} \circ \pi_H = \pi_{HK,K} \circ T_K^{-1} \circ \pi_K,$$

then $\tilde{T}(\theta(G/L)) = \theta(G/L)$ and so $T = \theta^{-1} \circ \tilde{T} \circ \theta$ gives a bijection of G/L onto itself.

Proof. (i) Here we only need to use the result that proper continuous bijections between locally compact spaces are homeomorphisms. Note that proper maps are closed.

(ii), (iii), and (iv) are routine verifications almost identical to those in CG.

Now let $\bar{\mu}$ be a Radon measure on G/L, and μ its Baire contraction. These are necessarily σ-finite since G/L is σ-compact, and every Baire measure (by definition finite on compact sets) is necessarily the restriction of a unique Radon measure. As in CG we denote the Baire and Borel sets of a space X by $\mathscr{B}_X{}^0$ and $\mathscr{B}_X$, respectively. We wish to show that

$$\mathscr{B}^0 = \mathscr{B}^0_{G/L} = \sum\left(\bigcup_{\mathfrak{H}} \pi_H^{-1}(\mathscr{B}_{G/LH})\right).$$

Note that if C is compact, U open, $C \subset U$ in G/L, then there exists $H \in \mathfrak{H}$, $\tilde{U}_H$ open in G/LH such that if $U_H = \pi_H^{-1}(\tilde{U}_H)$, then $C \subset U_H \subset U$, so if C is a $\mathscr{G}_\delta$ then $C \in \sum(\bigcup_{\mathfrak{H}} \pi_H^{-1}(\mathscr{B}_{G/LH}))$ and hence so does every set in $\mathscr{B}^0$. The reverse inclusion is trivial. Actually, since $\mathfrak{H}$ is σ-directed we have

$$\mathscr{B}^0 = \bigcup_{\mathfrak{H}} \pi_H^{-1}(\mathscr{B}_{G/LH}).$$

Similarly if $G/L = \operatorname{proj\,lim}_{j \in J} G/LK_j$, where $K_j \in \mathfrak{K}$, then

$$\mathscr{B}^0 = \sum\left(\bigcup_{J} \pi_{K_j}^{-1}(\mathscr{B}^0_{G/LK_j})\right).$$

The above proof is different from that in CG. Recall also that $\mathscr{B}^0_{G/LH} = \mathscr{B}_{G/LH}$ if $H \in \mathfrak{H}$. We put $\mu_K = \pi_K(\mu)$.

We define $\mathscr{G}^H$, $\mathscr{G}^K$, $E_{G/LK}$, $\mathscr{E}^K$, $\bar{\mathscr{B}}^0_{G/LK}$, $\bar{\mathscr{B}}_{G/LK}$, $\tilde{\mathscr{B}}^0_{G/LK}$, $\tilde{\mathscr{G}}^K$ as in CG except that G/L, G/LK replace G, G/K, respectively. In particular, note that $E_{G/LK}$, $\mathscr{E}^K$ are isomorphic and that

$$\tilde{\mathscr{B}}^0_{G/LK} = \bigcup\{\pi_{H,K}^{-1}(\bar{\mathscr{B}}_{G/LH}) : H \in \mathfrak{H}, H \supset K\}.$$

LEMMA 3. *Let $H, K \in \mathfrak{K}$ and satisfy $L(H \cap K) = LH \cap LK$.*

(i) *Then $\pi_{H,H\cap K}^{-1}(\mathscr{B}^0_{G/LH})$ and $\pi_{K,H\cap K}^{-1}(\mathscr{B}^0_{G/LK})$ generate $\mathscr{B}^0_{G/L(H\cap K)}$. Hence $\mathscr{G}^H$, $\mathscr{G}^K$ generate $\mathscr{G}^{H\cap K}$.*

(ii) *If further* $H, K \in \mathfrak{H}$, *then* $\bar{\mathscr{B}}_{G/L(H \cap K)}$ *is the completion of the σ-algebra generated by* $\pi_{H, H \cap K}^{-1}(\bar{\mathscr{B}}_{G/LH})$ *and* $\pi_{K, H \cap K}^{-1}(\bar{\mathscr{B}}_{G/LK})$.

(iii) *If* $H \in \mathfrak{H}$, *(but not necessarily* K*), then*

$$\begin{aligned} G/L(H \cap K) &= \text{proj lim}\{G/L(H \cap \bar{H}) : \bar{H} \supset K, \bar{H} \in \mathfrak{H}\} \\ &= \text{proj lim}\{G/LH \cap L\bar{H} : \bar{H} \supset K, \bar{H} \in \mathfrak{H}\}, \end{aligned}$$

$$\mathscr{B}^0_{G/L(H \cap K)} = \bigcup\{\pi^{-1}_{H \cap \bar{H}, H \cap K}(\mathscr{B}_{G/L(H \cap \bar{H})}) : \bar{H} \supset K, \bar{H} \in \mathfrak{H}\},$$

$$\tilde{\mathscr{B}}^0_{G/L(H \cap K)} = \bigcup\{\pi^{-1}_{H \cap \bar{H}, H \cap K}(\bar{\mathscr{B}}_{G/L(H \cap \bar{H})}) : \bar{H} \supset K, \bar{H} \in \mathfrak{H}\}.$$

Proof. (i) Essentially as in Lemma 2(i), we see, since $L(H \cap K) = LH \cap LK$ that the natural map $\theta : G/L(H \cap K) \to G/LH \times G/LK$ is injective, continuous, and proper and so a homeomorphism. The rest of (i) follows as in CG, again using the fact that $L(H \cap K) = LH \cap LK$.

(ii) is proved as in CG.

(iii) As in CG, given $H_0 \supset H \cap K$, there exists $\bar{H} \in \mathfrak{H}$ such that $\bar{H} \supset K$ and $H_0 \supset H \cap \bar{H}$. Hence

$$L(H \cap K) = \bigcap\{L(H \cap \bar{H} : \bar{H} \in \mathfrak{H}, \bar{H} \supset K\}.$$

Also $LK = \bigcap\{L\bar{H} : \bar{H} \in \mathfrak{H}, \bar{H} \supset K\}$ and so

$$LH \cap LK = \bigcap\{LH \cap L\bar{H} : \bar{H} \in \mathfrak{H}, \bar{H} \supset K\}.$$

Since $LH \cap LK = L(H \cap K)$ by assumption, the conclusions of (iii) follow immediately.

3. Measurable Transformations on Projective Limits: Invariance

As in CG we consider only those invertible measurable transformations T of a measure space $(X, \mathscr{S}, \mu)$ which satisfy in addition $\mu(Z) = 0$ if and only if $\mu(TZ) = 0$, i.e., which preserve sets of measure zero. These are exactly the invertible, measurable point transformations which induce measure algebra automorphisms. We define the notion of invariance as in CG.

Definition. (a) Let ϕ be an automorphism of E. $K \in \mathfrak{K}$ is said to be *invariant* under ϕ, if $\phi(\mathscr{E}^K) = \mathscr{E}^K$; ϕ then induces an automorphism ϕ_K of $E_{G/LK}$.

(b) Let T be a $\tilde{\mathscr{B}}^0$ measurable point transformation of G/L. $K \in \mathfrak{K}$ is said to be *invariant* under T if $T(\tilde{\mathscr{G}}^K) = \tilde{\mathscr{G}}^K$.

Note. If H and K are invariant under ϕ, respectively, T, and $LH \cap LK = L(H \cap K)$, then Lemma 3 shows that $H \cap K$ is invariant under ϕ, respectively, T. (We do not know if $H \cap K$ is invariant in general). Also if $\{K_i : i \in I\}$ are totally ordered and invariant under ϕ, then so is $\bigcap_{i \in I} K_i$, (see the proof of Lemma 9).

LEMMA 4. (a) *If ϕ is an automorphism of the measure algebra E of $(G/L, \mu)$, then for any $H \in \mathfrak{H}$, there exists $\bar{H} \in \mathfrak{H}$, with $H < \bar{H}$, i.e., $\bar{H} \subset H$ and $\bar{H}$ invariant under ϕ.*

(b) *If T is an invertible $\tilde{\mathscr{B}}^0$ measurable point transformation of $(G/L, \mu)$ then for any $H \in \mathfrak{H}$, there exists $\bar{H} \in \mathfrak{H}$, with $H < \bar{H}$ and $\bar{H}$ invariant under T.*

COROLLARY. (a) *If $K \in \mathfrak{K}$ and ϕ_K is an automorphism of $E_{G/LK}$, then for any $H \in \mathfrak{H}$, $H \supset K$, there exists $\bar{H} \in \mathfrak{H}$ with $H \supset \bar{H} \supset K$ and $\bar{H}$ invariant under ϕ_K.*

(b) *If $K \in \mathfrak{K}$ and T_K is a $\tilde{\mathscr{B}}^0_{G/LK}$ measurable, invertible point transformation of G/LK then for any $H \in \mathfrak{H}$, $H \supset K$, there exists $\bar{H} \in \mathfrak{H}$, with $H \supset \bar{H} \supset K$ and $\bar{H}$ invariant under T_K.*

Proof. The proofs of both the lemma and the corollary are identical with those in CG, since, for $H \in \mathfrak{H}$, G/LH has a countable basis and so $E_{G/LH}$, $\mathscr{B}_{G/LH}$ are countably generated, and $\mathscr{B}_{G/LH}$ has a countable, separating sequence of generators.

4. TWO GROUP THEORETIC PROPOSITIONS

In this section we give our two key propositions which tell us that given $K \in \mathfrak{K}$ there are "enough" groups $H \in \mathfrak{H}$ with $LH \cap LK = L(H \cap K)$.

PROPOSITION A. *Given $K \in \mathfrak{K}$, suppose there exists $H_0 \in \mathfrak{H}$ such that $H_0 \cap K = \{e\}$. Then there exists $\bar{H} \in \mathfrak{H}$, $\bar{H} \subset H_0$ such that $L = L(\bar{H} \cap K) = L\bar{H} \cap LK$. If in addition ϕ (respectively, T) is an automorphism (respectively, point transformation) of E (respectively G/L) we may choose $\bar{H}$ to be invariant under ϕ (respectively, T).*

Proof. Since distinct elements of K lie in distinct cosets of H_0, the continuous map $G \to G/H_0$ is injective when restricted to K and so is a homeomorphism of K in G/H_0, hence K is metrisable (since $H_0 \in \mathfrak{H}$) and hence so is LK/L (which is homeomorphic to $K/L \cap K$). LK/L is also compact. Now $\bigcap_{H \in \mathfrak{H}} LH \cap LK = L$. So $\{LH \cap LK/L : H \in \mathfrak{H}, H \subset H_0\}$ is a decreasing directed family of closed subsets of the compact metrisable space LK/L whose intersection is the singleton $\{L\}$. So there exists a decreasing *sequence* H_n in this family such that

$$\bigcap_{n=1}^{\infty} LH_n \cap LK/L = \{L\}.$$

Put $\bar{H} = \bigcap_{n=1}^{\infty} H_n$ then $L\bar{H} \cap LK \subset L$ and so equals L. Since any $\bar{\bar{H}} \subset \bar{H}$ also has the property

$$L = L(\bar{\bar{H}} \cap K) = L\bar{\bar{H}} \cap LK,$$

Lemma 4(a) or (b) shows that we may assume $\bar{H}$ invariant under a given ϕ or T.

PROPOSITION B. *Let ϕ be an automorphism of E, $K \in \mathfrak{K}$, such that $K \not\subset L$, so that $LK \neq L$ and there exists $H_0 \in \mathfrak{H}$ with $LK \not\subset LH_0$. Then there exists $\bar{H} \in \mathfrak{H}, \bar{H} \subset H_0, \bar{H}$ invariant under ϕ, such that $LK \not\subset L\bar{H}$ (and so $L(\bar{H} \cap K) \neq LK$) and such that $L(\bar{H} \cap K) = L\bar{H} \cap LK$. A similar result holds if ϕ is replaced by a point transformation.*

Proof. We may assume H_0 is invariant under ϕ. Put $L' = L(H_0 \cap K)$. Consider $G/H_0 \cap K$; now

$$(H_0/H_0 \cap K) \cap (K/H_0 \cap K) = \{e_{G/H_0 \cap K}\}.$$

So, by Proposition A, there exists $\tilde{H}$, a compact normal subgroup of $G/H_0 \cap K$ with metrisable quotient, such that $\tilde{H} \subset H_0/H_0 \cap K$ and such that

$$(L'/H_0 \cap K)\tilde{H} \cap (L'/H_0 \cap K)(K/H_0 \cap K) = L'/H_0 \cap K.$$

Put $H_1 = \pi_{H_0 \cap K}^{-1}(\tilde{H})$, then $G/H_1 = G/H_0 \cap K/H_1/H_0 \cap K = G/H_0 \cap K/\tilde{H}$. So $H_1 \in \mathfrak{H}$; also $H_0 \supset H_1 \supset H_0 \cap K$. Now we have,

$$L'H_1 \cap L'K = L'$$

(taking inverse images under $\pi_{H_0 \cap K}$) or

$$LH_1 \cap LK = L(H_0 \cap K) = L(H_1 \cap K)$$

since $H_0 \supset H_1 \supset H_0 \cap K$.

If H_1 is not invariant under ϕ, choose $H_2 \in \mathfrak{H}$ invariant under ϕ, $H_2 \subset H_1$ and so

$$LH_2 \cap LK \subset L(H_1 \cap K).$$

We choose inductively a sequence $H_n \in \mathfrak{H}$ as follows. H_{2n-1} is chosen as above so that $H_{2n-1} \subset H_{2n-2}$ and $LH_{2n-1} \cap LK = L(H_{2n-1} \cap K) =$

$L(H_{2n-2} \cap K)$. H_{2n} is chosen so that $H_{2n} \subset H_{2n-1}$ and H_{2n} is invariant under ϕ, hence $LH_{2n} \cap LK \subset L(H_{2n-1} \cap K)$. If at any stage an H_n has both properties H_n invariant under ϕ and $LH_n \cap LK = L(H_n \cap K)$, the process stops. If not put $\bar{H} = \bigcap_{n=0}^{\infty} H_n = \bigcap_{n=1}^{\infty} H_{2n-1} = \bigcap_{n=1}^{\infty} H_{2n}$. $\bar{H} \in \mathfrak{H}$ and is invariant under ϕ. Further

$$\begin{aligned} L\bar{H} \cap LK &= (\bigcap LH_{2n-1}) \cap LK = \bigcap(LH_{2n-1} \cap LK) \\ &= \bigcap L(H_{2n-1} \cap K) = L\bigcap (H_{2n-1} \cap K) = L((\bigcap H_{2n-1}) \cap K). \\ &= L(\bar{H} \cap K). \end{aligned}$$

Note that if K itself is invariant under ϕ, then so is $\bar{H} \cap K$.

5. Metrisable Homogeneous Spaces

In this section we assume that G is locally compact, σ-compact, and metrisable and so has a denumerable basis. Our first result is von Neumann's theorem in the form we need it.

Lemma 5. *If X is a locally compact, σ-compact metric space, and μ is a σ-finite Radon measure on X, then every automorphism of the measure algebra of $(X, \mathscr{B}_X, \mu)$ is induced by an invertible Borel point transformation of X.*

In addition to the results on Lusin spaces used in CG, all of which are to be found in [12], we use the following: A continuous bijection between Lusin spaces has a Borel measurable inverse. (This is a special case [12, Theorem 5, Corollary 3, p. 103].) More generally a Borel measurable bijection between Lusin spaces has a Borel measurable inverse (see [12, Lemma 16, p. 107]).

Now let $H \in \mathfrak{H}$, $\pi_H: G/L \to G/LH$. For each $x \in G/LH$, $\pi_H^{-1}(x)$ is homeomorphic to LH/L or $H/L \cap H$ which is a compact homogeneous space and so either finite or perfect and of cardinal c. If also $K \in \mathfrak{H}$, $K \subset H$, then by a theorem of Feldman and Greenleaf ([5], Theorem 1 and following remarks) G/LK which is $G/K/LK/K$ has a Borel section (=transversal) in G/K. Similarly G/LH has a Borel section R in G/K. If $\psi_{K,L}$ is the projection $G/K \to G/LK$, $\psi_{K,L}$ restricted to R is injective. Put $S = \psi_{K,L}(R)$. Then S is clearly a section of G/LH in G/LK. But R is Borel in G, which is Polish, so R is Lusin, hence S, which is the continuous bijective image of R, is Lusin and so Borel in G/LK. Thus S is a Borel section of G/LH in G/LK.

Lemma 6. *Let G be a locally compact, σ-compact metrisable group, L a closed subgroup, $H \in \mathfrak{H}$ such that $\pi_H^{-1}(x)$ has cardinal c for every $x \in G/LH$,*

μ a σ-finite Radon measure on G/L. Then there exists a Borel set $Z \subset G/L$ with $\mu(Z) = 0$ such that Z meets every fiber $\pi_H^{-1}(x)$ in a set of cardinal c.

Proof. We may assume without loss of generality that μ is finite. As above G/LH has a Borel section R in G whose image S in G/L under $\psi_L: G \to G/L$ is Borel, further ψ_L is a Borel isomorphism of the Lusin spaces R and S. The fibers $\pi^{-1}(x)$ are all homeomorphic to the compact homogeneous space LH/L which by assumption, is of cardinal c. Every element g of G has a unique representation $g = rk$, $r \in R$, $k \in LH$, further the bijective map $R \times LH \to G$ given by $(r, k) \to rk$ is continuous and so its inverse is Borel measurable, since $R \times LH$ is Lusin and G is Polish. Thus, again using the property that R is a section, the map $\bar{\theta}: R \times LH/L \to G/L$ is a continuous bijection, again between Lusin spaces, and so $\bar{\theta}^{-1}$ is again Borel measurable. We thus have a bijection $\theta: S \times LH/L \to G/L$, which is the composition of the maps $S \times LH/L \to R \times LH/L$ and $\bar{\theta}$. Since these are both Borel isomorphisms, so is θ. Since S is Lusin we may take on it a stronger Polish topology, which is Borel equivalent to it, call S with this topology S_0. We thus have a Borel isomorphism θ_0 between the Polish spaces $S_0 \times LH/L$ and G/L, in which the projection onto S_0 in the first space corresponds to projection onto G/LH in the second. Put $v = \theta_0^{-1}(\mu)$. It is clearly enough to prove that for the Borel measure v on $S_0 \times LH/L$, there exists a Borel set Z_0, with $v(Z_0) = 0$ such that $Z_0 \cap (\{s\} \times LH/L)$ has cardinal c for every $s \in S_0$. But since LH/L has cardinal c by assumption the proof of the statement is virtually identical with that of Lemma B in [3]. (In fact only minor verbal changes are needed, as is remarked in [3, p. 101]. The hypothesis that the Polish space replacing J has cardinal c is inadvertently omitted in this remark.) This completes the proof of Lemma 6, since $Z = \theta(Z_0)$ has the required properties.

LEMMA 7. *Let G be a locally compact, σ-compact, metrisable group, L a closed subgroup, $H \in \mathfrak{H}$, μ a σ-finite Radon measure on G/L. Let ϕ be an automorphism of the measure algebra E of $(G/L, \mathscr{B}_{G/L}, \mu)$, such that H is invariant under ϕ and so ϕ induces an automorphism ϕ_H of $E_{G/LH}$. Let T_H be an invertible, completion Borel measurable point transformation of G/LH inducing ϕ_H. Then there exists an invertible completion Borel point transformation T of G/L, inducing ϕ and such that $\pi_H \circ T = T_H \circ \pi_H$.*

Proof. We may assume without loss of generality that μ is finite. We note that the fibers $\pi_H^{-1}(x)$, $x \in G/LH$ are all homeomorphic to LH/L and so finite or of cardinal c. The proof of the lemma is, from there on, identical with that in CG. Note that we do not really need to assume μ Radon. Any Borel measure will do. The equivalent finite measure will automatically be Radon.

6. Final Lemmas and Theorems

LEMMA 8. *Let G be a locally compact, σ-compact group, L a closed subgroup,* $\mathfrak{H}$, $\mathfrak{K}$ *as before,* μ *a σ-finite Radon measure on* G/L, ϕ *an automorphism of the measure algebra E of* $(G/L, \mathscr{B}^0, \mu)$. *Suppose* $K \in \mathfrak{K}$ *is invariant under* ϕ, *so that* ϕ *induces an automorphism* ϕ_K *of* $E_{G/LK}$. *Suppose further that* ϕ_K *is induced by an invertible* $\tilde{\mathscr{B}}^0_{G/LK}$ *measurable point transformation* T_K *of* G/LK. *Suppose finally that there exists* $H_0 \in \mathfrak{H}$ *such that* $K \cap H_0 = \{e\}$. *Then there exists an invertible* $\tilde{\mathscr{B}}^0$ *measurable point transformation T of* G/L *which induces* ϕ *and which is such that* $\pi_K \circ T = T_K \circ \pi_K$.

Proof. We may assume without loss of generality that μ is finite. By Proposition A, there exists $H \in \mathfrak{H}$, invariant under ϕ such that $K \cap H = \{e\}$ and such that $LK \cap LH = L$. As in CG, there exists $H_1 \in \mathfrak{H}$ such that $HK \supset H_1 \supset K$ and H_1 is invariant under T_K and so also under ϕ_K and under ϕ. Put $H_2 = H \cap H_1$.

Since $H \cap K = \{e\}$, every element of HK has a *unique* representation hk, $h \in H$, $k \in K$. So since $HK \supset H_1 \supset K$, we have $H_1 = \tilde{H}K$, where $\tilde{H} \subset H$ and so $\tilde{H}$ necessarily equals $H \cap H_1 = H_2$. Thus $H_1 = H_2K$, (we note that this is proved in CG in a more complicated way). We show that

$$L(H_2K) \cap LH = L(H_2K \cap H) = LH_2,$$

i.e.,

$$LH_1 \cap LH = L(H_1 \cap H).$$

If $x \in L(H_2K) \cap LH$, then $x = lkh_2 = l'h$, where $l, l' \in L$, $k \in K$, $h_2 \in H_2$, $h \in H$. So

$$lk = l'hh_2^{-1} = l^*$$

since $LK \cap LH = L$. So $lkh_2 = l^*h_2$, i.e., $x = lkh_2 \in LH_2$. The reverse inclusion is trivial. Note that the *only* property of H_1 used here is that $HK \supset H_1 \supset K$. We thus get that $H_2 = H \cap H_1$ is invariant under ϕ; further $H_2 \cap K = \{e\}$ and $LH_2 \cap LK = L$.

Now the proof proceeds as in CG. From T_K we obtain a point transformation T_{H_1} inducing ϕ_{H_1}, and using Lemma 7, extend it to a point transformation T_{H_2} of G/LH_2 inducing ϕ_{H_2}. We have already shown that $KH_2 = H_1$; as in CG we prove that the maps T_K and T_{H_2} are consistent when projected down to G/LH_1 and so, by Lemma 2(iv) define an invertible map T of $G/L(H_2 \cap K) = G/L$.

It remains only to show that T, T^{-1} are $\tilde{\mathscr{B}}^0$ measurable and that T induces ϕ. First suppose that $\bar{H} \in \mathfrak{H}$, $\bar{H} \supset K$, $\bar{H}$ invariant under T_K and $L\bar{H} \cap LH_2 = L(\bar{H} \cap H_2)$. Then, as in CG, using Lemma 3, we show that for any $X \in \tilde{\mathscr{G}}^{\bar{H} \cap H_2}$,

TX, $T^{-1}X \in \tilde{\mathscr{G}}^{\bar{H} \cap H_2}$ and $\widehat{TX} = \phi \hat{X}$ (where $\hat{X}$ denotes, as in CG, the equivalence class to which X belongs in the measure algebra concerned).

Now let $\bar{H} \in \mathfrak{H}$, $\bar{H} \supset K$ and invariant under T_K. Since $H_1 \cap \bar{H} \in \mathfrak{H}$ and $H_1 \cap \bar{H} \supset K$, there exists $\bar{\bar{H}}$, such that $H_1 \cap \bar{H} \supset \bar{\bar{H}} \supset K$ and $\bar{\bar{H}}$ is invariant under T_K. Since $K \subset \bar{\bar{H}} \subset H_1 = H_2K$, using the fact that $H_2 \cap K = \{e\}$ and $LH_2 \cap LK = L$, we have by the earlier argument noted, with H replaced by H_2 that

$$L\bar{\bar{H}} \cap LH_2 = L(\bar{\bar{H}} \cap H_2).$$

Now $\tilde{\mathscr{G}}^{\bar{H} \cap H_2} \subset \tilde{\mathscr{G}}^{\bar{\bar{H}} \cap H_2}$, and combining this with the arguments in CG we get that

$$\tilde{\mathscr{B}}^0 = \bigcup \{\tilde{\mathscr{G}}^{\bar{\bar{H}} \cap H_2} : \bar{\bar{H}} \in \mathfrak{H}, \bar{\bar{H}} \supset K, \bar{\bar{H}} \text{ invariant under } T_K \text{ and}$$
$$L\bar{\bar{H}} \cap LH_2 = L(\bar{\bar{H}} \cap H_2)\},$$

which completes the proof of the lemma.

COROLLARY. *Suppose G/L, μ, ϕ, K, T_K satisfy the hypotheses of the lemma. Suppose $H \in \mathfrak{H}$ with $LH \not\supset LK$, so that $LH \cap LK \neq LK$. By Proposition* B, *there exists $\bar{H} \in \mathfrak{H}$ such that $\bar{H}$ is invariant under ϕ, $LK \not\subset L\bar{H}$ and*

$$L\bar{H} \cap LK = L(\bar{H} \cap K) \neq LK$$

so that $\bar{H} \cap K$ is invariant under ϕ, and ϕ induces an automorphism $\phi_{\bar{H} \cap K}$ of $E_{G/L(\bar{H} \cap K)}$. Then there exists an invertible, $\tilde{\mathscr{B}}^0_{G/L(\bar{H} \cap K)}$ measurable, point transformation $T_{\bar{H} \cap K}$ of $G/L(\bar{H} \cap K)$, which induces $\phi_{\bar{H} \cap K}$ and is such that $\pi_{K, \bar{H} \cap K} \circ T_{\bar{H} \cap K} = T_K \circ \pi_{K, \bar{H} \cap K}$.

Proof. As in CG, we apply the lemma with $G/\bar{H} \cap K$ in place of G, $L(\bar{H} \cap K)/\bar{H} \cap K = L\bar{H} \cap LK/\bar{H} \cap K$ in place of L, $\pi_{\bar{H} \cap K}(\mu)$ in place of μ, $\phi_{\bar{H} \cap K}$ in place of ϕ, $K/\bar{H} \cap K$ in place of K, ϕ_K and T_K unchanged, $\bar{H}/\bar{H} \cap K$ in place of $\bar{H}$.

LEMMA 9. *Let G be a locally compact, σ-compact group, L a closed subgroup, $\mathfrak{H}$, $\mathfrak{K}$ as before, μ a σ-finite Baire measure on G/L, ϕ an automorphism of the measure algebra E of $(G/L, \mathscr{B}^0, \mu)$. Let $\{K_j : j \in J\}$ be a totally ordered set of groups in $\mathfrak{K}$ such that*

(i) *each K_j is invariant under ϕ with ϕ_{K_j} the induced automorphism of E_{G/LK_j};*

(ii) *each ϕ_{K_j} is induced by a $\tilde{\mathscr{B}}^0_{G/LK_j}$ measurable, invertible point transformation T_{K_j} of G/LK_j;*

(iii) *if* $j_1, j_2 \in J$ *with* $K_{j_1} \supset K_{j_2}$, *then*

$$\pi_{K_{j_1},K_{j_2}} \circ T_{K_{j_2}} = T_{K_{j_1}} \circ \pi_{K_{j_1},K_{j_2}}.$$

If $K = \bigcap_{j\in J} K_j$, *then* K *is invariant under* ϕ *and if* ϕ_K *is the induced automorphism of* $E_{G/LK}$, *then there exists a* $\tilde{\mathscr{B}}^0_{G/LK}$ *measurable, invertible point transformation* T_K *of* G/LK *inducing* ϕ_K *and such that for all* $j \in J$,

$$\pi_{K_j,K} \circ T_K = T_{K_j} \circ \pi_{K_j,K}.$$

Proof. We may without loss of generality assume μ is finite. We have already noted that

$$G/LK = \operatorname{proj}\lim_{j\in J} G/LK_j$$

and that $\mathscr{B}^0_{G/LK} = \sum(\bigcup_{j\in J} \pi^{-1}_{K_j,K}\mathscr{B}^0_{G/LK_j})$,

$$\mathscr{G}^K = \sum\Bigl(\bigcup_{j\in J} \mathscr{G}^{K_j}\Bigr),$$

$$\mathscr{E}^K = \sum\Bigl(\bigcup_{j\in J} \mathscr{E}^{K_j}\Bigr).$$

This shows that K is invariant. By the same argument as in CG we obtain the invertible point mapping T_K of G/LK such that $\pi_{K_j,K} \circ T_K = T_{K_j} \circ \pi_{K_j,K}$ for all $j \in J$. It therefore remains to show that T_K, T_K^{-1} are $\tilde{\mathscr{B}}^0_{G/LK}$ measurable and that T_K induces ϕ_K. Note that we have already shown that $\bigcap_{j\in J} HK_j = HK$, and so if $H \supset K$, $H = \bigcap_{j\in J} HK_j$.

As in CG we first consider the case when J is countable. J then contains a countable cofinal sequence, $\tilde{J} = \{j_1, j_2, \ldots\}$. [For $J = \bigcup_{n=1}^{\infty} F_n$, F_n finite, $F_n \subset F_{n+1}$, let j_n be the largest j in F_n]. So if $H \in \mathfrak{H}$, $H \supset K$, then $H = \bigcap_{n=1}^{\infty} HK_{j_n}$. Since $HK_{j_n} \in \mathfrak{H}$ by Lemma 4, Corollary (b), there exists $H'_{j_n} \in \mathfrak{H}$, invariant under $T_{K_{j_n}}$ with $HK_{j_n} \supset H'_{j_n} \supset K_{j_n}$. By replacing HK_{j_n} by $HK_{j_n} \cap H'_{j_{n-1}} \supset K_{j_n}$ inductively, we may choose H'_{j_n} so that in addition $H'_{j_{n+1}} \subset H'_{j_n}$. Now we put $H' = \bigcap_{n=1}^{\infty} H'_{j_n}$. $H' \in \mathfrak{H}$, and we have $\mathscr{B}_{G/LH'} = \sum(\bigcup_{n=1}^{\infty} \pi^{-1}_{H'_{j_n},H'}(\mathscr{B}_{G/LH'_{j_n}}))$. Now the proof for the case J countable proceeds as in CG. [Note the first part of the proof in CG does not work as the use of Lemma 3 here cannot be justified. But the proof given above is simpler even in the compact groups case.]

The proof of the general case proceeds exactly as in CG, except of course, that $\pi_{HK_{j_0},H}$ is a homeomorphism because it is a proper continuous bijection between two locally compact spaces.

We now state (and prove) our main results for the σ-compact case.

THEOREM 1. *Let M be a locally compact homogeneous space acted upon by a locally compact, σ-compact group G, so that M is σ-compact and homeomorphic to G/L for some closed subgroup L of G. Let μ be a σ-finite Baire measure on M, which, since it is finite on compact sets is necessarily the restriction to the Baire sets $\mathscr{B}^0$ (of M or G/L) of a Radon measure $\bar{\mu}$ on M. Then every automorphism ϕ of the measure algebra E of $(M, \mathscr{B}^0, \mu)$ (or $(G/L, \mathscr{B}^0, \mu)$) is induced by an invertible, completion Baire measurable point transformation T of M; in fact T and T^{-1} are $\bar{\mathscr{B}}^0$ measurable.*

Proof. The proof is essentially identical with the proof of Theorem 1 in CG. One sets up the family $\mathscr{F}$ of ordered pairs (K, T_K) as in CG, one shows that it is inductive by Lemma 9 and nonempty by Lemmas 4(a) and 5; hence it has a maximal element, (K, T_K). If $LK = L$, then $G/LK = G/L$ and we are done. If not, by Lemma 1, there exists $H \in \mathfrak{H}$ such that $LH \not\supset LK$, and by Lemma 8 Corollary (recall that vital use of Proposition B is made here), there exists $\bar{H} \in \mathfrak{H}$ and $(K \cap \bar{H}, T_{K \cap \bar{H}})$ in $\mathscr{F}$ and different from (K, T_K). This contradicts the maximality of (K, T_K) and concludes the proof of the theorem.

As in CG, combining Theorem 1 with the theorem of Lamperti ([8], Theorem 3.1) we get

THEOREM 2. *Let M be a locally compact homogeneous space, acted upon by a locally compact, σ-compact group G, μ a σ-finite Baire measure on M, necessarily the restriction of a Radon measure $\bar{\mu}$ to the Baire sets $\mathscr{B}^0$. Let U be an invertible isometry of $L^p(M, \mathscr{B}^0, \mu)$, $1 \leqslant p < \infty$, $p \neq 2$, or a positive invertible isometry of $L^2(M, \mathscr{B}^0, \mu)$. [Note that $L^p(M, \mathscr{B}^0, \mu)$ coincides with $L^p(M, \mathscr{B}, \bar{\mu})$]. Then there exists an invertible, completion Baire measurable, point transformation T of M such that $(Uf)(x) = f(T^{-1}x)\alpha(x)$, with $|\alpha(x)|^p = \omega_T(x)$ [for $p = 2$, $\alpha(x) = \omega_T^{1/2}(x)$], where $\omega_T(x)$ is defined, for all $X \in \mathscr{B}^0$, by*

$$\mu(T^{-1}X) = \int_X \omega_T(x)\mu(dx).$$

7. THE NON-σ-COMPACT CASE

We now investigate to what extent the above theorems generalize to the case when G is not assumed σ-compact. It is then well known that M need not be homeomorphic to G/L. If M *is* homeomorphic to G/L for some closed subgroup L, our theorems can be proved. It is known that M is homeomorphic to G/L if and only if for one (and hence all) $a \in M$, the map $G \to M$ given by $g \to g(a)$ is open. This is elementary and is essentially shown in the proof (not the statement) of Theorem 20 in [11, Chapter III, Section 24, pp. 148–149]. We then have $L = G_a = \{g : g(a) = a\}$, and the map $g \to g(a)$ corresponds to $g \to \psi_L(g)$, where ψ_L is the natural projection $G \to G/L$. We also need the following elementary and well-known lemma (and its trivial corollary).

LEMMA. *If ψ_L is the natural projection $G \to G/L$, and K is compact in G/L, then there exists K_0, compact in G, such that $\psi_L(K_0) = K$.*

Proof. Let W be a compact neighbourhood of the identity in G. Then the interiors of $\{\psi_L(gW) : g \in G\}$ cover K and so there exists a finite sub-covering $\psi_L(g_1 W), \ldots, \psi_L(g_n W)$; $K_0 = \psi_L^{-1}(K) \cap \bigcup_{j=1}^n g_j W$ is the required compact set.

COROLLARY. *If E is σ-compact in G/L, there exists a σ-compact set E_0 in G, such that $\psi_L(E_0) = E$.*

In our theorem the Radon measure $\bar{\mu}$ will be regarded as the *essential* measure, this is virtually always the more important of the two measures involved: virtually every theorem on Radon measures, except for Fubini's theorem, is proved for the essential measure (see Schwartz [12, Part I, Chapter I]). Recall that a Radon measure on a locally compact space gives a unique measure on the σ relatively compact Borel sets. More precisely let $\mathscr{B}_R$ be the σ-ring generated by the compact sets in G/L, $\mathscr{B}$ the σ-algebra generated by the closed sets, ${\mathscr{B}_R}^0$ the σ-ring generated by the compact $\mathscr{G}_\delta$ sets (equivalently the least σ-ring with respect to which all real valued continuous functions with compact support are measurable), $\mathscr{B}^0$ the least σ-algebra with respect to which all real valued continuous bounded functions are measurable. $\mathscr{B}_R$, $\mathscr{B}$, ${\mathscr{B}_R}^0$, $\mathscr{B}^0$ are, respectively, the restricted Borel, Borel, restricted Baire, and Baire subsets of G/L. On $\mathscr{B}_R$ and so also on ${\mathscr{B}_R}^0$, the essential and the larger measure coincide. Let $\overline{\mathscr{B}}_{R,\bar{\mu}}$, and $\overline{\mathscr{B}}^0_{R,\mu}$ be their respective completions with respect to $\bar{\mu}$, μ. The 'completions' of $\mathscr{B}$, $\mathscr{B}^0$ with respect to the essential measure $\bar{\mu}$ and its restriction μ to $\mathscr{B}^0$, are defined as follows. We say $X \in \overline{\mathscr{B}}_{\bar{\mu}}$, if $X \cap Y \in \overline{\mathscr{B}}_{R,\bar{\mu}}$ for all $Y \in \overline{\mathscr{B}}_{R,\bar{\mu}}$ and then

$$\bar{\mu}(X) = \sup\{\bar{\mu}(X \cap Y) : Y \in \overline{\mathscr{B}}_{R,\bar{\mu}}\}.$$

Similarly $X \in {\overline{\mathscr{B}}_\mu}^0$ if $X \cap Y \in \overline{\mathscr{B}}^0_{R,\mu}$ for all $Y \in \overline{\mathscr{B}}^0_{R,\mu}$, it follows that $X \in \overline{\mathscr{B}}_{\bar{\mu}}$. We also then have $\bar{\mu}(X) = \mu(X) = \sup\{\mu(X \cap Y) : Y \in \overline{\mathscr{B}}^0_{R,\mu}\}$. As usual we write μ for $\bar{\mu} | {\overline{\mathscr{B}}_\mu}^0$. Note that if we extend $\bar{\mu}$ on $\mathscr{B}_R$ (respectively, ${\mathscr{B}_R}^0$) by the usual Carathéodory extension procedure, we get as measurable sets exactly $\overline{\mathscr{B}}_{\bar{\mu}}$ (respectively, ${\overline{\mathscr{B}}_\mu}^0$) but of course with the larger measure (see Srinivasan

[13]). We let E_R denote the measure ring of $\overline{\mathscr{B}}_{R,\bar{\mu}}$, or $\overline{\mathscr{B}}^0_{R,\mu}$ and E the measure algebra of $\overline{\mathscr{B}}_{\bar{\mu}}$, or $\overline{\mathscr{B}}_{\mu}{}^0$; of course $E_R \subset E$. E is a (lattice) complete algebra, since $\bar{\mu}$ is essential measure. Recall also that a measure space and its completion have the same measure algebra.

We only state and prove our generalization of Theorem 1, that of Theorem 2 follows trivially.

THEOREM 3. *Let G be a locally compact group, L a closed subgroup, $\bar{\mu}$ a Radon essential measure on the locally compact homogeneous space, G/L, μ its restriction to the Baire sets $\mathscr{B}^0$ of G/L. Then every automorphism ϕ of the measure algebra E of $(G/L, \mathscr{B}^0, \mu)$ is induced by an invertible completion Baire measurable (i.e., $\overline{\mathscr{B}}_{\mu}{}^0$ measurable) point transformation T of G/L.*

Proof. There exists a σ-compact full, i.e., open, subgroup F of G([6, Theorem B, p. 251]); F is open, so being a subgroup it is also closed, and hence is Baire, as well as σ-compact. For each $x \in G/L$, let $F(x)$ denote the orbit of F in G/L under the map $g \to g(x)$ of G in G/L, (i.e., $F(x) = \{y : y = f(x) \text{ for some } f \in F\}$), then $y \in F(x)$ if and only if $x \in F(y)$ and so $F(x) = F(y)$. Thus G/L is a disjoint union of such orbits, we let $\{F_i : i \in I\}$ be the class of disjoint orbits, each $F_i = F(x)$ for some $x \in G/L$. Since each map $g \to g(x)$ is open, and of course, continuous $G \to G/L$ (see the remark, at the beginning of this section), it follows that each F_i, the image of F under such a map, is open and σ-compact. Since $F_k = G/L - \bigcup\{F_i : i \in I, i \neq k\}$, each F_k is also closed and hence Baire, thus each F_i is in $\mathscr{B}_R{}^0$. Since the F_i are all open, Baire, there exists $i \in I$ such that $\mu(F_i) = \bar{\mu}(F_i) > 0$, else $\bar{\mu}(G/L) = \bar{\mu}(\bigcup F_i) = 0$. Let

$$\mathscr{D}_0 = \{i : \mu(F_i) = 0\}.$$

For each $i \notin \mathscr{D}_0$, we define the saturation class $\operatorname{sat}(F_i)$ of F_i as follows. $\hat{A}$ will denote the equivalence class in the measure algebra E of G/L to which A belongs. Let

$$\mathscr{I}_{i,1} = \{j \in I : \mu(\phi^n(\hat{F}_i) \wedge \hat{F}_j) > 0 \text{ for some } n \in \mathbf{Z}\}.$$

If $\mathscr{I}_{i,1}$ to $\mathscr{I}_{i,h}$ are defined let

$$\mathscr{I}_{i,h+1} = \{j \in I : \mu(\phi^n(\hat{F}_l) \wedge (\hat{F}_j) > 0 \text{ for some } l \in \mathscr{I}_{i,h}, n \in \mathbf{Z}\}.$$

$i \in \mathscr{I}_{i,h}$ for all $h = 1, 2, 3, \ldots$; also $\mathscr{I}_{i,h} \subset \mathscr{I}_{i,h+1}$. Further $\mathscr{I}_{i,h}$ is countable. For it is enough to show that $\mathscr{I}_{i,1}$ is countable and this follows since F_i is σ-compact and so of σ-finite μ measure. Put $\mathscr{I}_i = \bigcup_{h=1}^{\infty} \mathscr{I}_{i,h}$ and $\operatorname{sat}(F_i) = \bigcup_{j \in \mathscr{I}_i} F_j$. $\mathscr{I}_i$ is countable and so $\operatorname{sat}(F_i)$ is σ-compact, Baire, open and so locally compact. Further $\phi(\widehat{\operatorname{sat}(F_i)}) = \widehat{\operatorname{sat}(F_i)}$. Given any set $X \in \overline{\mathscr{B}}^0_{R,\mu}$, $\hat{X}$

meets at most countable many $\hat{F}_i$, and hence by our last remark so does $\phi(\hat{X})$, hence $\phi(\hat{X}) \in E_R$, i.e., $\phi(E_R) \subset E_R$, and since the same holds for ϕ^{-1}, $\phi(E_R) = E_R$. Finally $F_j \subset \text{sat}(F_i)$ if and only if $F_i \subset \text{sat}(F_j)$, so $\{F_i : i \in I, i \notin \mathscr{D}_0\}$ splits into equivalence classes, the union of the F_i in any such class is the saturation of any one of them. Let I_0 be a selection of indices $i \in I$, one from each equivalence class. Clearly if $\phi | \widehat{\text{sat}(F_i)}$ is induced by an invertible completion Baire measurable transformation T_i for each $i \in I_0$, then the resultant $T = \bigoplus_{i \in I_0} T_i$, and its inverse T^{-1} are both measurable with respect to the completion of the σ-ring of σ-compact Baire sets, i.e., with respect to $\bar{\mathscr{B}}^0_{R,\mu}$, and T induces $\phi | E_R$. So fix some $i \in I_0$.

Since $\text{sat}(F_i)$ is σ-compact, there exists, by the corollary to the lemma at the beginning of this section, a σ-compact set $E_i \subset G$ such that $\psi_L(E_i) = \text{sat}(F_i)$. There exists a σ-compact, full, i.e., open, and so locally compact subgroup G_i of G, such that $G_i \supset E_i$ [6, Theorem B, p. 251]. If $M_i = \psi_L(G_i)$, then $\text{sat}(F_i) \subset M_i$, M_i is σ-compact and open in G/L and so locally compact, further M_i is the orbit of G_i under $g \to \psi_L(g)$. Thus M_i is homogeneous under the action of the locally compact, σ-compact group G_i. (M_i being open, and a countable union of compact sets, is Baire in G/L, though we do not need this.) Define a Baire measure ν on M_i by $\nu = j\mu_0$, where $\mu_0 = \mu | \text{sat}(F_i)$, j is the injection map, $\text{sat}(F_i) \to M_i$. Since $\nu(G_i - \text{sat}(F_i)) = 0$, the measure algebras of μ_0 and ν are isomorphic and so $\phi | \text{sat}(F_i)$ induces a measure algebra automorphism $\tilde{\phi}_i$ of ν. By Theorem 1 (applied with $M = M_i$, $G = G_i$) there exists an invertible completion Baire measurable point transformation $\tilde{T}_i$ of M_i inducing $\tilde{\phi}_i$. Let $\Omega = \bigcap_{n \in \mathbf{Z}} \tilde{T}_i^n(\text{sat}(F_i))$, then $\tilde{T}_i(\Omega) = \Omega$. Define $T_i x = \tilde{T}_i x$ if $x \in \Omega$, $T_i x = x$ if $x \in \text{sat}(F_i) - \Omega$. Then T_i is the required invertible completion Baire measurable transformation of $\text{sat}(F_i)$ inducing $\phi | \text{sat}(F_i)$; if $T = \bigoplus_{i \in I_0} T_i$, then T, T^{-1} are $\bar{\mathscr{B}}^0_{R,\mu}$ measurable and T induces $\phi | E_R$. It follows from the definition of $\bar{\mathscr{B}}_\mu{}^0$ and of essential measure, that T, T^{-1} map sets in $\bar{\mathscr{B}}_\mu{}^0$ into sets in $\bar{\mathscr{B}}_\mu{}^0$, and sets of μ measure zero into sets of μ measure zero. Thus T induces an automorphism $\tilde{\phi}$ of E which coincides with ϕ on E_R. But since μ is essential measure, E is a complete measure algebra; hence for any $\beta \in E$, $\beta = \bigvee\{\alpha : \alpha \in E_R, \alpha \leqslant \beta\}$, and so, since ϕ, $\tilde{\phi}$ are automorphisms, $\tilde{\phi}(\beta) = \bigvee\{\phi(\alpha) : \alpha \in E_R, \alpha \leqslant \beta\} = \phi(\beta)$. Thus $\phi = \tilde{\phi}$ and T induces ϕ, which completes the proof of the theorem.

COROLLARY 1. *If $L = \{e\}$, then $G/L = G$, so the theorem holds for groups.*

COROLLARY 2. *If $G/L = G$, and if $\bar{\mu}$ is Haar measure, then T, T^{-1} are $\bar{\mathscr{B}}_{\bar{\mu}}$ measurable.*

Proof. For Haar measure, another theorem of Kakutani and Kodaira ([7], or [6, Theorem I, p. 288] shows that $\bar{\mathscr{B}}^0_{R,\mu} = \bar{\mathscr{B}}_{R,\bar{\mu}}$ and $\bar{\mathscr{B}}_\mu{}^0 = \bar{\mathscr{B}}_{\bar{\mu}}$.

Remark. In view of the remarks at the beginning of this section, we could in Theorem 3, replace the assumption that the homogeneous space M considered is G/L, by the equivalent assumption that the map $G \to M$, given by $g \to g(a)$ for some $a \in M$, is open.

ACKNOWLEDGMENTS

The authors wish to express thanks to Amit Roy and Gopal Prasad for showing them various counterexamples.

REFERENCES

1. N. BOURBAKI, "Topologie Générale," 3rd ed., Actualités Sci. Ind. No. 1142, Chs. 1 and 2, Hermann, Paris, 1961.
2. J. R. CHOKSI, Automorphisms of Baire measures on generalized cubes, *Z. Wahrscheinlichkeitstheorie und Verw. Gebiete* **22** (1972), 195–204.
3. J. R. CHOKSI, Automorphisms of Baire measures on generalized cubes, II, *Z. Wahrscheinlichkeitstheorie und Verw. Gebiete* **23** (1972), 97–102.
4. J. R. CHOKSI, Measurable transformations on compact groups, *Trans. Amer. Math. Soc.* **184** (1973), 101–124.
5. J. FELDMAN AND F. P. GREENLEAF, Existence of Borel transversals on groups, *Pacific J. Math.* **25** (1968), 455–461.
6. P. R. HALMOS, "Measure Theory," Van Nostrand, Princeton, New Jersey, 1950.
7. S. KAKUTANI AND K. KODAIRA, Über das Haarsche Mass in der lokal bikompakten Gruppe, *Proc. Imperial Acad. Tokyo* **20** (1944), 444–450.
8. J. LAMPERTI, On the isometries of certain function spaces, *Pacific J. Math.* **8** (1958), 459–466.
9. D. MAHARAM, Automorphisms of products of measure spaces, *Proc. Amer. Math. Soc.* **9** (1958), 702–707.
10. R. PANZONE AND C. SEGOVIA, Measurable transformations on compact spaces and o.n. systems on compact groups, *Rev. Un. Mat. Argentina* **22** (1964), 83–102.
11. L. S. PONTRJAGIN, "Topological Groups," 2nd ed., GITTL, Moscow, 1954; English transl., Gordon & Breach, New York, 1966.
12. L. SCHWARTZ, "Radon Measures on Arbitrary Topological Spaces," Studies in Mathematics, Tata Institute of Fundamental Research, Oxford Univ. Press, London and New York, 1973.
13. T. P. SRINIVASAN, On measurable sets, *J. Indian Math. Soc.* **18** (1954), 1–8.
14. J. VON NEUMANN, Einige Sätze über die messbare Abbildungen, *Ann. of Math.* **33** (1932), 574–586.

AMS (MOS) 1970 subject classifications: primary 28A65, 28A60; secondary 22D40, 43A85.

STUDIES IN PROBABILITY AND ERGODIC THEORY
ADVANCES IN MATHEMATICS SUPPLEMENTARY STUDIES, VOL. 2

Ergodic Transformations of Lebesgue Spaces†

ANTHONY LO BELLO

Department of Mathematics, Allegheny College, Meadville, Pennsylvania

Let $(\mathcal{X}, \mathfrak{A}, \mu)$ be a nonatomic Lebesgue measure space of total measure 1. Let $\mathcal{Z}$ be the set of integers, $\mathfrak{B}$ the σ-field of all subsets of $\mathcal{Z}$, and ν the measure on $(\mathcal{Z}, \mathfrak{B})$ whose value at any set $A \in \mathfrak{B}$ is the cardinal number of A.

THEOREM. *If R is an ergodic, measure preserving transformation of $(\mathcal{X}, \mathfrak{A}, \mu)$, then there is an ergodic, measure preserving transformation T on $(\mathcal{X} \times \mathcal{Z}, \mathfrak{A} \times \mathfrak{B}, \mu \times \nu)$ and a measurable function $K: \mathcal{X} \times \mathcal{Z} \to \mathcal{Z}$ such that for almost all $(x, n) \in \mathcal{X} \times \mathcal{Z}$,*

$$T(x, n) = (R(x), K(x, n)).$$

Let $\mathcal{E}_{\mathrm{III}}(\mathcal{X}, \mathfrak{A}, \mu)$ be the set of all ergodic transformations on $(\mathcal{X}, \mathfrak{A}, \mu)$ that admit no invariant measure equivalent to μ.

THEOREM. *If $R \in \mathcal{E}_{\mathrm{III}}(\mathcal{X}, \mathfrak{A}, \mu)$, then there is an ergodic transformation T on $\mathcal{X} \times \mathcal{Z}$ and a measurable function $K: \mathcal{X} \times \mathcal{Z} \to \mathcal{Z}$ such that for almost all $(x, n) \in \mathcal{X} \times \mathcal{Z}$*

(1) *$T(x, n) = (R(x), K(x, n))$ and*

(2) *T preserves a measure $\pi \sim \mu \times \nu$ if and only if R admits an admissible pair (S, λ).*

1. INTRODUCTION AND PRELIMINARIES

Hajian *et al.* [1] developed a systematic method of constructing on a nonatomic, σ-finite measure space ergodic transformations that admit no σ-finite invariant measure. More recently, the same authors undertook the investigation of a generalized notion of induced transformation [2]. The first result of this paper (Theorem 1) is the construction of an ergodic measure preserving transformation (e.m.p.t.) on a product space. The problem is related to certain ideas of Dye [3] and is closely connected with the work done in the articles of Hajian, Ito, and Kakutani cited above.

Throughout this paper $(\mathcal{X}, \mathfrak{A}, \mu)$ will denote a nonatomic Lebesgue measure space of total measure 1. $(\mathcal{R}, \mathfrak{A}, \mu)$ will denote a nonatomic Lebesgue measure space with $\mu(\mathcal{R}) = \infty$. $\mathcal{Z}$ is the set of integers, $\mathfrak{B}$ is the σ-field of all

† The results in this paper are included in the author's doctoral dissertation written at Yale University under the direction of Professor S. Kakutani.

ISBN 0-12-599102-9

subsets of $\mathscr{Z}$, and ν is the counting measure on $(\mathscr{Z}, \mathfrak{B})$. $\mathscr{N}$ is the set of positive integers. We do not define the elementary notions of ergodic theory and refer the reader to the introductions of [2] and [4] for the notation and the definition of the terms which those articles have in common with this paper. Finally, all statements below are made modulo sets of measure 0.

In Lemma 1 we rely decisively on Section 3 in [1] to prove our main result for the special case of a von Neumann transformation. Lemma 2 is part (i) of Theorem 1 in [4]. We accomplish the proof of Theorem 1 by a method depending on Lemmas 1 and 2. Theorem 2 is a generalization of Theorem 1 to the infinite σ-finite case, and in Theorem 3 we characterize those elements of $\mathscr{E}_{\mathrm{III}}(\mathscr{X}, \mathfrak{A}, \mu)$ that admit a result of the type shown in Theorems 1 and 2. In Remarks 1 and 2 we summarize briefly the applications of the theorems of this paper to the work done in [1] and [2], respectively. In conclusion, all proofs are given in complete detail in [5].

2. Main Results

LEMMA 1. *If R is a von Neumann transformation on* $(\mathscr{X}, \mathfrak{A}, \mu)$, *then there is an e.m.p.t. T on* $\mathscr{X} \times \mathscr{Z}$ *and a measurable function* $k: \mathscr{X} \to \mathscr{Z}$ *such that*

$$T(x, n) = (R(x), n - k(x)). \tag{1}$$

Proof. Define $k: \mathscr{X} \to \mathscr{Z}$ by

$$k(x) = n - 2 \qquad \text{if} \quad x \in D(n, 2^{n-1})$$

where $\{D(n, i): n \in \mathscr{N}, i = 1, \ldots, 2^n\}$ is as in the definition of von Neumann transformation in [4]. Define T by (1). The proof that T is an e.m.p.t. follows the argument given in Section 3 in [1] for the case $\alpha = 1$. Complete details may also be found in [5].

LEMMA 2. *If R is an e.m.p.t. on* $(\mathscr{X}, \mathfrak{A}, \mu)$, *then there is a von Neumann transformation* $S \in [R]^+$.

Proof. See Theorem 1 in [4].

THEOREM 1. *If R is an e.m.p.t. on* $(\mathscr{X}, \mathfrak{A}, \mu)$, *there is an e.m.p.t. T on* $\mathscr{X} \times \mathscr{Z}$ *and a measurable function* $K: \mathscr{X} \times \mathscr{Z} \to \mathscr{Z}$ *such that*

$$T(x, n) = (R(x), K(x, n)).$$

Proof. By Lemma 2, there is a von Neumann transformation $\hat{R} \in [R]^+$, and by Lemma 1 there is an e.m.p.t. $\hat{T}$ on $\mathscr{X} \times \mathscr{Z}$ and a measurable function

$k:\mathscr{X} \to \mathscr{Z}$ such that

$$\hat{T}(x, n) = (\hat{R}(x), n - k(x)).$$

We also have

$$\mathscr{X} = \bigcup_{i \in \mathscr{N}} A_i \qquad \text{(disjoint)}$$

where

$$A_i = \{x \in \mathscr{X} : \hat{R}(x) = R^i(x)\}.$$

Let $\{r_i : i \in \mathscr{N}\}$ be a set of real numbers such that $0 < r_i < 1$ and $r_{i-1} < r_i$. Put

$$\tilde{\mathscr{X}} = \mathscr{X} \times \mathscr{Z} \cup \bigcup_{m \in \mathscr{Z}} \bigcup_{n \geqslant 2} \bigcup_{j=1}^{n-1} (R^j A_n, m + r_j).$$

If we set $r_0 = 0$, then we have

$$\tilde{\mathscr{X}} = \bigcup_{m \in \mathscr{Z}} \bigcup_{n \in \mathscr{N}} \bigcup_{j=0}^{n-1} (R^j A_n, m + r_j),$$

and $(\mathscr{X} \times \mathscr{Z}, \mathfrak{A} \times \mathfrak{B}, \mu \times \nu)$ induces a measure space structure on $\tilde{\mathscr{X}}$ in the natural way. We observe that each element $\xi \in \tilde{\mathscr{X}}$ may be expressed uniquely by a pair $(R^j(x), m + r_j)$ for some $m \in \mathscr{Z}$, $x \in A_n$, $n \in \mathscr{N}$, and j, $0 \leqslant j < n$. We define a transformation $\tilde{T}$ on $\tilde{\mathscr{X}}$, putting, for $x \in A_n$,

$$\tilde{T}(R^j(x), m + r_j) = \begin{cases} (R^{j+1}(x), m + r_{j+1}) & \text{if } \ j < n - 1 \\ (R^{j+1}(x), m - k(x)) & \text{if } \ j = n - 1. \end{cases}$$

$\tilde{T}$ is clearly a measure preserving transformation (m.p.t.) of $\tilde{\mathscr{X}}$. We also have

$$\tilde{T}_{\mathscr{X} \times \mathscr{Z}} = \hat{T}$$

where $\tilde{T}_{\mathscr{X} \times \mathscr{Z}}$ is the transformation induced by $\tilde{T}$ on $\mathscr{X} \times \mathscr{Z}$, and

$$\bigcup_{i \in \mathscr{Z}} \tilde{T}^i(\mathscr{X} \times \mathscr{Z}) = \tilde{\mathscr{X}}.$$

We therefore conclude that $\tilde{T}$ is an e.m.p.t. on $\tilde{\mathscr{X}}$.

We now construct a m.p.t. $W:\tilde{\mathscr{X}} \to \mathscr{X} \times \mathscr{Z}$ such that for $\xi \in \tilde{\mathscr{X}}$, $\xi = (R^j(x), m + r_j)$,

$$W(\xi) = (R^j(x), p)$$

where p is a function of x, m, and j. When we have such a W, we may then define

$$T = W\tilde{T}W^{-1}$$

and thereby complete the demonstration of the theorem.

To construct W, we first find a dissipative m.p.t. V on $\tilde{\mathscr{X}}$ with section $\mathscr{X} \times 0$ such that for each $\xi \in \tilde{\mathscr{X}}$, $\xi = (R^j(x), m + r_j)$,

$$V(\xi) = (R^j(x), q)$$

where q is a function of x, m, and j. Once we have found such a V, we may define W by

$$W(\xi) = (R^j(x), -i)$$

where

$$\xi = (R^j(x), m + r_j)$$

and

$$V^i(\xi) \in \mathscr{X} \times 0.$$

We get this dissipative m.p.t. V on $\tilde{\mathscr{X}}$ as follows. For each $m \in \mathscr{N}$, put

$$C_m = \{x \in \mathscr{X} : x \text{ belongs to exactly } m \text{ different members of } \{R^j(A_n) : n \in \mathscr{N}, j = 0, \dots, n-1\}\}.$$

Then C_m is measurable for each m, and we put

$$C_\infty = \mathscr{X} - \bigcup_{m \in \mathscr{N}} C_m.$$

Since R is ergodic, either $\mu(C_\infty) = 0$ or $\mu(C_\infty) = 1$. Define a measurable function U on $\tilde{\mathscr{X}}$ by

$$U(R^j(x), m + r_j) = (R^j(x), m + r_k),$$

if there is a $k > j$, such that there is an $n' \in \mathscr{N}$ with $1 \leqslant k \leqslant n' - 1$, and a $y \in A_{n'}$ such that $R^j(x) = R^k(y)$ and k is the smallest such integer $> j$, and

$$U(R^j(x), m + r_j) = (R^j(x), m + 1) \qquad \text{otherwise.}$$

If $\mu(C_\infty) = 0$, let $V = U$. If $\mu(C_\infty) = 1$, let ψ be a bijection of $\mathscr{Z} \times \mathscr{N} \cup 0$ onto $\mathscr{Z}$ such that $\psi(0, 0) = 0$. For $(R^j(x), m + r_j) \in \tilde{\mathscr{X}}$ there is an $i \in \mathscr{N} \cup 0$ such that $(R^j(x), m + r_j) = U^i(R^j(x), m)$. We define V at $(R^j(x), m + r_j)$ by

$$V(R^j(x), m + r_j) = (R^j(x), \psi^{-1}(\psi(m, i) + 1)).$$

This completes the proof of the theorem. A simpler and more direct proof is available and will be contained in a forthcoming paper of Hajian and Ito.

THEOREM 2. *If R is an e.m.p.t. on $(\mathscr{R}, \mathfrak{A}, \mu)$, then there is an e.m.p.t. T on $\mathscr{R} \times \mathscr{Z}$ and a measurable function $K : \mathscr{R} \times \mathscr{Z} \to \mathscr{Z}$ such that*

$$T(x, n) = (R(x), K(x, n)).$$

Proof. Let $\mathscr{X}$ be a subspace of $\mathscr{R}$ with $\mu(\mathscr{X}) = 1$. Consider the e.m.p.t. $R_{\mathscr{X}}$ induced by R on $\mathscr{X}$. By Theorem 1, there is an e.m.p.t. T_1 on $\mathscr{X} \times \mathscr{Z}$ and a measurable function $K_1 : \mathscr{X} \times \mathscr{Z} \to \mathscr{Z}$ such that

$$T_1(x, n) = (R_{\mathscr{X}}(x), K_1(x, n)) \qquad \text{for} \quad (x, n) \in \mathscr{X} \times \mathscr{Z}.$$

Define a measurable function $K : \mathscr{R} \times \mathscr{Z} \to \mathscr{Z}$ by

$$\begin{aligned} K(x, n) &= K_1(x, n) \qquad &&\text{if} \quad x \in \mathscr{X} \\ K(x, n) &= n &&\text{if} \quad x \notin \mathscr{X}. \end{aligned}$$

Now define T on $\mathscr{R} \times \mathscr{Z}$ by

$$T(x, n) = (R(x), K(x, n)).$$

It is clear that $T_{\mathscr{X} \times \mathscr{Z}} = T_1$. Since T_1 is ergodic, we conclude the proof by observing that

$$\bigcup_{i \in \mathscr{Z}} T^i(\mathscr{X} \times \mathscr{Z}) = \mathscr{R} \times \mathscr{Z}.$$

DEFINITION. $R \in \mathscr{E}_{\text{III}}(\mathscr{X}, \mathfrak{A}, \mu)$ admits an admissible pair (S, λ) if

$$S \in [R], \qquad S \text{ is ergodic}, \qquad \text{and} \qquad S \text{ preserves } \lambda, \lambda \sim \mu.$$

THEOREM 3. *Let $R \in \mathscr{E}_{\text{III}}(\mathscr{X}, \mathfrak{A}, \mu)$. There is an ergodic transformation T on $\mathscr{X} \times \mathscr{Z}$ and a measurable function $K : \mathscr{X} \times \mathscr{Z} \to \mathscr{Z}$ such that*

$$T(x, n) = (R(x), K(x, n)) \tag{2}$$

and

$$T \quad \text{preserves a measure} \quad \sigma \sim \mu \times \nu \tag{3}$$

if and only if R admits an admissible pair (S, λ).

Proof. If there exist ergodic T and measurable K satisfying (2) and (3), then we put

$$S = T_{\mathscr{X} \times 0} \qquad \text{(induced transformation)}$$

and

$$\lambda = \text{measure induced by} \quad \sigma \quad \text{on} \quad \mathscr{X} \times 0.$$

(S, λ) is an admissible pair for R.

Conversely, suppose (S, λ) is an admissible pair for R. By Lemma 7 in [2], there is an ergodic M on $(\mathscr{X}, \mathfrak{A}, \mu)$ such that

$$M \in [R]^+, \qquad S \in [M], \qquad \text{and} \qquad M \text{ preserves} \quad \lambda.$$

By Theorem 1, we know there is an e.m.p.t. $\tilde{T}$ on $(\mathscr{X} \times \mathscr{Z}, \mathfrak{A} \times \mathfrak{B}, \pi = \lambda \times \nu)$ and a measurable function $\tilde{K}$ such that

$$\tilde{T}(x, n) = (M(x), \tilde{K}(x, n)).$$

We now use the method of Theorem 1 to obtain an e.m.p.t. T on $(\mathscr{X} \times \mathscr{Z}, \mathfrak{A} \times \mathfrak{B}, \pi)$ and a measurable function K such that

$$T(x, n) = (R(x), K(x, n)).$$

We do this by extending π to a measure ρ on

$$\tilde{\mathscr{X}} = \mathscr{X} \times \mathscr{Z} \cup \bigcup_{m \in \mathscr{Z}} \bigcup_{n \geqslant 2} \bigcup_{j=1}^{n-1} (R^j A_n, m + r_j)$$

by putting

$$\rho(B) = \pi(R^{-j}B) \qquad \text{if} \quad B \subset (R^j A_n, m + r_j).$$

We then define $\tilde{\tilde{T}}$ on $\tilde{\mathscr{X}}$ by setting, for $x \in A_n$,

$$\tilde{\tilde{T}}(R^j(x), m + r_j) = \begin{cases} (R^{j+1}(x), m + r_{j+1}) & \text{if} \quad j < n - 1 \\ (R^{j+1}(x), \tilde{K}(x, n)) & \text{if} \quad j = n - 1. \end{cases}$$

$\tilde{\tilde{T}}$ preserves ρ on $\tilde{\mathscr{X}}$, and we have

$$\tilde{\tilde{T}}_{\mathscr{X} \times \mathscr{Z}} = \tilde{T}$$

and

$$\bigcup_{i \in \mathscr{Z}} \tilde{\tilde{T}}^i(\mathscr{X} \times \mathscr{Z}) = \tilde{\mathscr{X}},$$

so $\tilde{\tilde{T}}$ is an e.m.p.t. on $\tilde{\mathscr{X}}$.

We now construct the bijective (though no longer measure preserving) transformation $W: \tilde{\mathscr{X}} \to \mathscr{X} \times \mathscr{Z}$, using the technique of Theorem 1. Then $T = W\tilde{\tilde{T}}W^{-1}$ preserves ρW^{-1} on $\mathscr{X} \times \mathscr{Z}$ and ρW^{-1} is equivalent to π. T is ergodic and has the property

$$T(x, n) = (R(x), K(x, n))$$

for some measurable function K. This completes the proof of Theorem 3.

Remark 1. It is interesting to observe that Theorem 1 takes the following form in the terminology in [2].

If R is an e.m.p.t. on $(\mathscr{X}, \mathfrak{A}, \mu)$ then there is an e.m.p.t. T and a dissipative m.p.t. Q on $(\mathscr{X} \times \mathscr{Z}, \mathfrak{A} \times \mathfrak{B}, \mu \times \nu)$ such that the pair (T, Q) induces R on the Q-section $\mathscr{X} \times 0$.

Remark 2. Theorems 1, 2, and 3 above show that the method of Theorem 1 in [1] enables us to construct exactly those ergodic transformations R of $(\mathscr{X}, \mathfrak{A}, \mu)$ that admit an admissible pair and no others.

Remark 3. Theorem 3 provides a considerably different proof of the main theorem in [2] free from an extra assumption made in that paper.

References

1. A. Hajian, Y. Ito, and S. Kakutani, Invariant measures and orbits of dissipative transformations, *Advances in Math.* **9** (1972), 52–65.
2. A. Hajian, Y. Ito, and S. Kakutani, Orbits, sections, and induced transformations, *Israel J. Math.* **18**, No. 2 (1974), 97–115.
3. H. Dye, On groups of measure preserving transformations, I, *Amer. J. Math.* **81** (1959), 119–159.
4. A. Hajian, Y. Ito, and S. Kakutani, Full groups and a theorem of Dye, *Advances in Math.* **17** (1975), 48–59.
5. A. Lo Bello, Ergodic transformations of Lebesgue spaces, Ph.D. Thesis, Yale Univ., 1975.

AMS (MOS) 1970 subject classification: primary, 28A65.

A
B
C 8
D 9
E 0
F 1
G 2
H 3
I 4
J 5